Atlas of Exfoliative Cytopathology

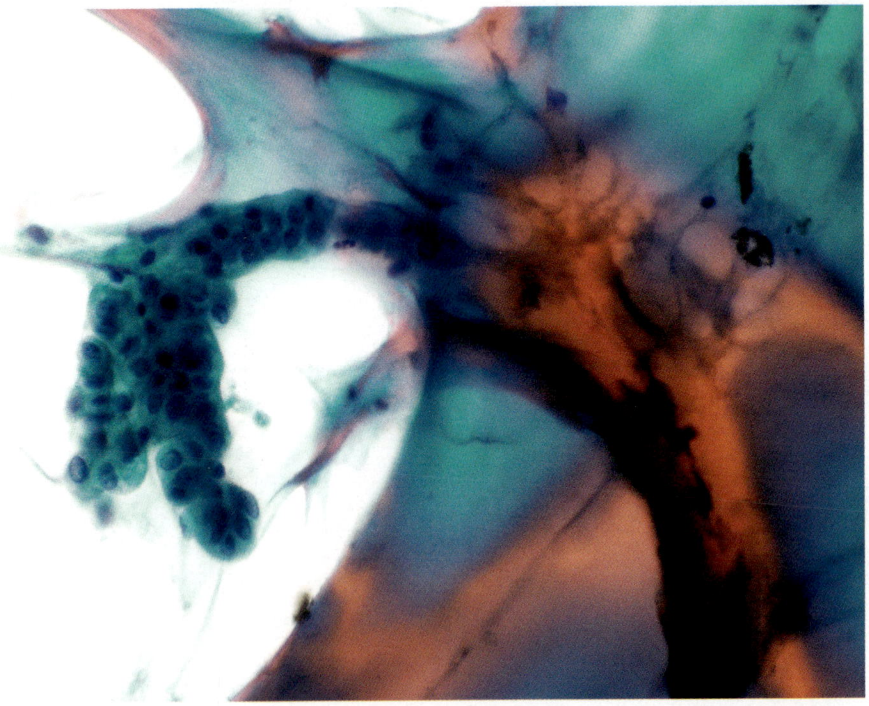

Syed Z. Ali, MD, FRCPath, FIAC
Professor of Pathology and Radiology
Director of Cytopathology
The Johns Hopkins University School of Medicine
Baltimore, Maryland

Christopher J. VandenBussche, MD, PhD
Assistant Professor of Pathology
Associate Director of Cytopathology
The Johns Hopkins University School of Medicine
Baltimore, Maryland

Cheng-Ying Ho, MD, PhD
Assistant Professor of Pathology
University of Maryland School of Medicine
Baltimore, Maryland

Doreen N. Palsgrove, MD
Postdoctoral Fellow
Departments of Pathology and Oncology
The Johns Hopkins Hospital
Baltimore, Maryland

Derek B. Allison, MD
Chief Resident, Department of Pathology
The Johns Hopkins Hospital
Baltimore, Maryland

Ashley Cimino-Mathews, MD
Associate Professor of Pathology and Oncology
Director of the Division of Breast Pathology
The Johns Hopkins University School of Medicine
Baltimore, Maryland

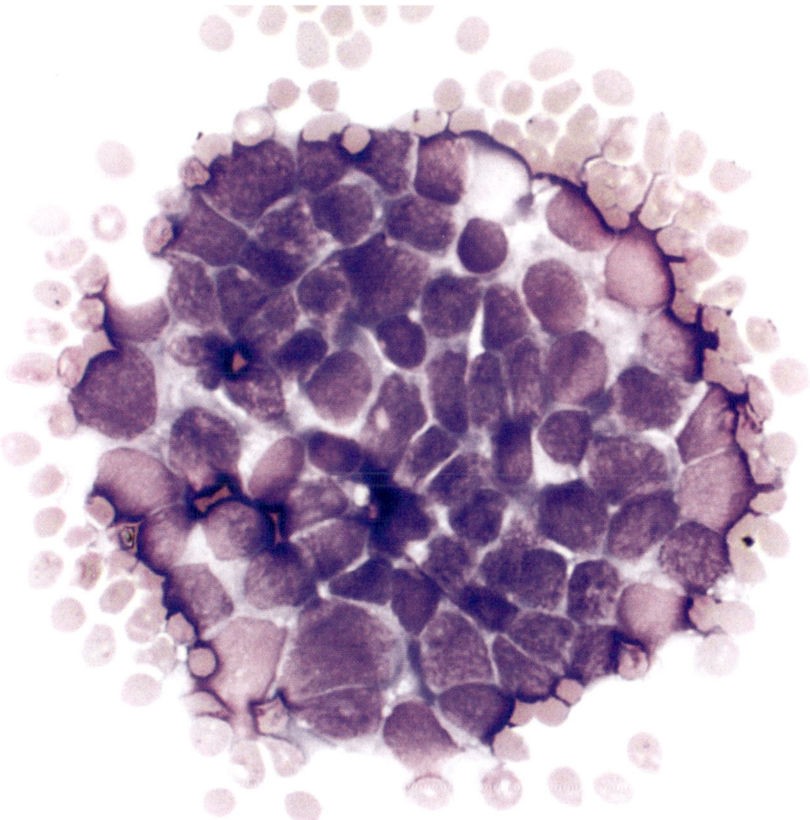

Atlas of Exfoliative Cytopathology

With Histopathologic Correlations

demosMEDICAL
An Imprint of Springer Publishing

Visit our website at www.springerpub.com

ISBN: 9781620701102
ebook ISBN: 9781617052873

Acquisitions Editor: David D'Addona
Compositor: diacriTech

Medicine is an ever-changing science. Research and clinical experience are continually expanding our knowledge, in particular our understanding of proper treatment and drug therapy. The authors, editors, and publisher have made every effort to ensure that all information in this book is in accordance with the state of knowledge at the time of production of the book. Nevertheless, the authors, editors, and publisher are not responsible for errors or omissions or for any consequences from application of the information in this book and make no warranty, expressed or implied, with respect to the contents of the publication. Every reader should examine carefully the package inserts accompanying each drug and should carefully check whether the dosage schedules mentioned therein or the contraindications stated by the manufacturer differ from the statements made in this book. Such examination is particularly important with drugs that are either rarely used or have been newly released on the market.

Library of Congress Cataloging-in-Publication Data
Names: Ali, Syed Z., author. | VandenBussche, Christopher J., author. | Ho, Cheng-Ying (Of Children's National Medical Center), author.
Title: Atlas of exfoliative cytopathology with histopathologic correlations / Syed Z. Ali, Christopher J. VandenBussche, Cheng-Ying Ho.
Description: New York : Demos Medical, [2018] | Includes index.
Identifiers: LCCN 2017045222| ISBN 9781620701102 | ISBN 9781617052873 (e-book)
Subjects: | MESH: Serous Membrane—cytology | Serous Membrane—pathology | Body Fluid Compartments—diagnostic imaging | Neoplastic Cells, Circulating—pathology | Cytodiagnosis–methods | Histological Techniques—methods | Atlases
Classification: LCC QP90.5 | NLM QS 517 | DDC 612/.01522—dc23
LC record available at https://lccn.loc.gov/2017045222

Printed in the United States of America by Sheridan Books, Inc.
17 18 19 20 21 / 5 4 3 2 1

To my mentors, Drs. Steve I. Hajdu, Yener S. Erozan, and Dorothy L. Rosenthal.—Syed Z. Ali

To my father, Tom, and his parents, Albert and Gertrude.—Christopher J. VandenBussche

To my lovely daughter, Josephine Louise.—Cheng-Ying Ho

To Andy, for his unwavering love and support.—Doreen N. Palsgrove

To my father, who gave me a passion for science; my mother, who always supported me; and my wife and son, who inspire me daily.—Derek B. Allison

To my family, with love.—Ashley Cimino-Mathews

Contents

Foreword

Translating cellular morphology to a definite diagnosis that guides treatment decisions is at the very heart of cytopathology. The diagnostic precision of each individual cytopathologist grows with every new specimen and is enhanced by the continuous correlation with histology and clinical findings. The enormous spectrum of changes that we see under the microscope requires good guidance in order not to get lost in what might appear as visual and biological chaos. Although every case is unique, classification and grouping of recurring observations help us to find our way to the right diagnosis. In this regard, illustrated textbooks continue to play an important role as a solid reference and good guide through the professional life of every cytopathologist. Despite the rapid progress in ancillary methods and molecular testing that we are experiencing in cytopathology, superb morphological diagnosis remains the cornerstone for all further downstream analyses and clinical decisions. It is my true pleasure to write a foreword to this outstanding work, which is destined to serve as a trusty guide in a large and important field of non-gynecological cytopathology. *Atlas of Exfoliative Cytopathology With Histopathologic Correlations* covers the wide spectrum of cytological findings in a well-organized format. The volume is divided into six chapters. A treasure of more than 500 high-resolution images illustrates the cytomorphologic characteristics and the differential diagnostic challenges associated with exfoliative cytology across different organ systems. The exquisite figures catch the essentials of a specific condition and each of them is supplemented by a concise description. Beyond morphology, the authors have also incorporated contemporary immunohistochemistry and histological correlations that further enrich our understanding. Urinary tract and gynecologic cytopathology specimens have been excluded as two previous volumes in the series have already broadly covered those topics. This *Atlas* has grown out of the fruitful ground of the Johns Hopkins Division of Cytopathology, where Dr. Ali and his team of co-authors are internationally well known as brilliant teachers and cytopathologists dedicated to moving the field forward.

Although there is no shortage of textbooks, atlases and virtual image databases in the field of cytopathology, *Atlas of Exfoliative Cytopathology With Histopathologic Correlations* stands out as a most practical volume tailored to assist cytopathologists in their daily diagnostic work irrespective of their level of experience. Besides enjoying the perfume of newly printed book pages, many of us will appreciate having this kind of atlas within easy reach. The authors should be congratulated on this excellent contribution.

Lukas Bubendorf, MD
Professor and Head of Cytopathology
University of Basel, Switzerland

1

Benign and Reactive Serous Cavity Effusions and Washings

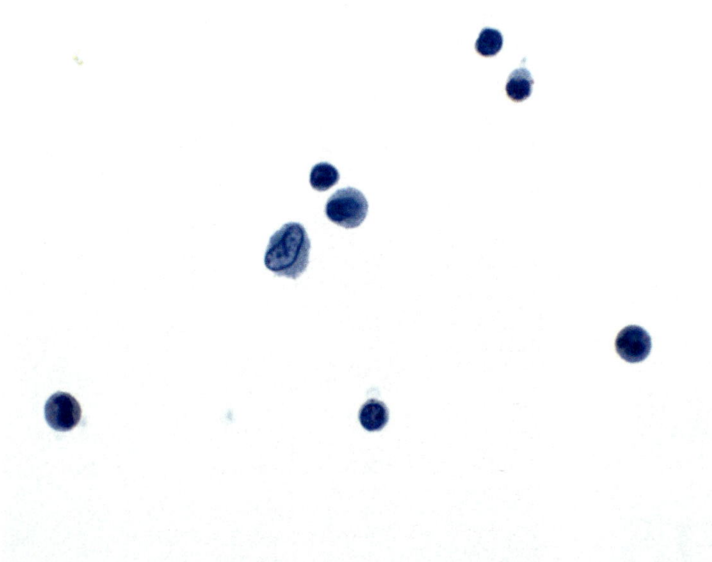

Figure 1.1a — Benign Histiocytes and Lymphocytes, Ascites Fluid [Pap stain; high power]. All serous cavity fluids typically contain three benign cellular components: mesothelial cells, histiocytes, and lymphocytes. A histiocyte can be seen in the center of this field, containing granular cytoplasm and an elongated, slightly curved nucleus. Several lymphocytes can be seen, with a thin rim of cytoplasm and relatively regular nuclear borders.

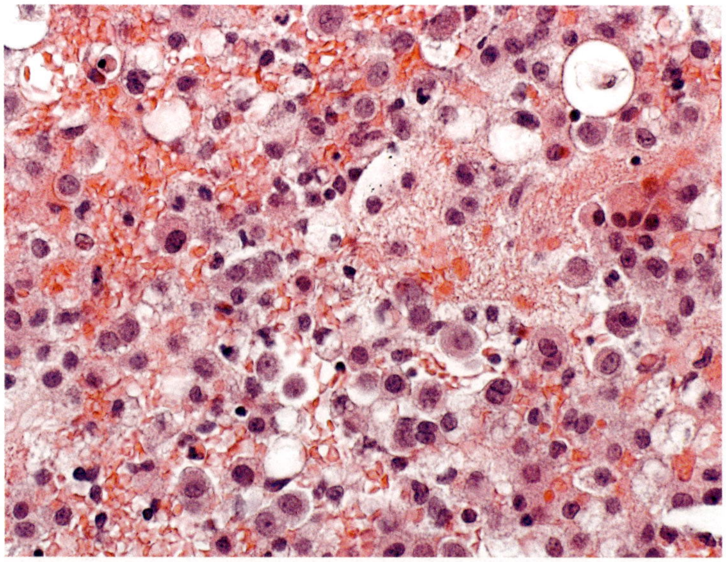

Figure 1.1b — Benign Ascites Fluid, Cell Block Section [H&E stain; medium power]. This field depicts numerous histiocytes admixed with occasional mesothelial cells and scattered lymphocytes. Mesothelial cells are easily identified by their round to oval, centrally located nuclei; smooth nuclear borders; and pale fuzzy cell borders. Reactive mesothelial cells can vary in size and exhibit enlarged central or paracentral nuclei with binucleation, irregular nuclear outlines, and prominent nucleoli. Coarsening of chromatin can also occur but is typically evenly distributed. Histiocytes tend to have smaller nuclei than mesothelial cells, with granular or vacuolated cytoplasm. Occasionally, cytoplasmic vacuolization and pseudo–signet ring type of changes can also be seen in degenerating mesothelial cells; however, in general, mesothelial cells tend to have denser or more opaque peripheral cytoplasm. A Mucicarmine stain would be negative in these cases. Lymphocytes should be small with round nuclei, inconspicuous nucleoli, and scant cytoplasm.

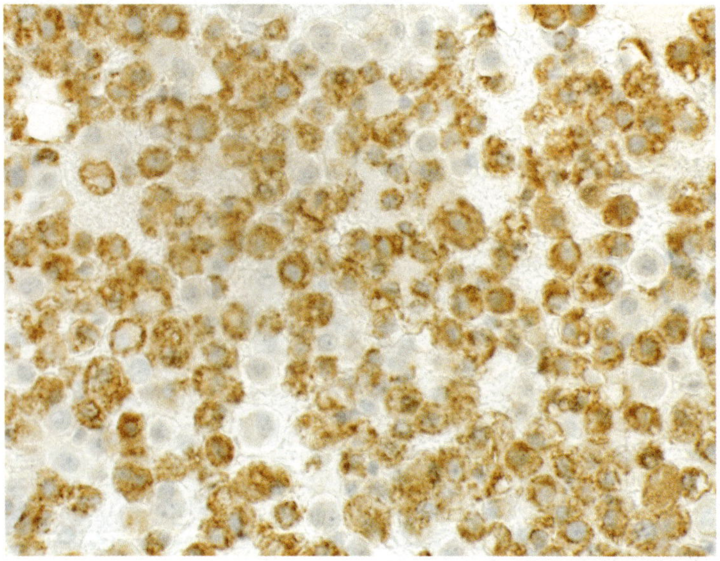

Figure 1.1c — Benign Ascites Fluid, Cell Block Section [CD68 immunohistochemical stain; medium power]. The histiocytes are diffusely immunoreactive for CD68, whereas the mesothelial cells are negative. Although primarily used as a histiocytic marker, CD68 can also be expressed in malignant melanoma, granular cell tumors, malignant fibrous histiocytoma, peripheral nerve sheath tumors, and rare carcinomas. CD163 is also a marker of monocyte and macrophages, with less nonspecific staining.

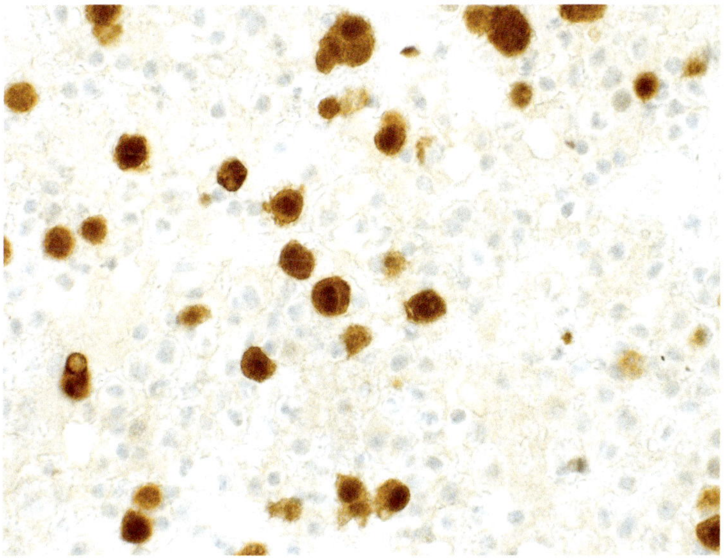

Figure 1.1d — Benign Ascites Fluid, Cell Block Section [Calretinin immunohistochemical stain; medium power]. A Calretinin immunostain is strongly and diffusely positive in the cytoplasm and nuclei of the mesothelial cells but is negative in background histiocytes and mononuclear cells. Calretinin is an intracytoplasmic Ca^{++}-binding protein typically expressed by mesothelial cells. Other positive mesothelial markers include WT-1, D2-40, and CK 5/6. Benign mesothelial cells may also stain with desmin, but are generally negative for epithelial membrane antigen (EMA).

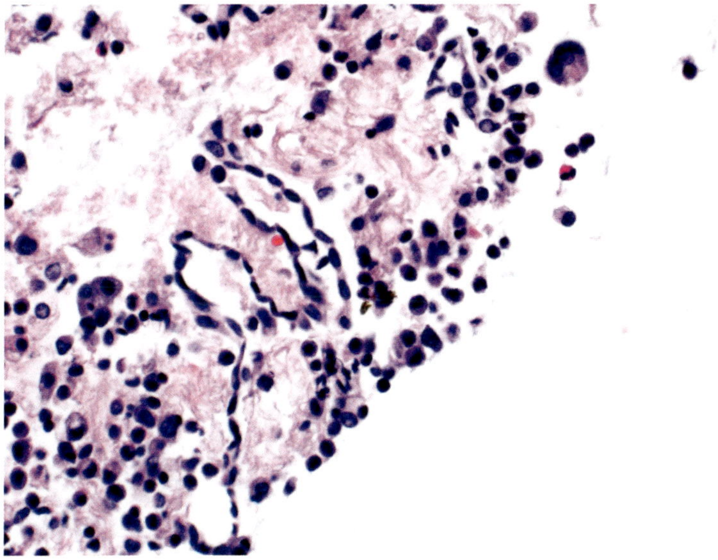

Figure 1.2a — Benign Ascites Fluid, Cell Block Section [H&E stain; medium power]. — Serous membranes consist of loose fibroconnective tissue lined by a monolayer of mesothelial cells (center), which readily exfoliate as thin strips and discrete cells, as seen here, or in small clusters. Mesothelial cells synthesize proinflammatory cytokines, growth factors, and extracellular matrix proteins to aid in serosal repair. They are also capable of phagocytosis, antigen presentation, and secrete plasminogen, which promotes clearance of fibrin deposits. When the mesothelium becomes damaged (e.g., during surgical procedures), fibrin accumulates, causing fibrous adhesions to form between opposing surfaces.

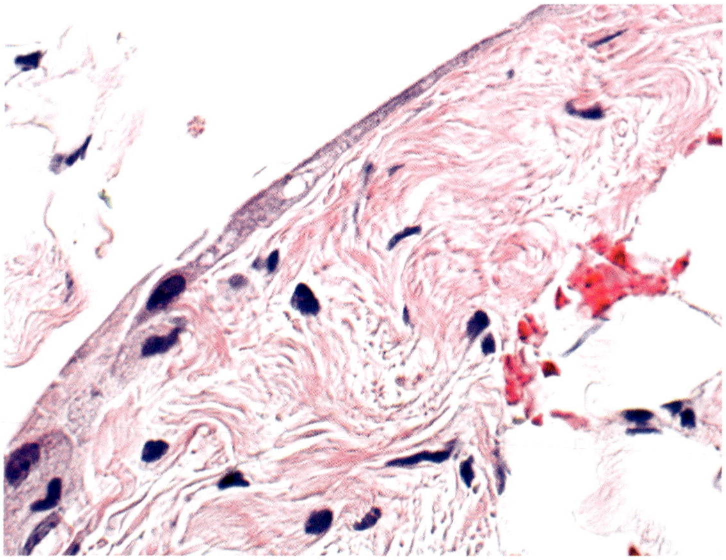

Figure 1.2b — Normal Mesothelial Lining, Biopsy [H&E stain; high power]. Mesothelium is a simple squamous epithelium that lines the body cavities (e.g., pleural, peritoneal, pericardial) and organs. In many surgical pathology biopsies and excisions, the normal mesothelial lining sloughs off during processing and is not seen histologically. When it is visible, the mesothelium lining, as seen in this field, consists of bland simple squamous cells with elongated, pale cytoplasm and bland nuclei. The function of the mesothelium is to provide a smooth surface lining to the body organs.

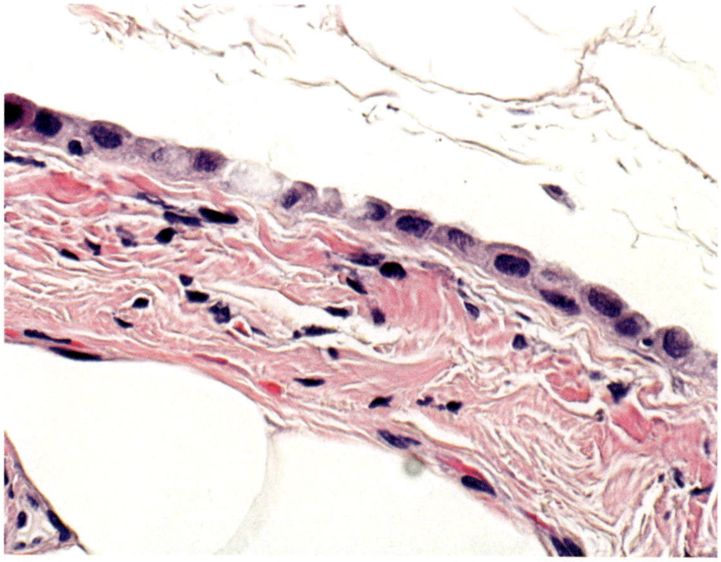

Figure 1.2c — Normal Mesothelial Lining, Biopsy [H&E stain; high power]. Under reactive conditions, the mesothelial cells can display a low cuboidal or hobnail appearance, as seen here. Reactive mesothelial hyperplasia can become quite florid and be difficult to distinguish from atypical or malignant mesothelial proliferations, particularly on the limited material of small biopsies. Mesotheliomas give rise to malignant mesotheliomas, which can occur in any mesothelial-lined site, including the pleura, the peritoneum, and the testicular tunica.

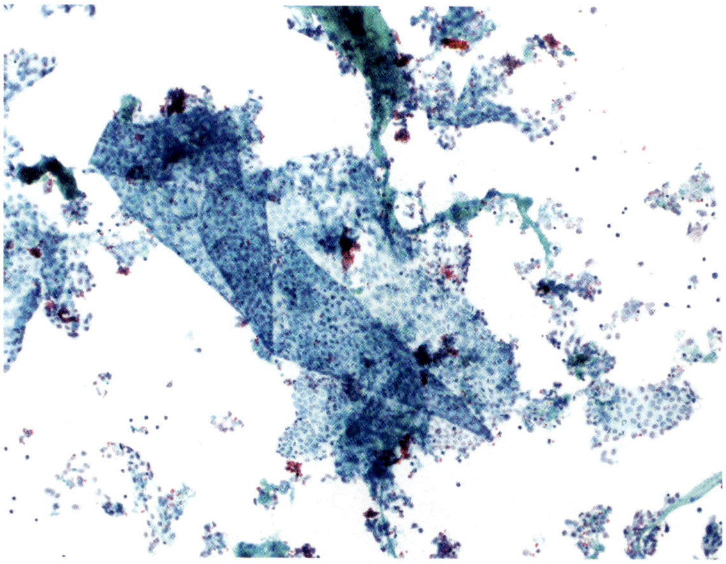

Figure 1.3a — Benign Mesothelial Cells, Pelvic Washing [Pap stain; low power]. It is normal to see mesothelial fragments in washing specimens, as the procedure disrupts the mesothelial lining, whereas nonwashing effusion specimens contain single mesothelial cells that have naturally exfoliated. Although the presence of mesothelial fragments in an effusion should raise some concern for a mesothelioma, they are more often seen secondary to inflammatory processes or benign proliferations. Furthermore, malignant mesothelioma would not have the bland cytomorphologic feature seen here or the evenly spaced cells organized within the fragments. Benign mesothelial sheets often appear in a monolayer that may fold, whereas mesothelial proliferations form three-dimensional structures and may have irregular branching architecture.

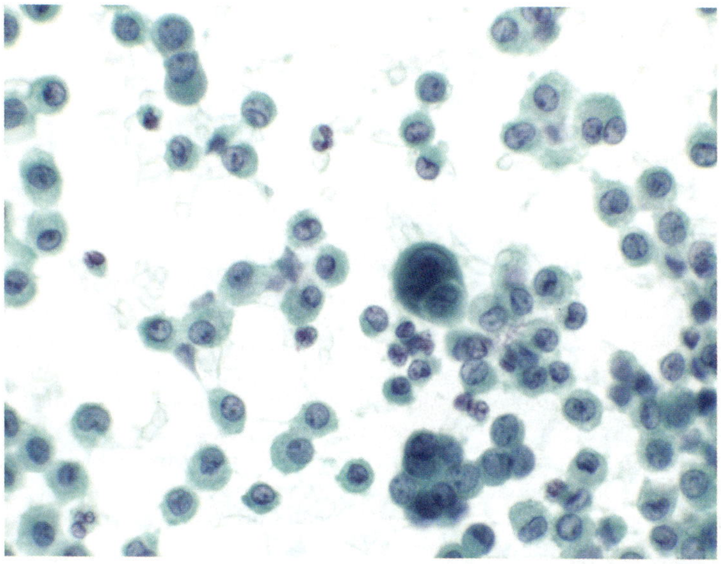

Figure 1.3b — Benign Mesothelial Cells, Pelvic Washing [Pap stain; high power]. Reactive mesothelial cells exhibiting occasional binucleation, cell-in-cell engulfment (center), and monolayer cell groups (bottom center) are common occurrences in serous effusions. Reactive mesothelial cell proliferations may, rarely, share some of the same morphologic features seen in malignant cells, usually in the presence of inflammation. The nuclei, however, are uniform and lack the degree of nuclear atypia characteristically seen in malignancies.

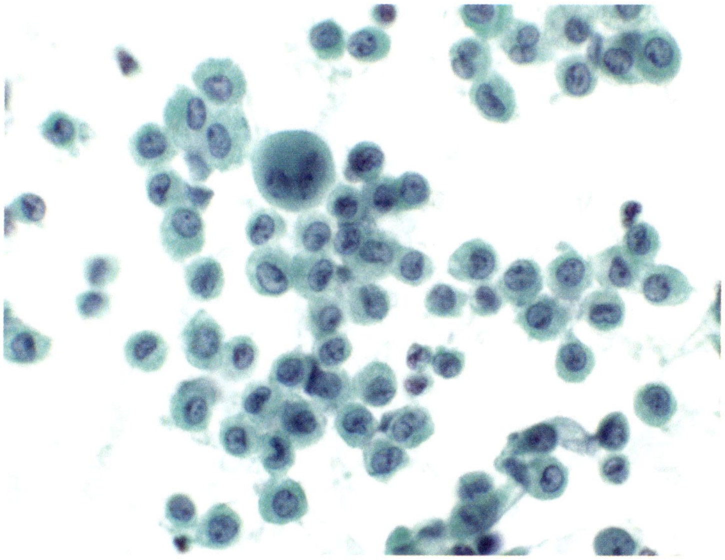

Figure 1.3c — Benign Mesothelial Cells, Pelvic Washing [Pap stain; high power]. Giant multinucleated mesothelial cells are not uncommonly seen in serous fluids but lack nuclear atypia and are usually accompanied by numerous smaller mesothelial cells, as seen in this micrograph. Unlike multinucleated histiocytes, they maintain smooth nuclear contours and well-defined membranes. Well-defined cytoplasmic vacuoles may be present, but the remaining cytoplasm should be dense with a fuzzy peripheral cell border.

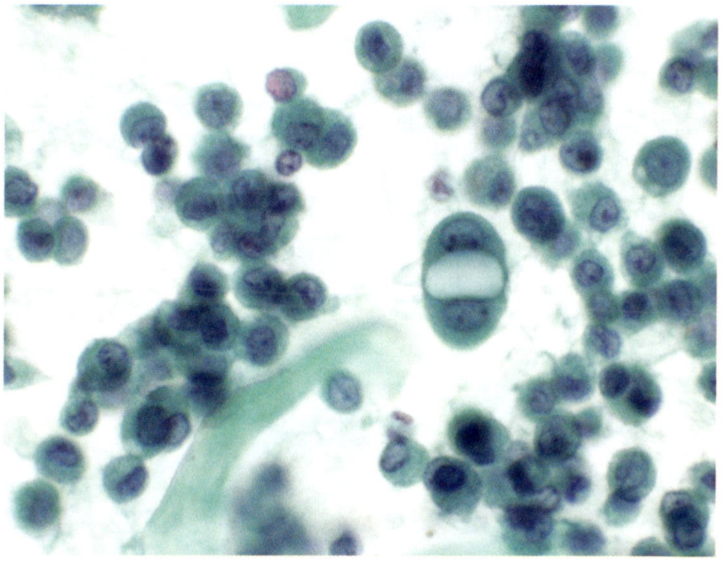

Figure 1.3d — Benign Mesothelial Cells, Pelvic Washing [Pap stain; high power]. Reactive mesothelial cells are frequently shed as doublets or triplets, separated by intercellular "windows" or clefts created by long and floppy microvilli between the opposing cellular surfaces. This field, however, depicts a mixture of reactive mesothelial cells with well-defined intracytoplasmic vacuoles. The binucleate cell in the central portion of the image contains a single intracytoplasmic vacuole that pushes but does not indent the nuclei.

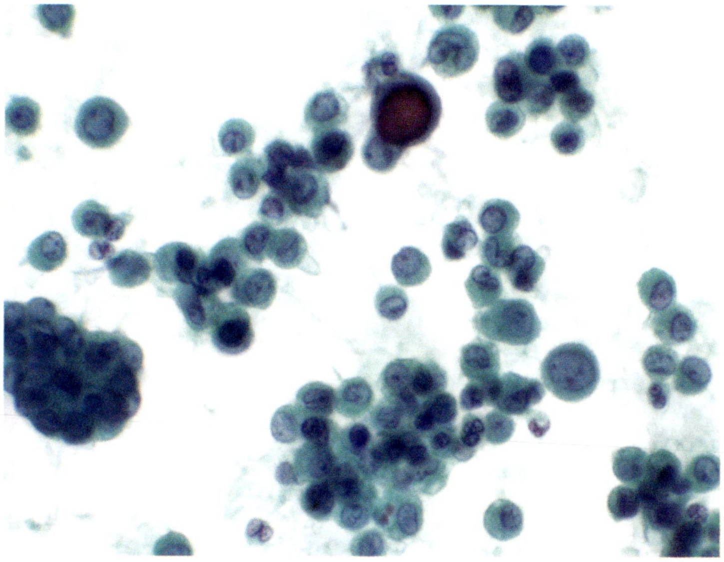

Figure 1.4a — Benign Mesothelial Cell Proliferation [Pap stain; high power]. Papillary fragments of mesothelium, consisting of a connective tissue core covered by a monolayer of mesothelial cells, can occasionally be seen in serous fluid, especially during a washing procedure. These tissue cores frequently infarct and calcify to form laminar concentric, acellular, basophilic concretions known as psammoma bodies. They are commonly seen in tumors such as ovarian serous cystadenomas, mesothelioma, and endometrial adenocarcinoma, but can also present in benign processes, including endosalpingiosis and benign mesothelial proliferations. To avoid a false-positive interpretation, a diagnosis of malignancy should not be based on the presence of psammoma bodies alone. Complex papillary fragments with broad stromal cores, particularly in the cell block, should raise suspicion of a neoplastic process.

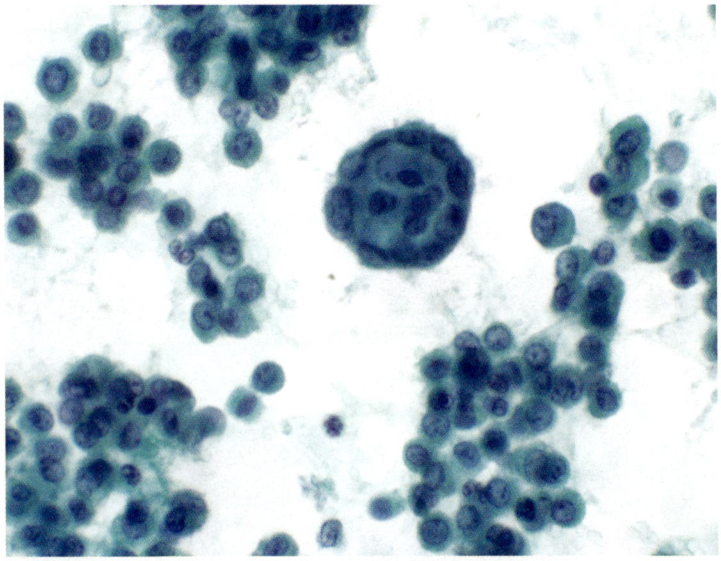

Figure 1.4b — Benign Mesothelial Cell Proliferation, Pelvic Washing [Pap stain; high power]. Reactive mesothelial proliferations occasionally form organized spherical cell balls with well-spaced cellular arrangements, as seen here. They are typically admixed with smaller clusters and single cells of mesothelium with similar uniform nuclear features. Unfortunately, these structures can also be seen in malignant mesothelioma and metastatic adenocarcinoma. In challenging cases of florid reactive mesothelial hyperplasia, a broad immunohistochemical panel may be necessary to exclude mesothelioma. Although no one immunostain is 100% sensitive or specific, loss of BAP1 staining (seen in up to 50% of nonsarcomatous mesotheliomas), strong membranous epithelial membrane antigen (EMA) staining, membranous glucose transporter 1 (GLUT1) expression, and cytoplasmic telomerase reverse transcriptase (TERT) expression would support a diagnosis of malignant mesothelioma.

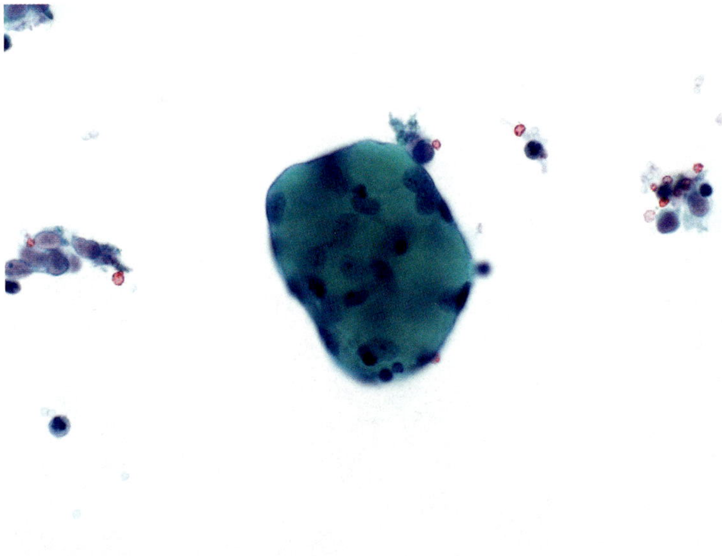

Figure 1.5a — Benign "Collagen Ball," Pelvic Washing [Pap stain; high power]. "Collagen balls" are spherical collections of green matrix material lined by attenuated bland mesothelial cells. They are typically seen in peritoneal washings and are thought to represent pinched off mesothelial-lined stromal projections from serosal surfaces. They are of no clinical significance and should not be misinterpreted as metastatic adenocarcinoma.

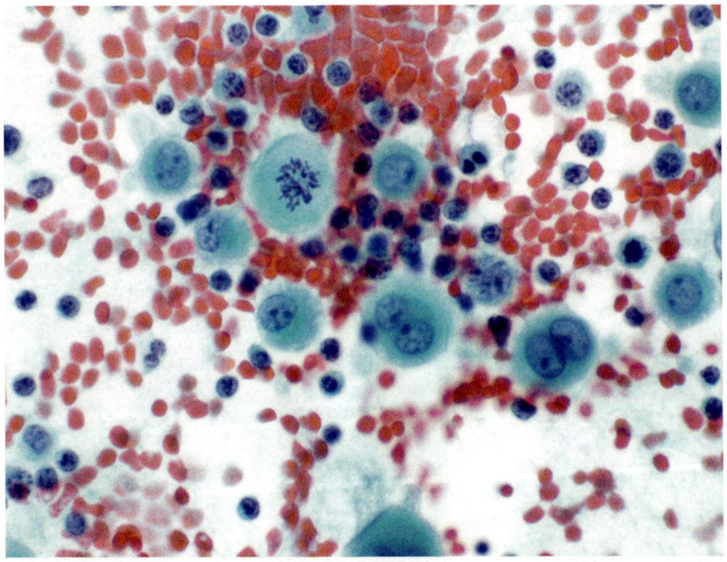

Figure 1.5b — Benign Mesothelial Cells, Paracentesis [Pap stain; high power]. Mitotic figures such as this are not an uncommon finding in benign serous cavity specimens. The cells seen with mitotic figures are reactive mesothelial cells. The presence of multiple mitotic figures may raise suspicion that a fluid specimen is involved by a malignancy and warrant additional workup using immunohistochemistry or examination of clinical data. Atypical mitotic figures—such as ring mitoses or aneuploidy mitotic figures—should not be seen in benign effusion specimens.

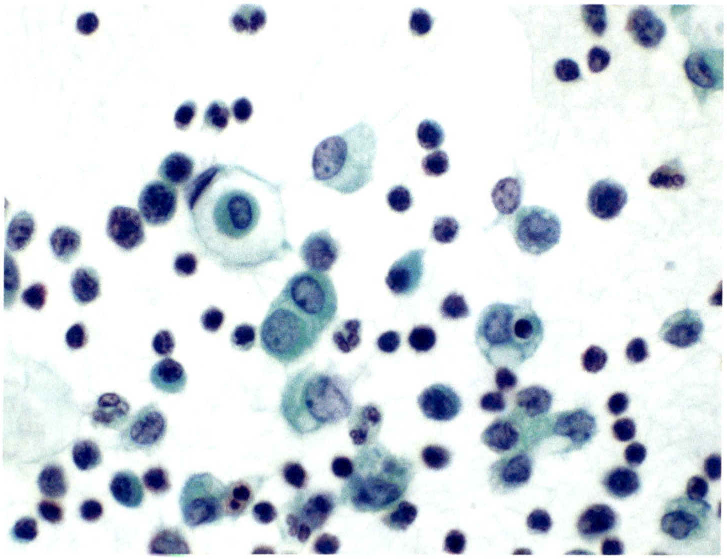

Figure 1.5c — Benign Mesothelial Cells, Paracentesis [Pap stain; high power]. This is a benign effusion with reactive mesothelial cells seen in the background of polymorphous lymphocytes. One mesothelial cell has become largely vacuolated, giving it a signet-ring-cell appearance. Further confounding the morphology is the presence of an engulfed cell within the vacuole. Although entosis, or cell cannibalism, is often seen in neoplastic processes, it is commonly demonstrated by benign mesothelial cells in serous cavity specimens.

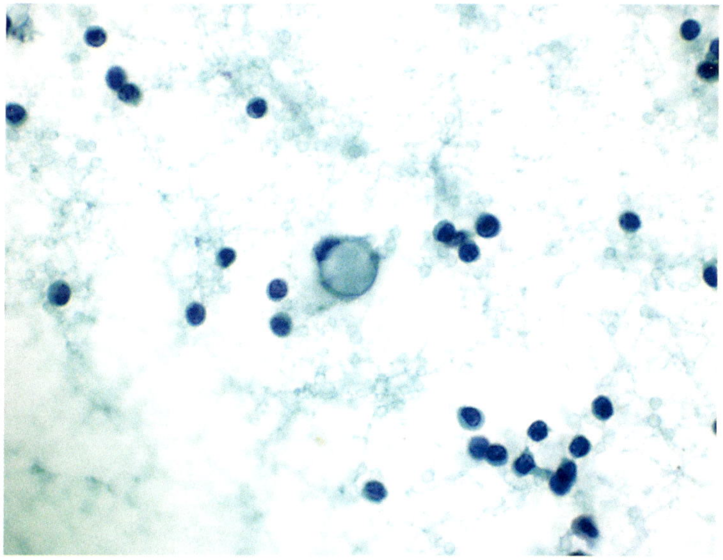

Figure 1.5d — Benign Mesothelial Cell, Paracentesis [Pap stain; high power]. A lone mesothelial cell is seen in the center of the field, surrounded by background polymorphous lymphocytes. The cell has abundant, vacuolated cytoplasm that gives the cell a signet-ring-cell appearance. This can raise concern for a metastatic adenocarcinoma, such as from the gastrointestinal tract. Rare vacuolated cells with nuclei of similar quality to those of nearby mesothelial cells can typically be dismissed. It can be difficult to distinguish benign vacuolization from mucin vacuoles.

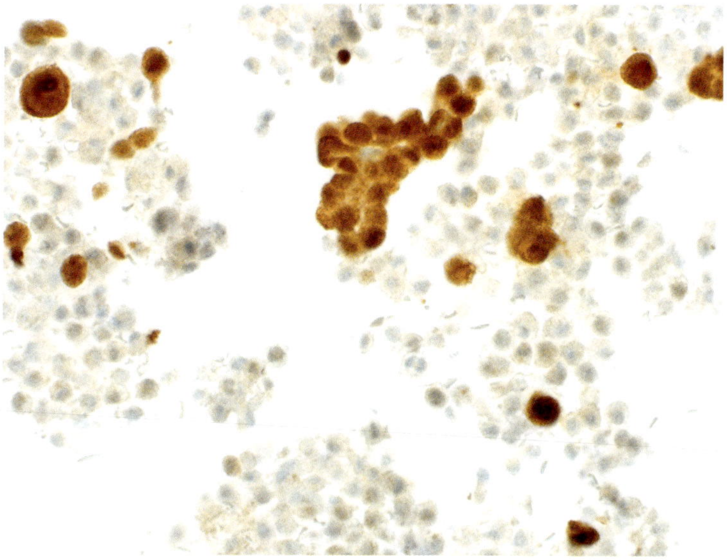

Figure 1.6a — Benign Mesothelial Cells, Ascites Cell Block Section [Calretinin immunohistochemical stain; high power]. Although calretinin is typically positive in mesothelial proliferations (intense nuclear and cytoplasmic/membranous immunoreactivity, as seen here) and negative in most cases of metastatic carcinomas, it should be interpreted only within the context of a larger immunohistochemical profile. Calretinin immunoreactivity has been reported in a subset of pulmonary adenocarcinomas, small cell carcinoma, squamous cell carcinoma, colorectal carcinoma, adrenocortical carcinoma, serous carcinoma, and endometrioid adenocarcinoma. Other tumors which may enter into the differential diagnosis that are also frequently calretinin positive include granulosa cell tumors, ovarian sex cord stromal tumors, extra-adrenal paragangliomas, and atypical mucinous neoplasms. Strong, diffuse staining is seen in almost all epithelioid mesotheliomas but in only 57% of sarcomatoid subtypes.

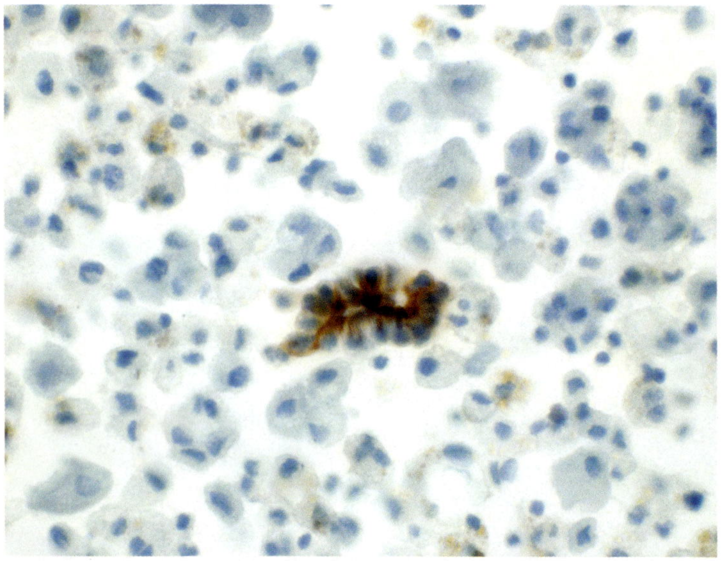

Figure 1.6b — Benign Mesothelial Cells, Ascites Cell Block Section [Ber-EP4 immunohistochemical stain; high power]. Ber-EP4 is an antibody against the epithelial cell adhesion molecule (EpCAM) that is expressed on the basolateral surface and in the cytoplasm of virtually all epithelial cells with the exception of most squamous epithelia, hepatocytes, renal proximal tubular cells, gastric parietal cells, and myoepithelial cells. Normal mesothelial cells are typically EpCAM negative but may show focal immunoreactivity, as shown here, when undergoing reactive changes.

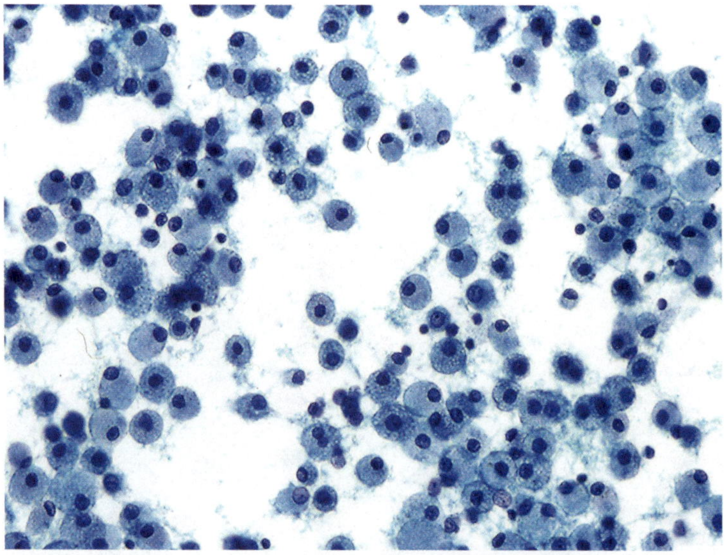

Figure 1.7a — Chylous Effusion, Pleural Fluid [Pap stain; medium power]. Chylous effusions are characteristically composed of numerous large, vacuolated histiocytes containing chylomicrons. Knowledge of the appearance of the fluid (turbid, milky) is also helpful. Benign causes of chylothorax may include tuberculous pleuritis, sarcoidosis, cirrhosis, lupus, venous thrombosis, mitral stenosis, heart failure, nephrosis, and protein-losing enteropathy.

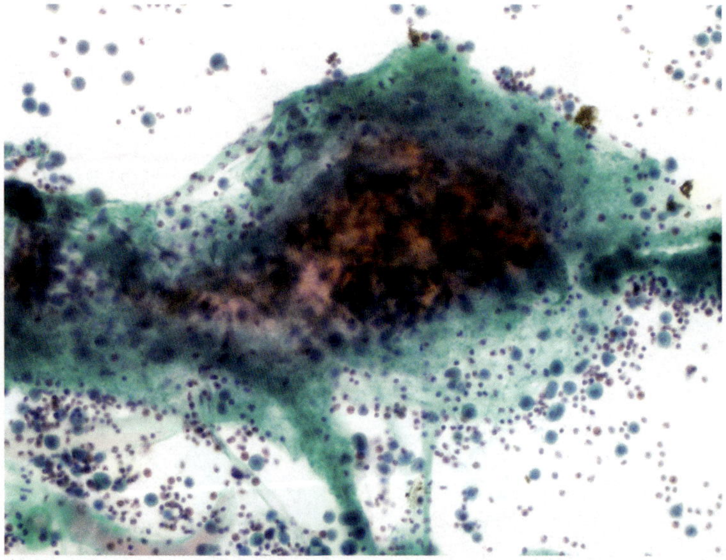

Figure 1.7b — Fibrinous Pericarditis, Pericardial Effusion [Pap stain; low power]. Fibrinous pericarditis is an exudative inflammatory process in which the epicardium contains acute inflammatory cells and a fibrinous exudate. Here, numerous inflammatory cells can be seen in the background, but the large collection of fibrinous material in the center of the field is an important finding and should be reported. Although cytology is predominantly the study of cells, any background material seen in cytology specimens can provide clues and, in some instances, are they key to a diagnosis (e.g., colloid, amyloid, matrix material).

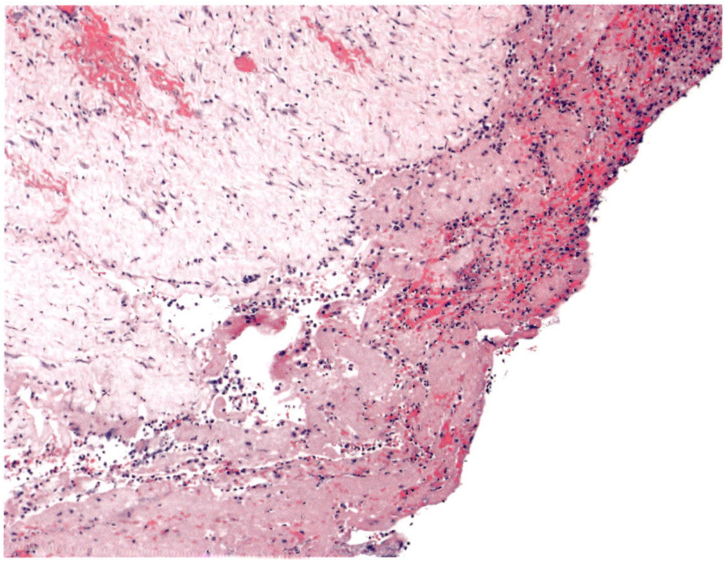

Figure 1.7c — Fibrinous Pericarditis, Pericardial Biopsy [H&E stain; low power]. The corresponding surgical specimen shows a layer of fibrin (red amorphous material) deposited on the pericardium. Numerous inflammatory cells are present within the fibrinous deposit, which provides an excellent morphologic correlate to the cytology specimen. Uremia may be one cause of fibrinous pericarditis; a pericardial friction rub is the classic clinical finding.

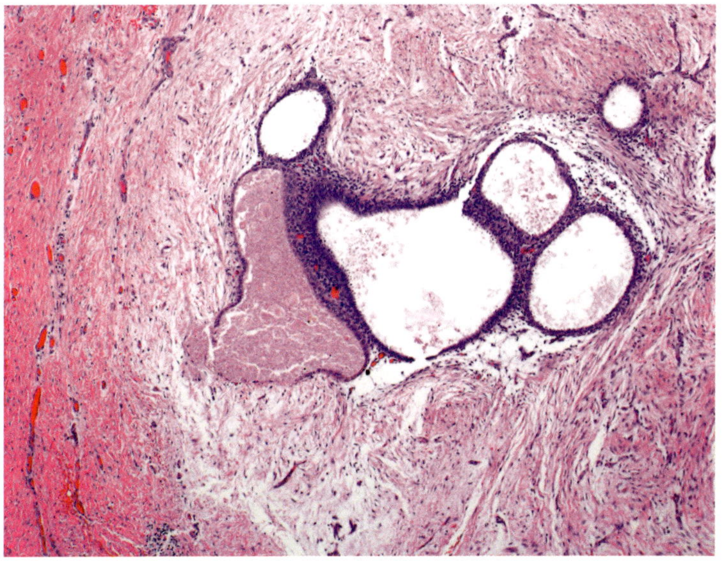

Figure 1.8a — Peritoneal Endometriosis, Biopsy [H&E stain; low power]. This abdominal wall excisional biopsy reveals numerous clusters of rounded glands within fibrous tissue and smooth muscle. The differential diagnosis of glands in the abdominal wall always includes metastatic adenocarcinoma, commonly from the gastrointestinal or gynecologic tract. However, the differential diagnosis should also include benign processes such as inclusion cysts, enteric duplication cysts, and Mullerianosis, including endometriosis, endosalpingiosis, and endocervicosis. Endometriosis can be an incidental finding or may present with associated symptoms such as pelvic pain or intestinal obstruction.

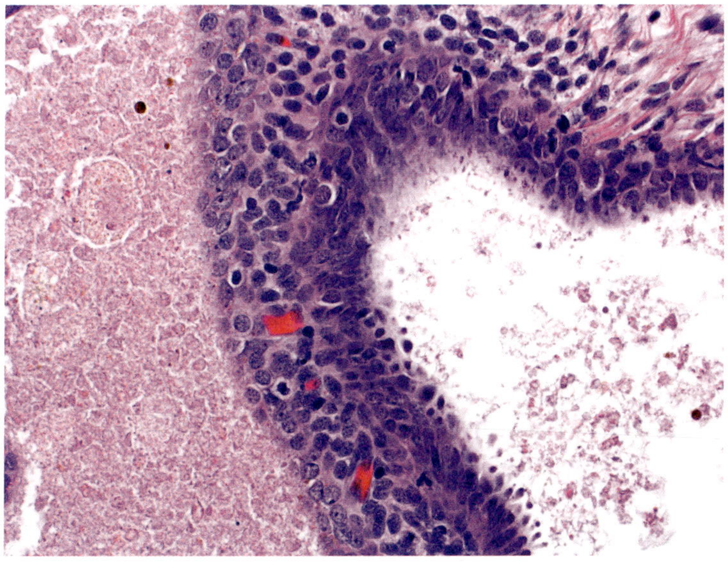

Figure 1.8b — Peritoneal Endometriosis, Biopsy [H&E stain; intermediate power]. High-power examination of these glands reveals a bland glandular lining with ciliated cells, an associated ovarian-type stroma, hemosiderin-laden macrophages, and hemosiderin-rich fluid. These are characteristic features of endometriosis. The presence of ciliated glands and the ovarian-type stroma distinguishes endometriosis from adenocarcinoma. The glandular epithelium is immunoreactive for estrogen receptor and negative for CDX2, and the stroma is reactive for CD10.

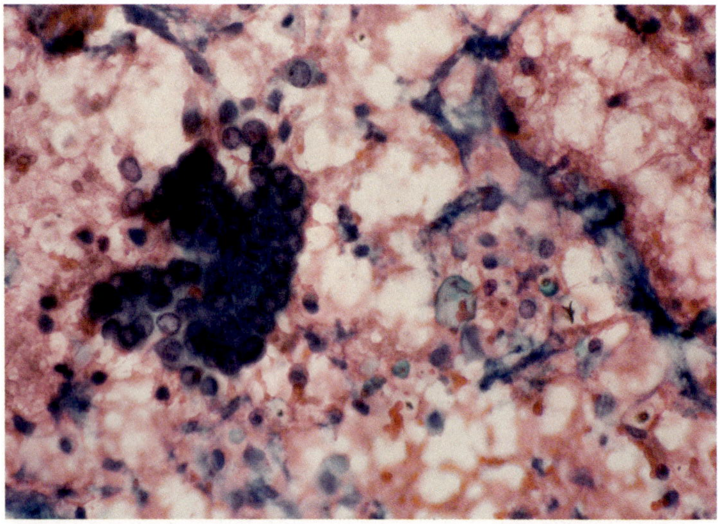

Figure 1.8c — Endometriosis, Pelvic Washing [Pap stain; high power]. Endometriosis is the ectopic growth of endometrial tissue (glands and stroma) that reacts to hormonal stimulation during menstruation. Normal endometrial cells tend to be arranged in sheets, tight balls with central stromal cells, or branching tubules, as shown here. The cells are ovoid to columnar with eccentrically placed, uniformly round nuclei, and regularly distributed chromatin. Endometrial stromal cells are usually not identified but, when present, have a histiocytic or spindle cell appearance. A definitive diagnosis requires the presence of endometrial epithelial cells and/or hemosiderin-laden macrophages. Stromal cells are rarely, if ever, seen.

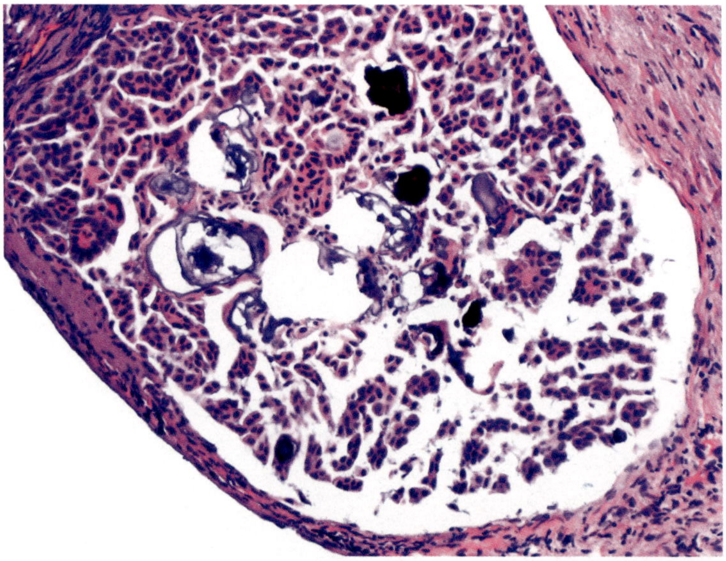

Figure 1.9a — Florid Endosalpingiosis, Abdominopelvic Washing Cell Block Section [H&E stain; medium power]. Endosalpingiosis can display variable degrees of micropapillary architecture, often within mesothelial or epithelial-lined cystic spaces surrounded by fibrous stroma. This is contrast to atypical proliferative serous (borderline) tumors (APSTs/SBTs), which typically have thicker tumor epithelium, fusion of adjacent papillae, and confluent micropapillary areas greater than 5 mm in diameter. However, in cases of limited tissue sampling, florid endosalpingiosis may be indistinguishable from noninvasive epithelial implants of APSTs. Thus a conservative diagnosis of low-grade serous proliferation may be warranted until more extensive tissue sampling is obtained to avoid inappropriate upstaging of patients with respect to risk for advanced stage disease and invasive implants.

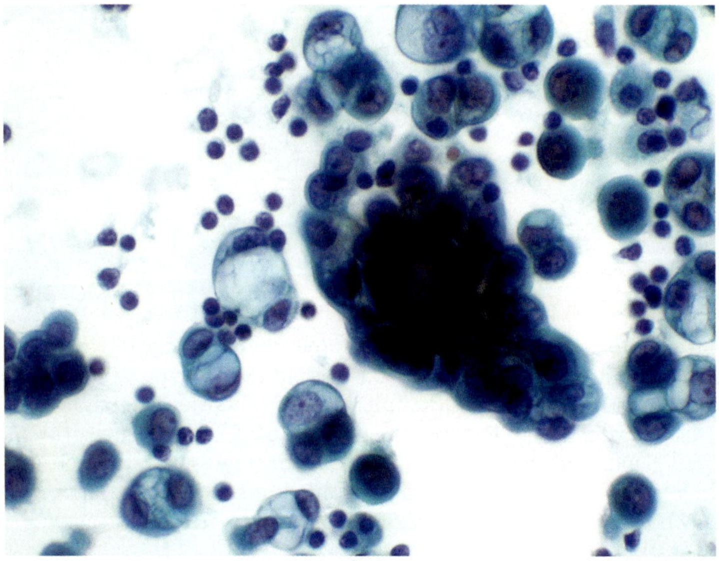

Figure 1.9b — Florid Endosalpingiosis, Abdominopelvic Washing [Pap stain; high power]. Psammoma bodies can be present in the center of small papillae, as shown here, or “naked” without associated epithelium. Naked psammoma bodies, which form secondary to degeneration of the overlying epithelium, often act as an irritant in the peritoneal cavity, inducing fibrosis that can lead to adhesions and bowel obstruction. Most cases of endosalpingiosis are incidental findings without associated clinical symptoms. Chronic inflammation, as seen in this field, is a frequent finding in patients with previous pelvic inflammatory disease and endosalpingiosis. In fact, it is well known that inflammation can stimulate the proliferation of tubal epithelium and may play a role in the pathogenesis of papillary tubal hyperplasia and endosalpingiosis.

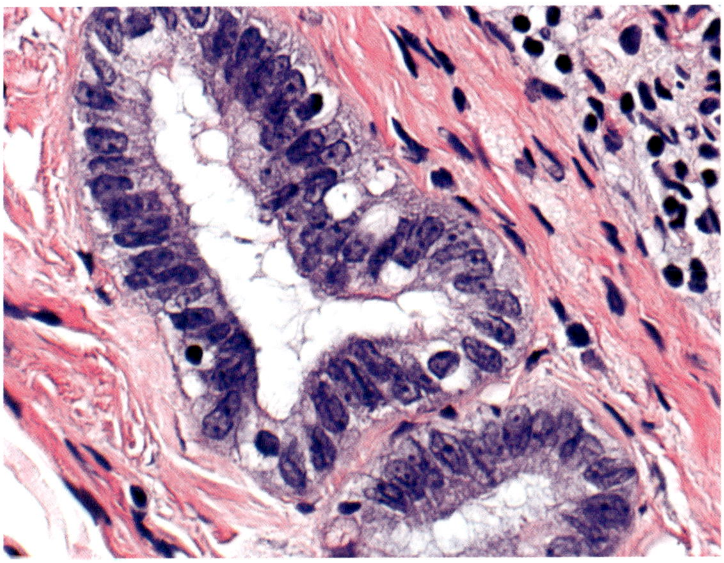

Figure 1.10a — Nodal Endosalpingiosis, Excisional Biopsy [H&E stain; high power]. This pelvic lymph node contains small glands in the lymph node capsule. The glands are characterized by bland columnar cells with cilia and scattered smaller cells with round nuclei and clear cytoplasm (i.e., "intercalated" or "peg" cells). The histology of these glands resembles the lining of the normal fallopian tube and is thus consistent with endosalpingiosis, which can involve the capsule or the parenchyma of lymph nodes, as well as the pelvic wall, abdominal wall, or organs. Supradiaphragmatic endosalpingiosis has also been reported, including in axillary lymph nodes. Endosalpingiosis in lymph nodes mimics metastatic well-differentiated adenocarcinoma; however, the presence of ciliated cells and intercalated cells are histologic clues that support classification as benign glands.

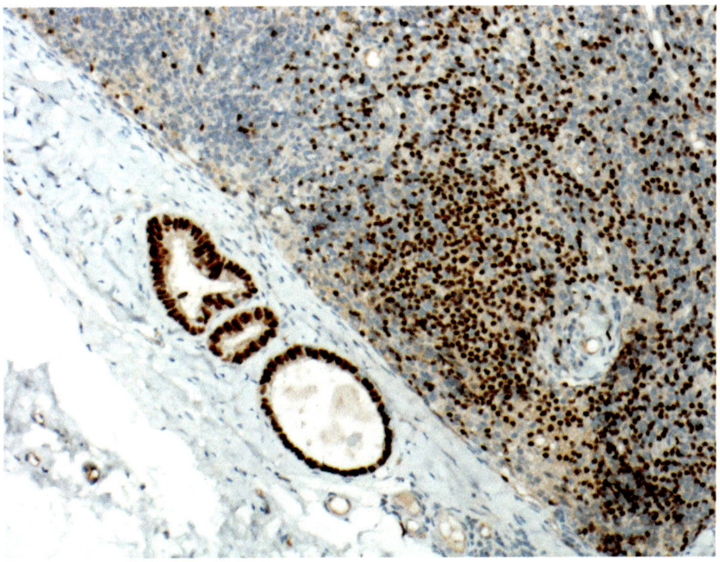

Figure 1.10b — Nodal Endosalpingiosis, Excisional Biopsy [PAX8 immunohistochemical stain; intermediate power]. The glands of endosalpingiosis label by immunohistochemistry like benign fallopian tube epithelium. The glands are immunoreactive for PAX8 (seen here), as well as WT-1 and estrogen receptor (ER). They lack myoepithelial cells and are negative for the intestinal marker CDX2; the breast markers gross cystic disease fluid protein (GCDFP), mammaglobin, and GATA-binding protein 3 (GATA3); and the lung marker thyroid transcription factor-1 (TTF-1). A targeted immunopanel can help support the diagnosis of endosalpingiosis if the histologic features are ambiguous. Furthermore, the capsular location favors a benign inclusion, rather than a metastatic adenocarcinoma, which preferentially involves the sinus.

2

Neoplastic Serous Cavity Effusions and Washings

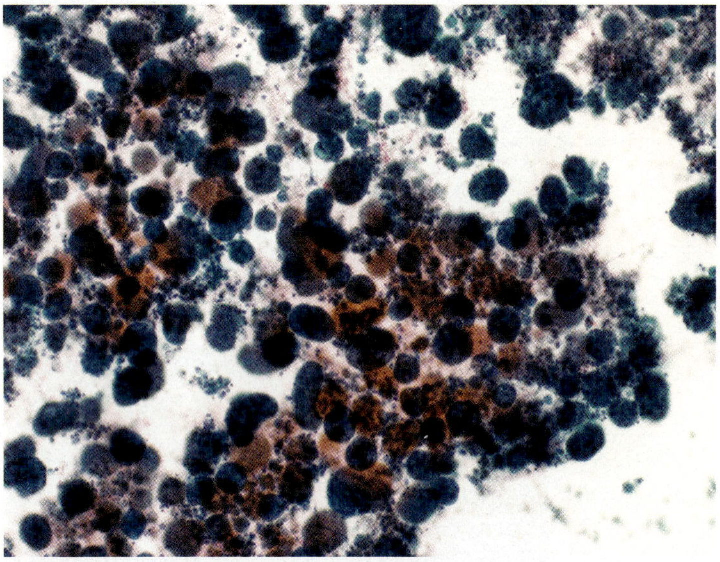

Figure 2.1a — Mesothelioma, Pleural Fluid [Pap stain; low power]. Numerous three-dimensional spherical cell fragments are dispersed throughout the field, giving the appearance of cannonballs. The fragments appear to have smooth edges, and the overall picture is suggestive of adenocarcinoma. However, this patient had mesothelioma. Mesothelioma seen in fluid specimens can have diverse morphologies and may mimic adenocarcinoma. Clinicoradiologic impression, as well as an initial immunohistochemical screening, is useful.

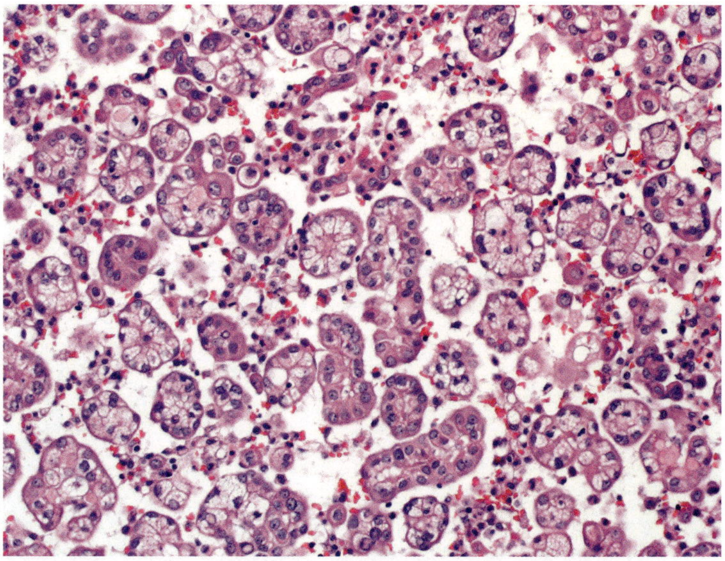

Figure 2.1b — Mesothelioma, Pleural Fluid Cell Block Section [H&E stain; medium power]. The cells here seem to form gland-like spaces and have foamy or vacuolated cytoplasm. Both of these features are suggestive of an adenocarcinoma rather than a mesothelioma. A few cells have abundant, pink cytoplasm, a finding associated with mesothelioma; however, cell block preparations are often unreliable because of artifacts in the preparation. Immunohistochemical studies on cell block material usually begin with epithelial markers (Ber-EP4 or MOC-31) and a mesothelial marker (calretinin).

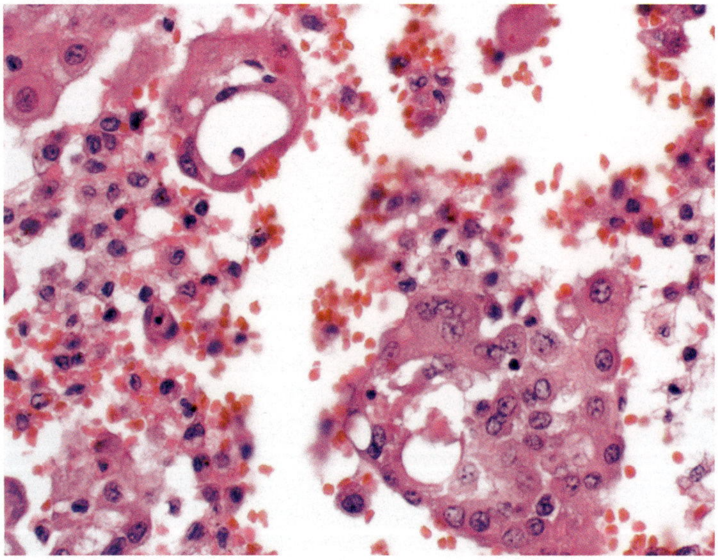

Figure 2.1c — Mesothelioma, Pleural Fluid Cell Block Section [H&E stain; high power]. This patient's mesothelioma involved the pleural fluid. On the cell block preparation, the cells have abundant pink cytoplasm. Despite this, the nuclei are pleomorphic and have irregular borders and coarse chromatin. Numerous gland-like spaces appear to be present, which may give the impression that this is an adenocarcinoma. In this case, the empty spaces appear to be large vacuoles within some mesothelioma cells. Although mesothelioma is rare, it should be on the differential of "large pink cells."

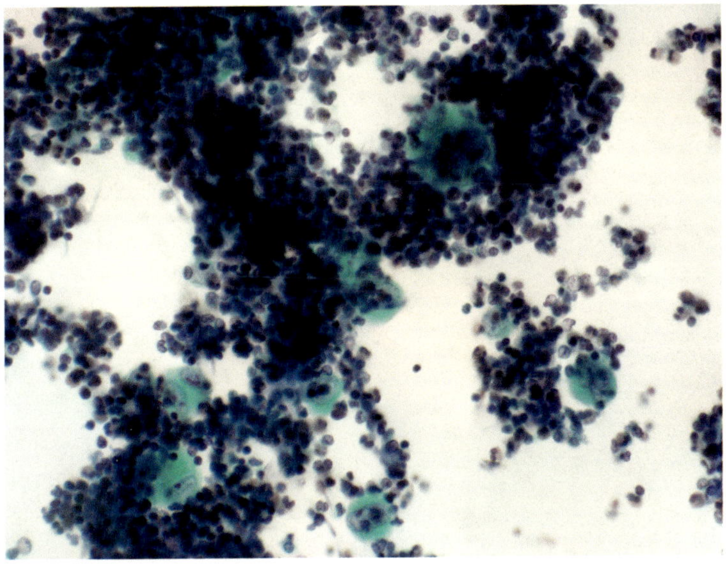

Figure 2.2a — Mesothelioma, Pleural Fluid [Pap stain; medium power]. The specimen is very cellular, and numerous inflammatory cells appear in the background. The presence of numerous giant cells is a startling finding. These cells almost appear as normal mesothelial cells, except they are many times larger than the surrounding normal cells. Some have binucleated forms and most have abundant cytoplasm. The malignant cell in the top right-hand field is multinucleated, has a high nuclear-to-cytoplasmic ratio, and has irregular nuclear borders. These singly dispersed cells could represent a poorly differentiated carcinoma or even a sarcoma; in this case, the patient had a mesothelioma.

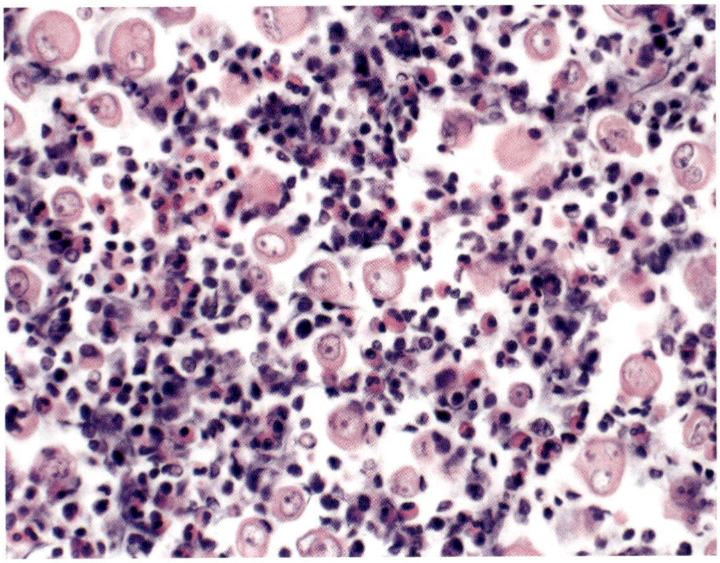

Figure 2.2b — Mesothelioma, Pleural Fluid Cell Block Section [H&E stain; high power]. Note the pink cytoplasm possessed by the mesothelioma cells. The cells are much larger than the background cells, some of which are likely benign reactive mesothelial cells. The mesothelioma cells have chromatin at the nuclear rim and in coarse granules, leaving much of their large nuclei empty and pale. In some instances, it may be difficult to distinguish whether a mesothelial cell proliferation is benign or malignant on cytology; in this specimen, the cells are unquestionably malignant and their origin is less certain based on cytomorphology alone.

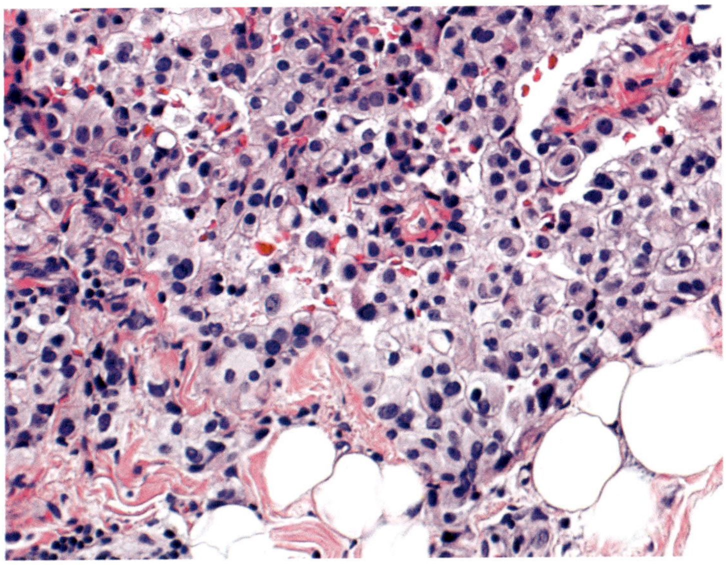

Figure 2.3a — Malignant Mesothelioma, Biopsy [H&E stain; intermediate power]. Malignant mesothelioma is a neoplastic proliferation of mesothelial cells and can arise at any site containing mesothelial lining. In the pleura, as seen here, asbestos exposure is a major risk factor for the development of mesothelioma. It can be very difficult to distinguish between reactive mesothelial hyperplasias and malignant mesotheliomas, particularly on superficial or limited samples in biopsy material. The presence of infiltration into fat, as seen here, is helpful in supporting the diagnosis of malignancy.

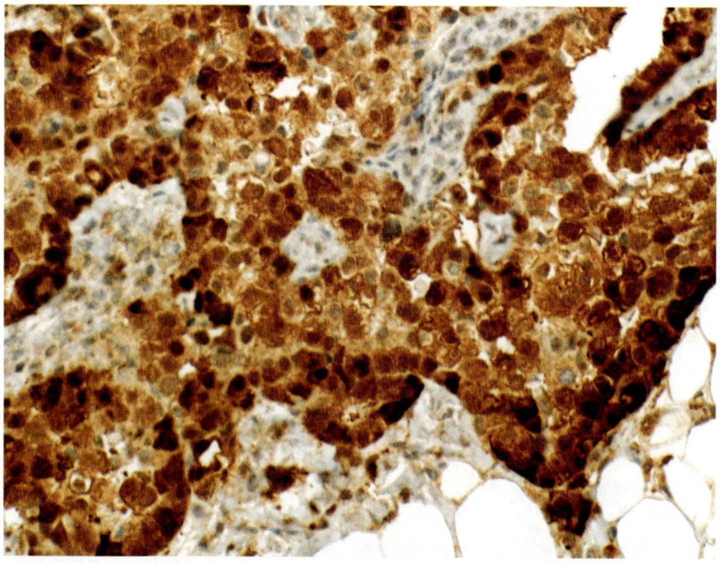

Figure 2.3b — Malignant Mesothelioma, Biopsy [Calretinin immunohistochemical stain; intermediate power]. Mesothelial cells and mesotheliomas are immunoreactive for calretinin, as illustrated here. Other immunostains that demonstrate positivity in mesothelial cells and mesotheliomas are cytokeratin 5/6 (CK5/6), WT-1, and D2-40. The differential diagnosis of pleural mesothelioma includes pulmonary non–small cell carcinoma. Mesotheliomas are typically negative for Ber-EP4, thyroid transcription factor-1 (TTF-1), and Napsin A, whereas pulmonary adenocarcinomas are typically positive. Mesotheliomas are typically negative for p40, whereas pulmonary squamous cell carcinomas are typically positive.

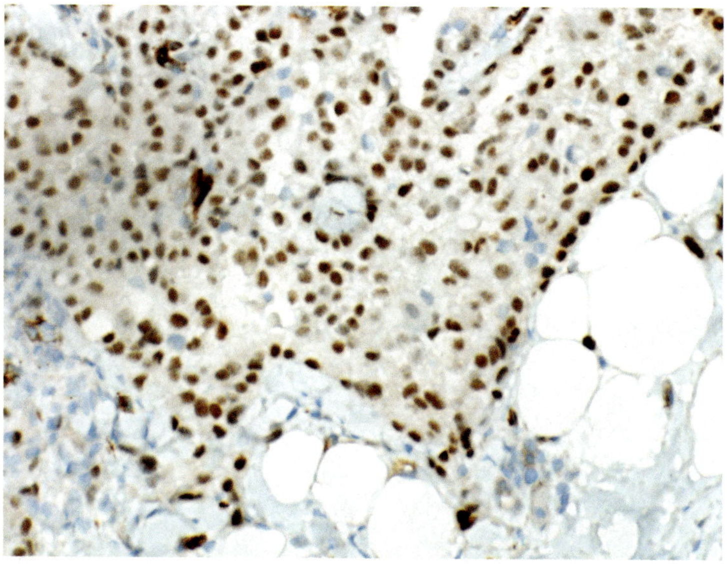

Figure 2.3c — Malignant Mesothelioma, Biopsy [WT-1 immunohistochemical stain; intermediate power]. This solid epithelial proliferation shows diffuse nuclear WT-1 labeling, supporting the diagnosis as a mesothelial proliferation. WT-1 labeling can be seen in both benign and malignant mesothelial proliferations, so the presence of other features (atypia, mitoses, and infiltration) must be used to support a diagnosis of mesothelioma. WT-1 immunoreactivity is also seen in gynecologic tract malignancies, such as high-grade serous carcinoma, such that WT-1 labeling can be a diagnostic pitfall in peritoneal papillary proliferations. Other immunostains that are positive in high-grade serous carcinomas are estrogen receptor (ER) and PAX8. Although PAX8 labeling is not seen in pleural mesotheliomas, focal labeling has been reported in peritoneal mesotheliomas.

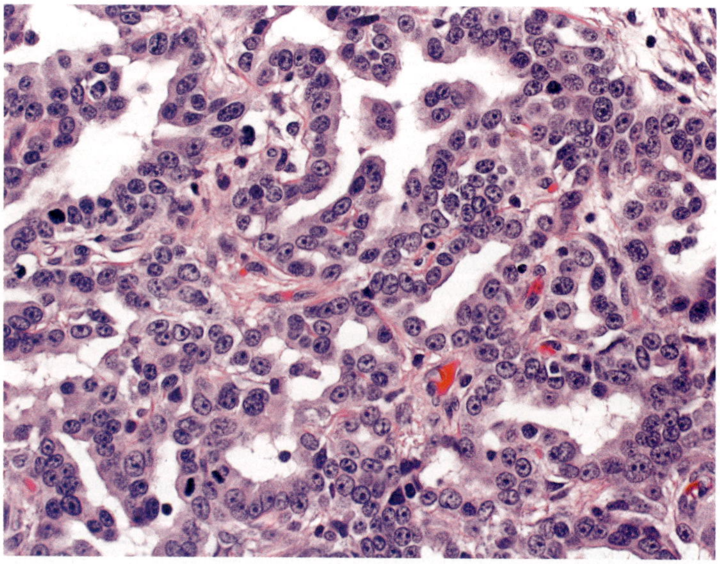

Figure 2.4a — Malignant Mesothelioma, Biopsy [H&E stain; intermediate power]. Mesotheliomas can display a wide range of morphologic appearances and are broadly separated into epithelioid, sarcomatoid/spindled, and mixed variants. Sarcomatoid, desmoplastic mesotheliomas can be difficult to characterize as they may show limited labeling for mesothelial markers. In this epithelioid variant seen here, the cells are low cuboidal and hobnailed and form gland-like and micropapillary structures. The morphology mimics a gland-forming adenocarcinoma. Of note, Mucicarmine reactivity should not be used to support a diagnosis of adenocarcinoma in this setting because Mucicarmine can label the hyaluronic acid produced by mesotheliomas.

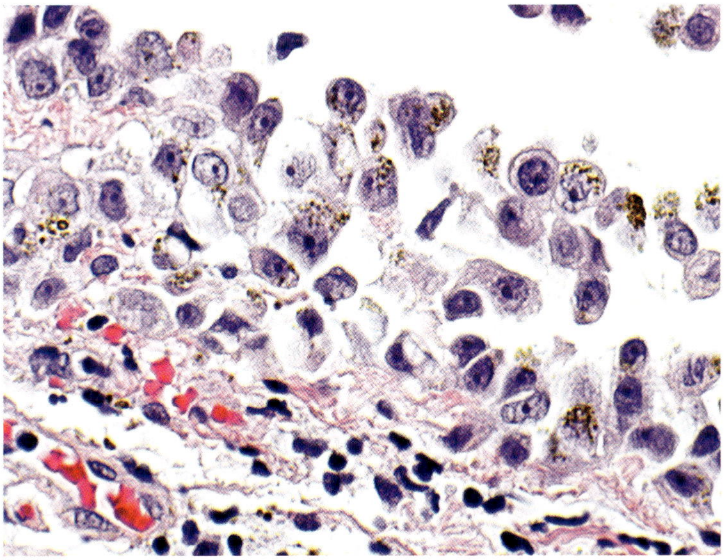

Figure 2.4b — Malignant Mesothelioma, Biopsy [H&E; high power]. This epithelioid malignant mesothelioma displays hobnail growth of markedly atypical cells with a high nuclear-to-cytoplasmic ratio and prominent nucleoli. Elsewhere, invasive growth is seen. There is extensive hemorrhage throughout the specimen, and the mesothelioma cells contain hemosiderin. This is a potential diagnostic pitfall with melanin-containing malignant melanoma. Evaluation of the entire specimen, use of targeted immunostains, and knowledge of clinical history can aid in reaching the correct diagnosis.

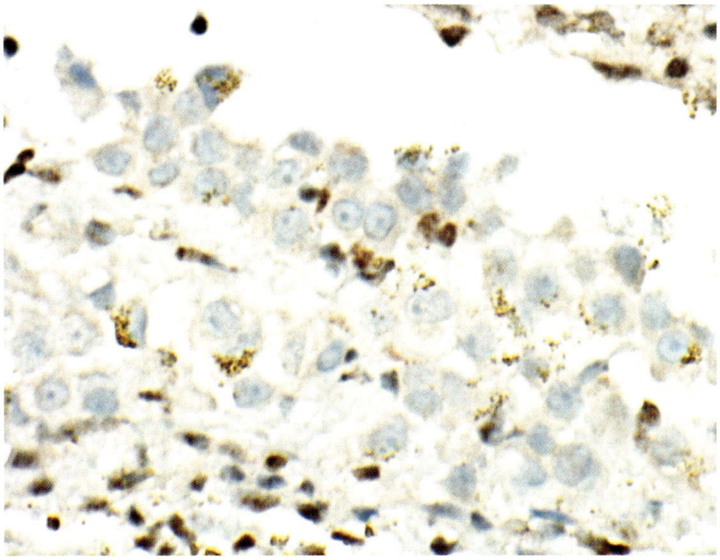

Figure 2.4c — Malignant Mesothelioma, Biopsy [BAP1 immunohistochemical stain; high power]. The majority of the immunostains that label malignant mesotheliomas also label benign mesothelial proliferations. However, loss of nuclear BAP1 labeling, as seen here, can be useful in distinguishing malignant mesothelioma from reactive mesothelial proliferations. Germline mutations in the *BAP1* (BRCA1-associated protein 1) gene are known to result in increased risk of melanoma and mesothelioma, and the BAP1 immunostain can identify tumors with homozygous loss of *BAP1*. Loss of BAP1 labeling is highly specific for pleural mesothelioma and is not seen in benign reactive mesothelial proliferations or in lung adenocarcinoma. Intact BAP1 labeling, however, does not exclude the diagnosis of malignancy. Intact BAP1 labeling in lymphocytes, adipocytes, or mesothelial cells should serve as a positive internal control.

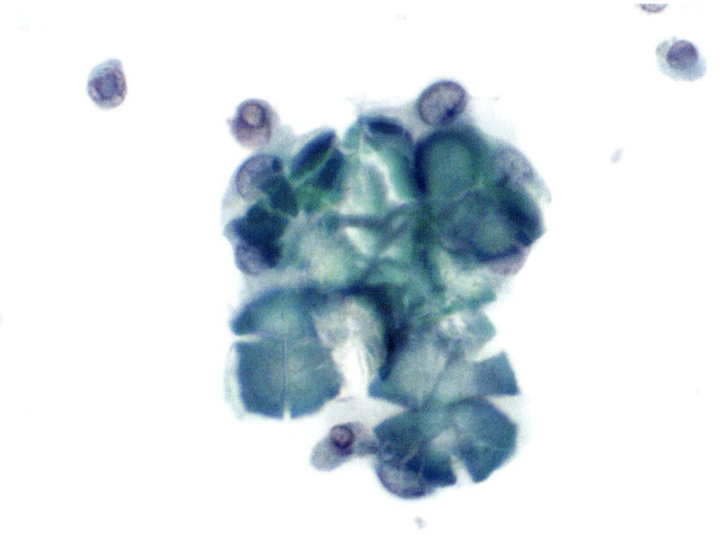

Figure 2.5a — Psammomatous Calcification Associated With Ovarian Cystadenoma, Pelvic Washing [Pap stain; high power]. Psammoma bodies can be seen in both benign and malignant processes occurring in the peritoneal cavity. In this case, the patient had surgery to remove an ovarian cystadenoma. Here, the calcified material appears to be cracking and is surrounded by bland-appearing epithelioid cells.

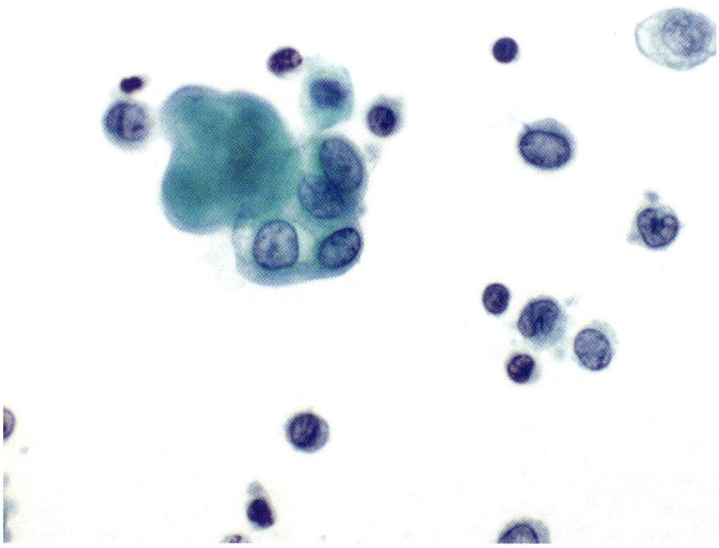

Figure 2.5b — Psammomatous Calcification Associated With Ovarian Cystadenoma, Pelvic Washing [Pap stain; high power]. This field contains psammomatous calcification, which appears as amorphous cyan material with concentric layering that is most apparent at the outmost rim. The psammoma body is loosely associated with bland-appearing epithelial cells, which have regular nuclear borders, pale chromatin, and abundant cytoplasm. Some benign processes associated with psammoma bodies include endosalpingiosis, ovarian cystadenoma and cystadenofibroma, endometriosis, and papillary mesothelial hyperplasia.

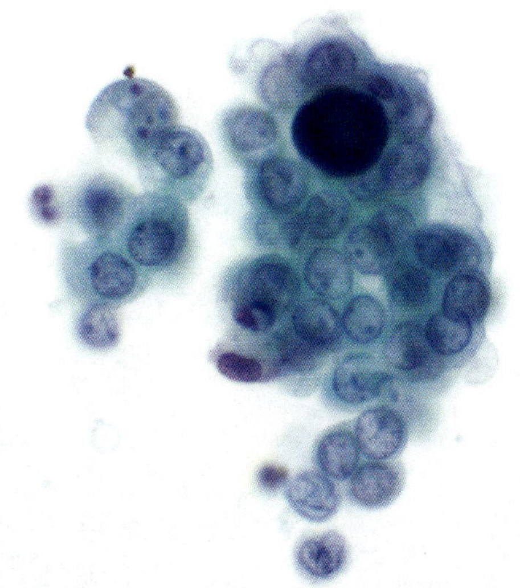

Figure 2.5c — Psammomatous Calcification Associated With Ovarian Cystadenoma, Pelvic Washing [Pap stain; high power]. Here is the classic appearance of a psammoma body—a round concretion of acellular material concentrically layered. The body is contained within a papillary-like fragment composed of monotonous epithelial cells. The epithelial cells have round and regular borders, a single small nucleolus, and a powdery chromatin pattern.

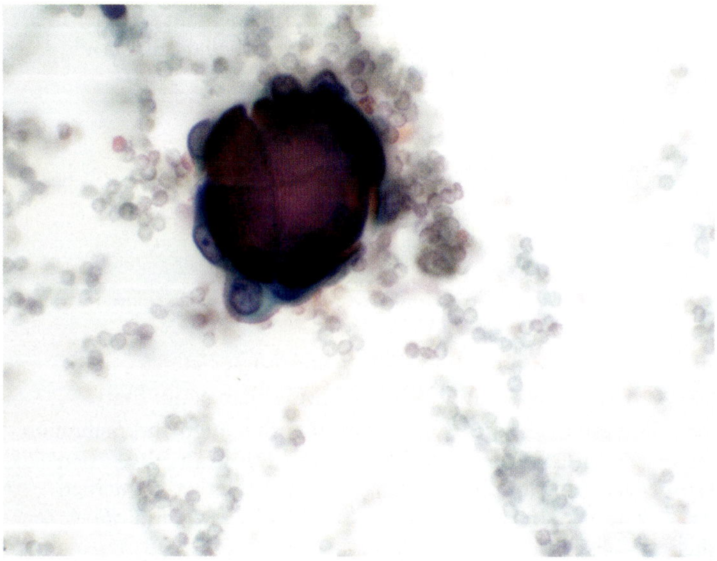

Figure 2.6a — Ovarian Atypical Proliferative Serous Tumor, Pelvic Washing [Pap stain; high power]. This patient had a pelvic washing during surgery for resection of an atypical proliferative serous tumor of the ovary. Implants may occur on the peritoneal cavity, and these implants may be invasive or noninvasive. Because invasion cannot be determined on a cytology specimen, this specimen can be noted as "involved by" the patient's ovarian tumor. In instances of ovarian lesions, one should not assume that the presence of tumor in a peritoneal specimen necessarily indicates a metastatic disease. One should take care in choosing proper wording of the diagnosis, as it may affect a patient's prognosis and/or treatment. Here, bland epithelial cells surround a psammoma body with "cracking artifact."

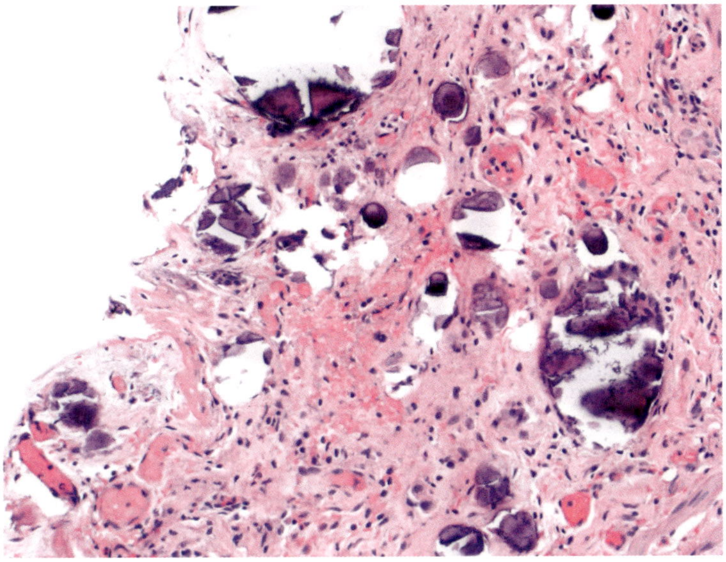

Figure 2.6b — Ovarian Atypical Proliferative Serous Tumor, Peritoneal Implant Biopsy [H&E stain; medium power]. This particular tumor demonstrated prominent psammomatous calcification. The larger bodies have cracked, likely during tissue sectioning, and fragments have been lost. Smaller bodies are intact and are layered in a concentric fashion. The epithelial component is overshadowed by the calcification.

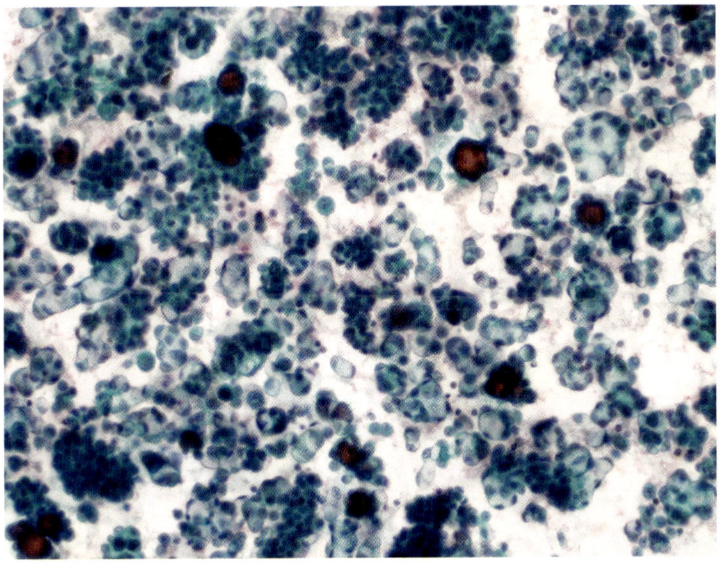

Figure 2.7a — Low-Grade Serous Primary Peritoneal Carcinoma, Pelvic Washing [Pap stain; medium power]. Numerous bland and monotonous epithelial cells form acinar and papillary structures. The neoplastic fragments have hobnailed edges, and the tumor cells are vacuolated in many places. Some areas show the formation of psammoma bodies within fragments, which appear homogenously red in this field. Although low-grade serous carcinoma occurs in the ovary, it can also be primary to the peritoneum, so the presence of such a picture does not necessarily represent a metastatic process. This illustrates the importance of caution when clinicoradiologic data are not available.

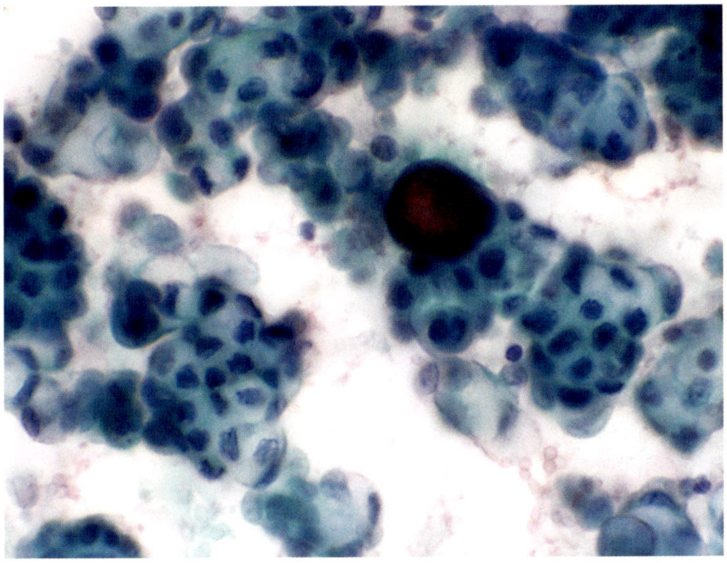

Figure 2.7b — Low-Grade Serous Primary Peritoneal Carcinoma, Pelvic Washing [Pap stain; high power]. A psammoma body is colored red and has an amorphous appearance. The surrounding cells are bland, with abundant cytoplasm, slightly irregular nuclear borders, and only slight hyperchromasia. The cells all have small nucleoli. Primary peritoneal carcinoma is a rare tumor; it is most often found in women, but rare instances of men with this disease have been reported.

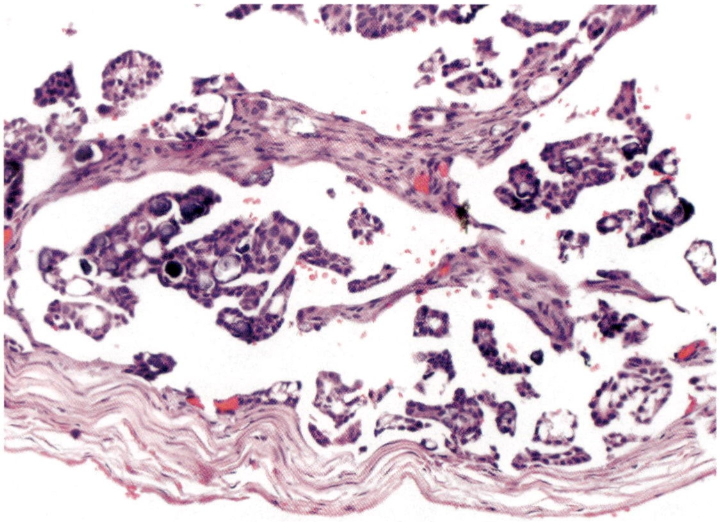

Figure 2.7c — Low-Grade Serous Primary Peritoneal Carcinoma, Peritoneal Biopsy [Pap stain; medium power]. Bland epithelial cells form papillary structures, and psammoma bodies are seen as concentric purple structures. The large clear space surrounding the neoplastic cells is secondary to the papillary growth pattern of this tumor. When seen on cell block material, the cells appear glandular (non-mesothelial) and stain positively with Ber-EP4 or MOC-31. This may lead one in the wrong direction, such as a search for the primary tumor if this is thought to represent a metastatic process.

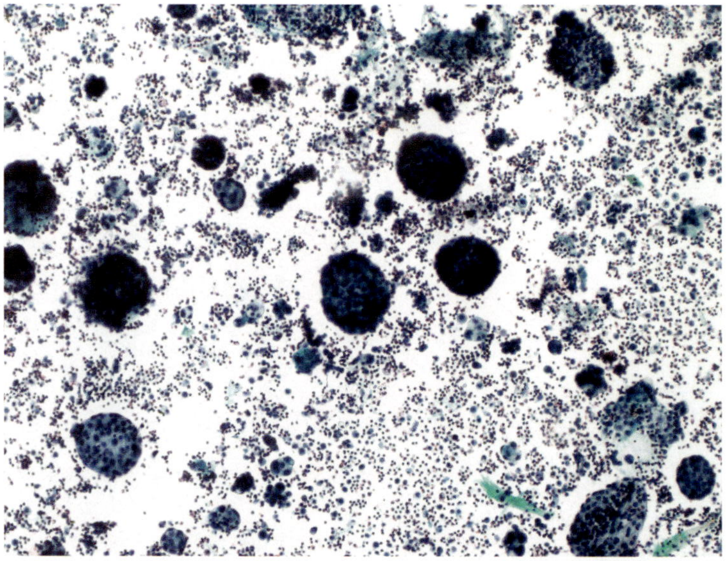

Figure 2.8a — Ovarian Clear Cell Carcinoma, Pelvic Washing [Pap stain; low power]. Although mesothelial fragments can be found in washing specimens, they typically form monolayered sheets that may fold. They may form small three-dimensional fragments with hobnailed borders. Here, the specimen is cellular, and cellular balls of varying sizes are seen. Some of these fragments have distinctly flat edges (lower left corner), indicating that they are unlikely to be mesothelial in origin. This appearance of cannonballs in fluid specimens has been seen with mesotheliomas and metastatic breast carcinomas but may be seen in other neoplasms as well. Here, the patient was diagnosed on ovarian resection with clear cell carcinoma.

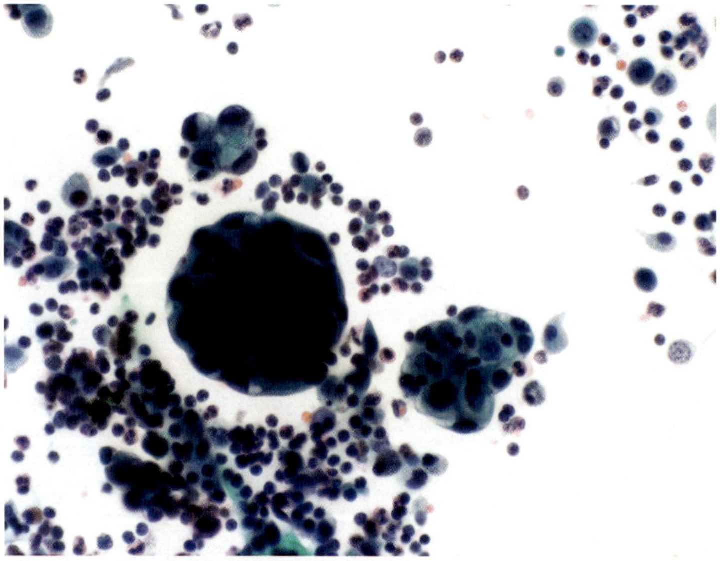

Figure 2.8b — Ovarian Clear Cell Carcinoma, Pelvic Washing [Pap stain; high power]. Note the relatively flat edge of the large cellular fragment in the center of the field. The adjacent fragment is less three-dimensional, and thus the cytomorphology of the cells can be better appreciated. The cells have abundant foamy cytoplasm, a correlate of the "clear cell" appearance the cells have on histology. It is difficult to deny that these are malignant cells, given the presence of severe hyperchromasia, enlarged nuclei, and irregular nuclear borders.

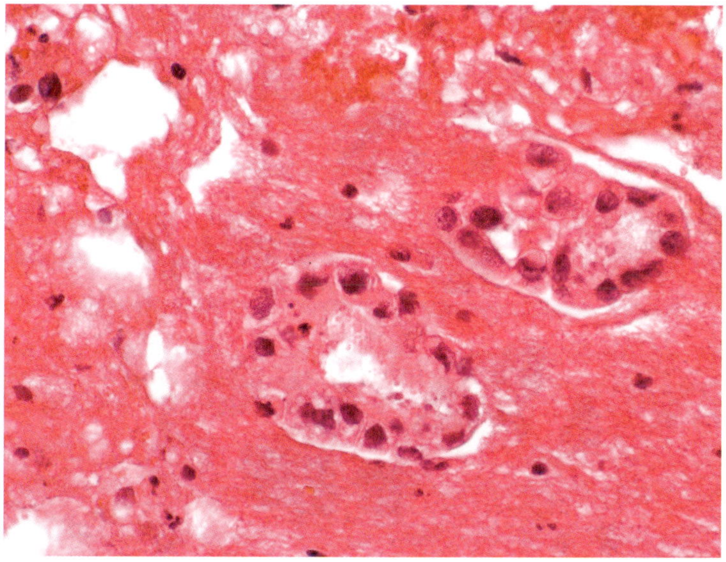

Figure 2.8c — Ovarian Clear Cell Carcinoma, Pelvic Washing Cell Block Section [H&E stain; high power]. Note the excellent cytomorphologic correlation between the cells seen here and those seen on the Pap-stained preparations. Although three-dimensional fragments often obscure their contents, cell block preparations section through the fragments and allow their contents to be examined. Here, the cells form gland-like structures with secretion into the spaces. This is a helpful clue that this represents an adenocarcinoma and not a mesothelioma.

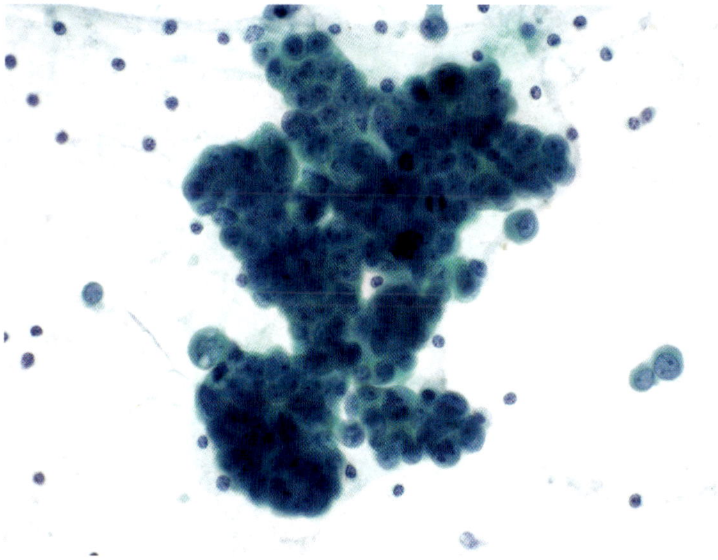

Figure 2.9a — High-Grade Serous Ovarian Carcinoma, Pelvic Washing [Pap stain; medium power]. This three-dimensional papillary fragment is composed of monotonous neoplastic cells with round nuclei, high nuclear-to-cytoplasmic ratios, and prominent nucleoli. The hobnailed edges, are suggestive of mesothelial cell origin. However, a few cells have large vacuoles in their cytoplasm, which is a feature seen in some ovarian and endometrial carcinomas.

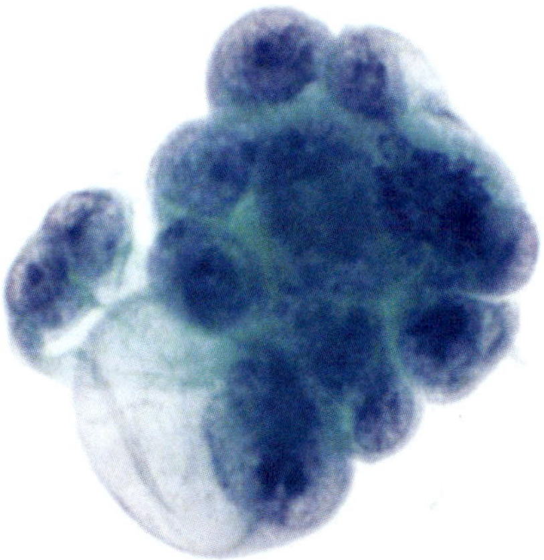

Figure 2.9b — High-Grade Serous Ovarian Carcinoma, Pelvic Washing [Pap stain; high power]. On close examination, the neoplastic cells are less uniform in size than they appeared on low power. Some nuclei are more than four times the size of adjacent nuclei, though all nuclei appear round with regular borders. The edges of this three-dimensional cluster have a hobnailed appearance; an aneuploid mitotic figure can be seen in one of the larger cells. The cell in the bottom left-hand corner of the cluster has a large clear vacuole. When present in the pelvic cavity of a female patient, this cytomorphology should lead one to suspect a serous carcinoma, which may be primary to the peritoneum or involving the cavity secondary to a gynecologic tract (ovarian, endometrial, or fallopian tube) primary.

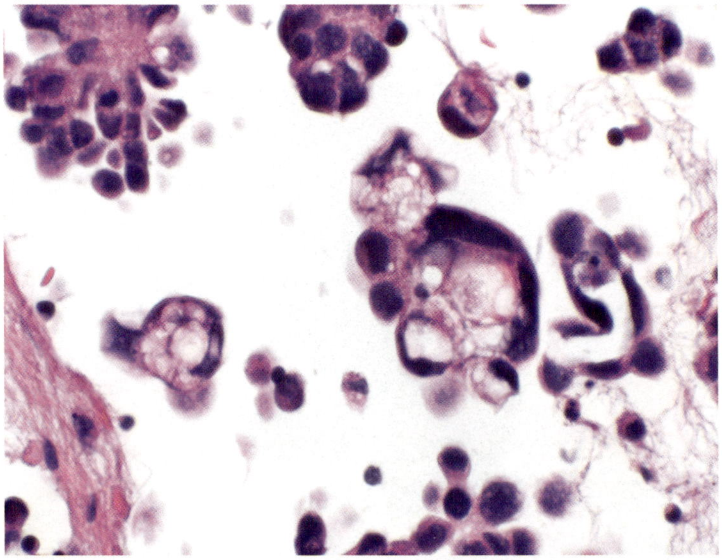

Figure 2.9c — High-Grade Serous Ovarian Carcinoma, Pelvic Washing Cell Block Section [H&E stain; high power]. Cell block material shows pleomorphic, dark nuclei, some of which have been flattened to one side from cytoplasmic vacuoles. Other fragments have a more papillary arrangement formed from neoplastic cells with dark nuclei and very little cytoplasm. The arrangement of small clusters, loosely cohesive single cells, and papillary formations are commonly seen in involved cell block material.

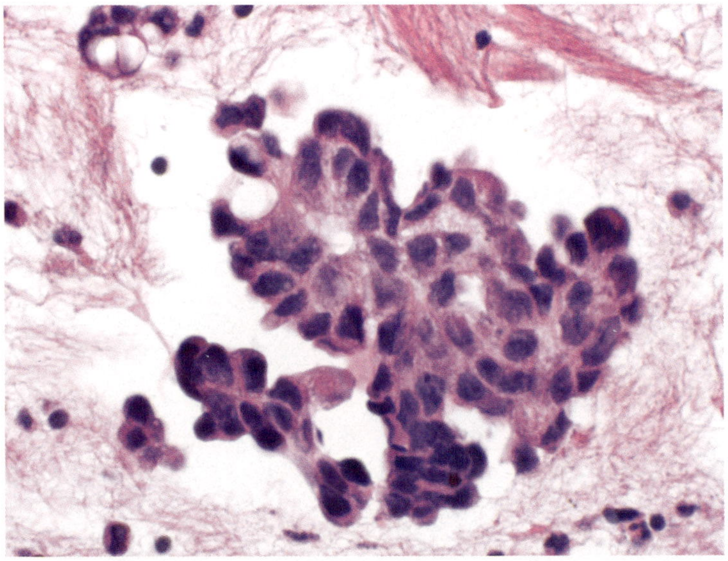

Figure 2.9d — High-Grade Serous Ovarian Carcinoma, Pelvic Washing Cell Block Section [H&E stain; high power]. The fragment contains hyperchromatic malignant cells with irregular shapes and nuclear borders. Papillary projections can be seen arising from the fragment. Prominent nucleoli can be seen in some of the tumor cells, and some cytoplasmic vacuoles are present. Although some of these features are shared with the Pap-stained preparations (nucleoli, cytoplasmic vacuoles, high nuclear-to-cytoplasmic ratios), other features are more prominent on the cell block preparation (papillary architecture, nuclear pleomorphism, hyperchromasia). High-grade ovarian serous carcinomas are positive for PAX8, p53, and WT-1 but may lose estrogen receptor (ER) expression.

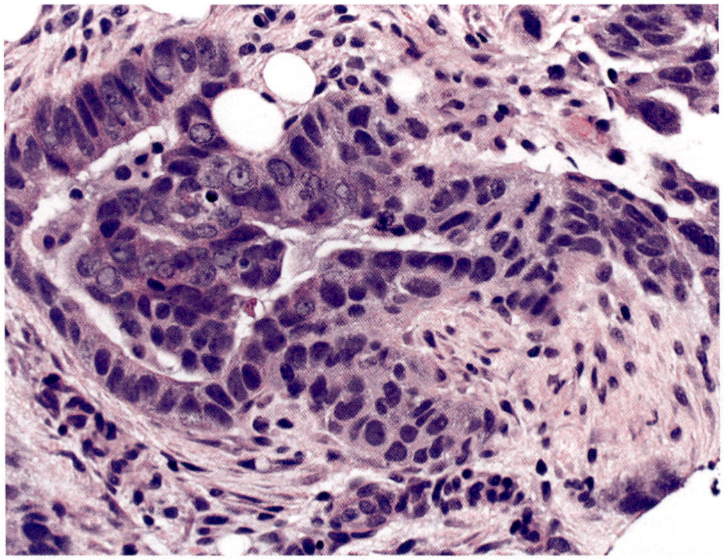

Figure 2.10a — High-Grade Serous Carcinoma, Biopsy [H&E stain; intermediate power]. This peritoneal biopsy in a middle-aged woman reveals a complex glandular structure with an associated desmoplastic stromal response. The glands in this field display micropapillary architecture with slit-like fenestrations, nuclear atypia, and several mitotic figures. Psammomatous calcifications are not a prominent feature of this biopsy but are frequently seen in association with serous carcinomas.

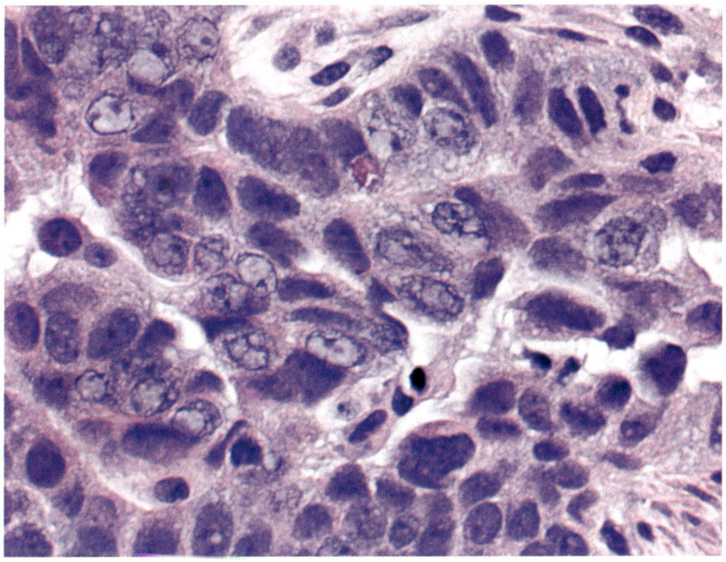

Figure 2.10b — High-Grade Serous Carcinoma, Biopsy [H&E stain; high power]. Higher-power examination of the atypical glands confirm the presence of a high nuclear-to-cytoplasmic ratio, nuclear hyperchromasia, and irregular nuclear contours. Serous carcinomas are classified as either low or high grade on the basis of nuclear atypia and the mitotic rate. Peritoneal, ovarian, and tubal serous carcinomas are posited to arise from serous tubal intraepithelial carcinoma.

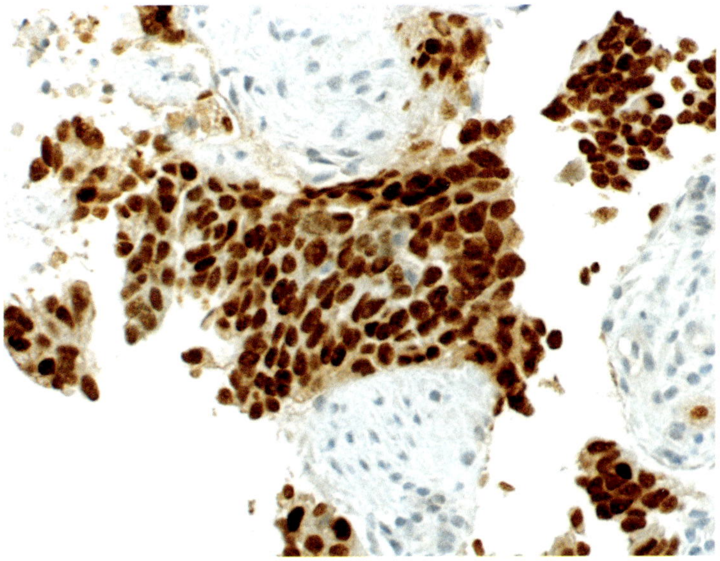

Figure 2.11 — High-Grade Serous Carcinoma, Biopsy [PAX8 immunohistochemical stain; intermediate power]. High-grade serous carcinomas show strong and diffuse nuclear immunoreactivity for PAX8. This helps distinguish them from metastatic breast carcinoma, which is negative for PAX8. However, PAX8 is not entirely specific for high-grade serous carcinoma, as Renal Cell Carcinoma (RCC) is also immunoreactive for PAX8. Of note, the differential diagnosis also includes peritoneal malignant mesothelioma, which can occasionally display focal PAX8 labeling. An immunopanel containing PAX8, WT-1, p53, p16, and estrogen receptor (ER) is helpful in confirming the diagnosis of high-grade serous carcinoma. Other useful immunostains may include GATA-binding protein 3 (GATA3; for breast carcinoma) and calretinin (for mesothelioma).

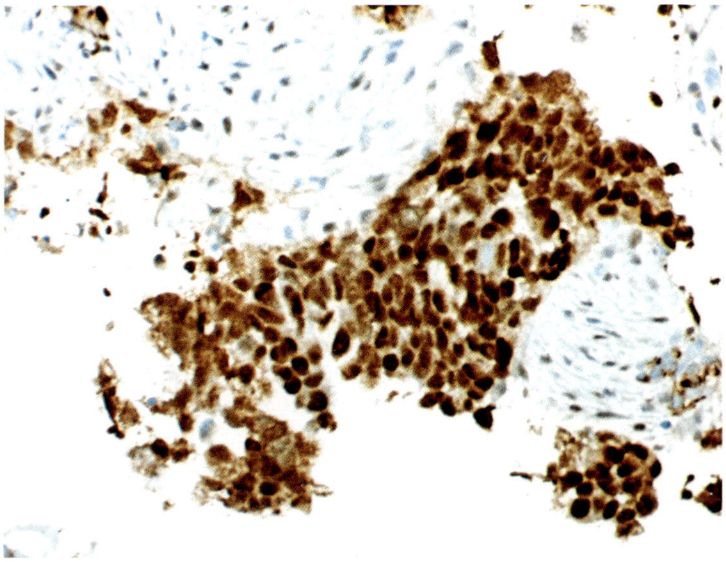

Figure 2.12a — High-Grade Serous Carcinoma, Biopsy [WT-1 immunohistochemical stain; intermediate power]. High-grade serous carcinomas show strong and diffuse nuclear immunoreactivity for WT-1. However, WT-1 is not specific for high-grade serous carcinoma, and strong and diffuse labeling is also seen in peritoneal malignant mesothelioma. WT-1 labeling has also been reported in a small subset of primary breast carcinomas. It is important to perform a targeted immunopanel, including PAX8, WT-1, p53, p16, estrogen receptor (ER), GATA-binding protein 3 (GATA3), and calretinin, in this setting.

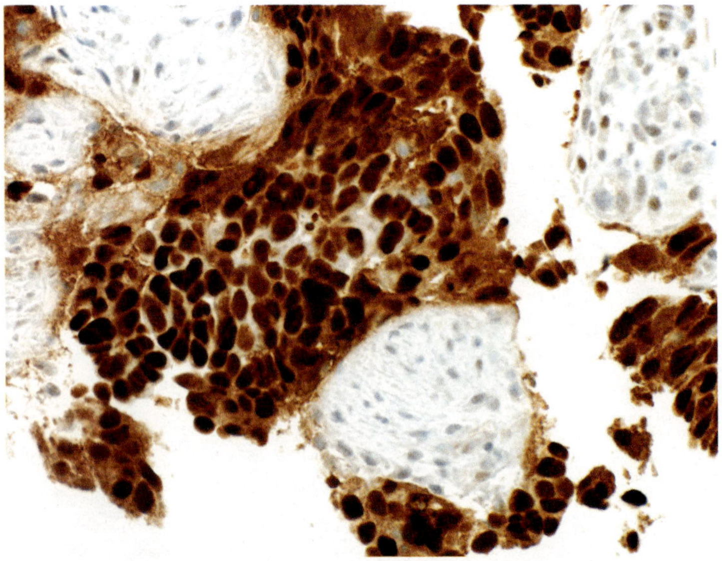

Figure 2.12b — High-Grade Serous Carcinoma, Biopsy [p53 immunohistochemical stain; intermediate power]. High-grade serous carcinomas show strong and diffuse nuclear p53 labeling. The gene *TP53* is a tumor suppressor gene, and most p53 immunohistochemistry antibodies recognize both wild-type and mutant p53 protein; however, because the wild-type p53 protein has a very short half-life, the strong and diffuse labeling in high-grade serous carcinomas reflects the accumulation of stable, mutant p53. p53 labeling is not specific for high-grade serous carcinoma and can be seen in other high-grade malignancies. However, most endometrioid carcinomas are negative for p53.

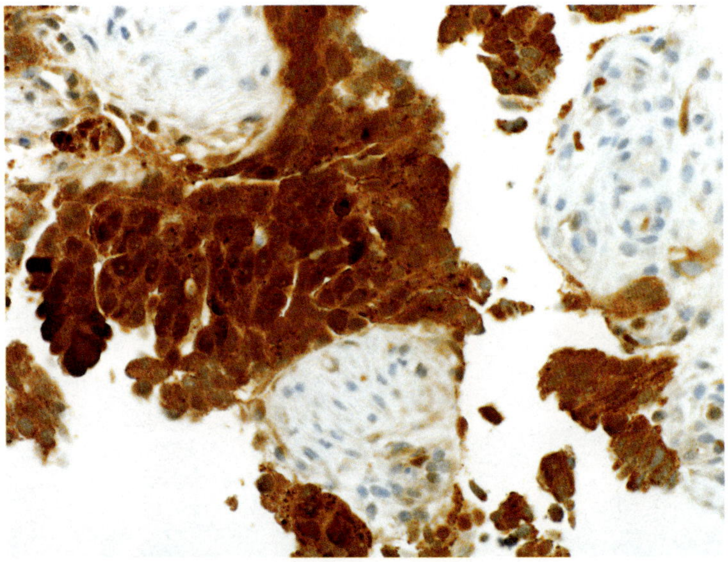

Figure 2.12c — High-Grade Serous Carcinoma, Biopsy [p16 immunohistochemical stain; intermediate power]. Immunohistochemistry for p16 shows strong and diffuse, nuclear and cytoplasmic labeling in this high-grade serous carcinoma. Strong and diffuse p16 immunoreactivity reflects increased p16 protein accumulation, which occurs via the release of a negative feedback loop when the retinoblastoma (Rb) tumor suppressor protein is inactivated. Rb inactivation can occur due to cancer cell mutations (as seen in small cell carcinomas) or secondary to viral integration by high-risk human papilloma viruses (HPVs) (as is seen in cervical squamous cell carcinomas).

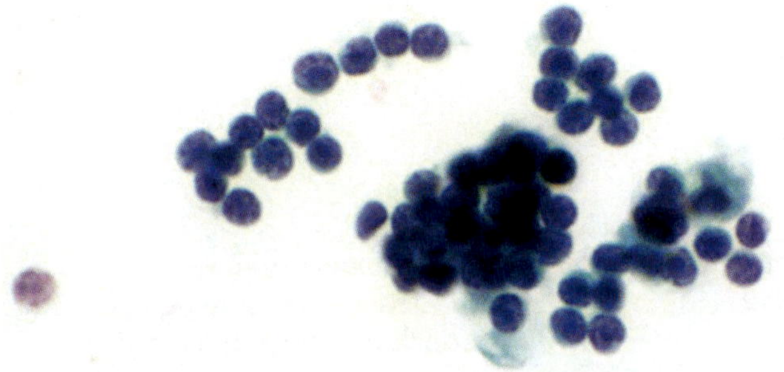

Figure 2.13a — Metastatic Small Cell Carcinoma of the Ovary—Hypercalcemic Type (SCCOHT), Ascites Fluid [Pap stain; high power]. SCCOHT typically consists of poorly cohesive, monomorphic small cells with scant cytoplasm, round to oval hyperchromatic nuclei, inconspicuous nucleoli, and occasional mitotic figures. The differential diagnosis includes lymphoma, small cell carcinoma of the ovary—pulmonary type, small cell lung carcinoma, Ewing's sarcoma, metastatic melanoma, and adult or juvenile granulosa cell tumors. A large cell variant of SCCOHT has also been described but typically contains cells with large nuclei, prominent nucleoli, and abundant eosinophilic cytoplasm, which may take on a hyaline-like quality.

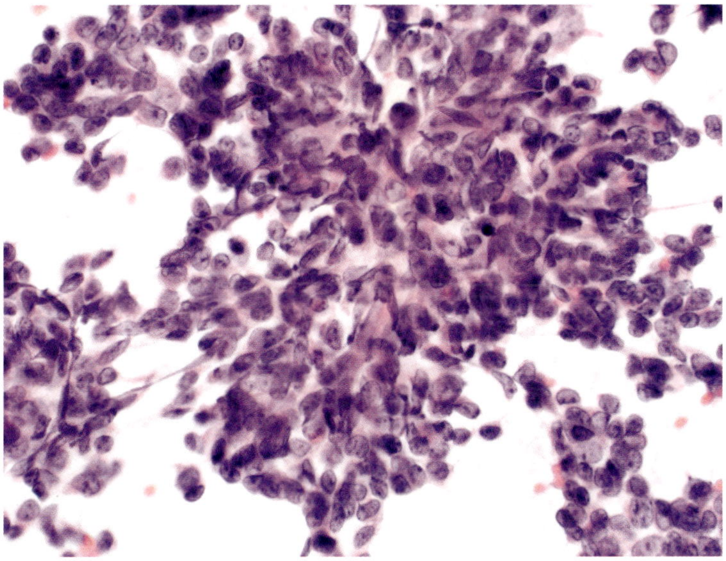

Figure 2.13b — Metastatic Small Cell Carcinoma of the Ovary — Hypercalcemic Type (SCCOHT), Touch Preparation [Diff-Quik stain; high power]. Tumors cells are small with scant cytoplasm, round to oval nuclei, and small nucleoli. Mitotic figures and necrosis are frequent. SCCOHT shows a unique immunoprofile, including expression of WT-1, epithelial membrane antigen (EMA), vimentin, and cytokeratins, and lack of expression of inhibin, chromogranin, thyroid transcription factor-1 (TTF-1), S-100 protein, and nuclear SMARCA4.

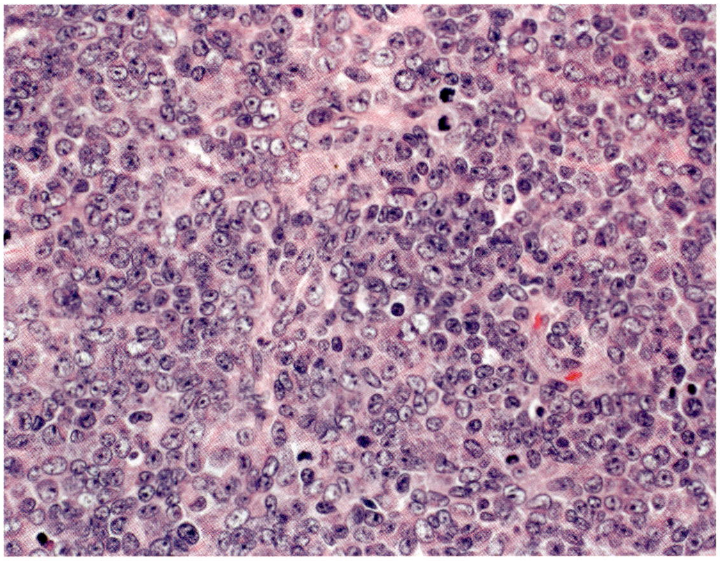

Figure 2.13c — Small Cell Carcinoma of the Ovary — Hypercalcemic Type (SCCOHT) Type, Resection [H&E stain; intermediate power]. This field shows a diffuse growth of small to medium cells with scant amounts of eosinophilic cytoplasm, frequent mitotic figures, and small but identifiable nucleoli. Tumor cells can also form small nests, cords, and trabeculae. Cystic follicle–like structures lined by tumor cells are present in a majority of cases. These spaces typically contain eosinophilic fluid. Rarely, small foci of tumor cells may demonstrate clear cytoplasm or intracellular mucin.

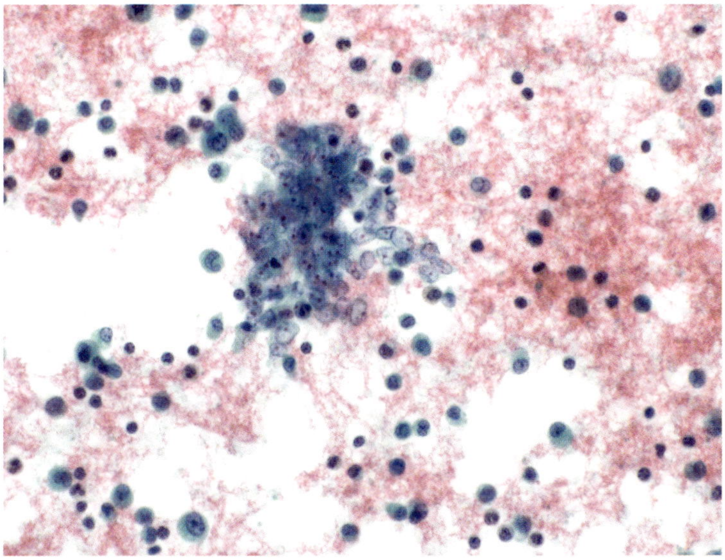

Figure 2.14a — Adult Granulosa Cell Tumor (AGCT), Pelvic Washing [Pap stain; high power]. AGCT is associated with hyperestrogenism and can cause precocious puberty when found in children. However, most patients are above the age of 15. The cytomorphologic features of AGCT have some overlap with papillary thyroid carcinoma, with elongated nuclei, nuclear grooves, and a pale chromatin pattern. The cells also can have distinct nucleoli and very little cytoplasm, as seen here. This effect may mimic small cell carcinoma.

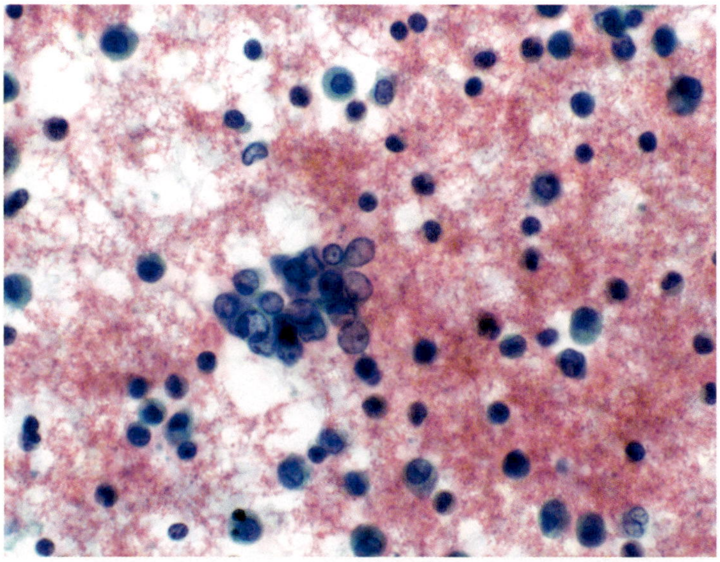

Figure 2.14b — Adult Granulosa Cell Tumor (AGCT), Pelvic Washing [Pap stain; high power]. In this field, the nuclei appear more round than elongated, which is more consistent with the histologic appearance of these cells. Note the condensation of chromatin along the nuclear borders, which are highly irregular. During pregnancy, cells may become luteinized and possess more cytoplasm.

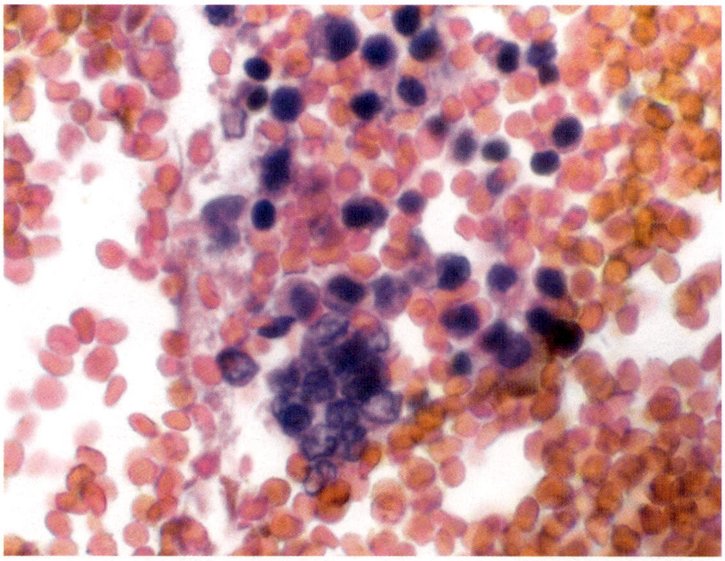

Figure 2.14c — Adult Granulosa Cell Tumor (AGCT), Pelvis Washing Cell Block Section [H&E stain; high power]. The cell block morphology is similar to that seen on the Pap stain, with irregular nuclear borders, condensed chromatin, and little cytoplasm. The cells are noticeably smaller than the mesothelial cells seen as individual cells surrounding the fragment of tumor cells. In this case, the patient had peritoneal involvement by her ovarian tumor.

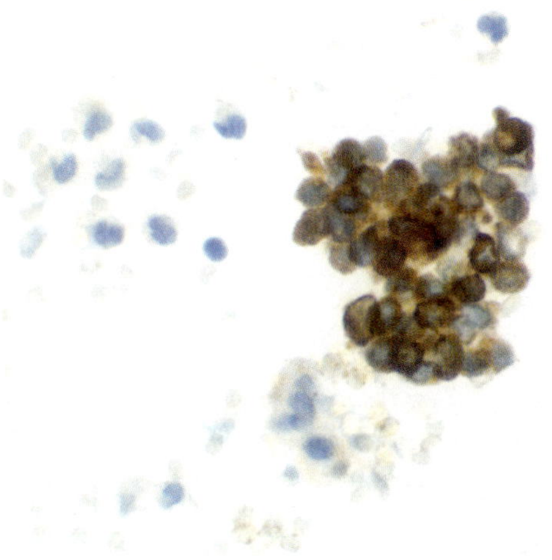

Figure 2.14d — Adult Granulosa Cell Tumor (AGCT), Pelvic Washing Cell Block Section [Inhibin immunohistochemical stain; high power]. Inhibin-alpha is a hormone marker for AGCT and can be used to prove involvement of a pelvic washing in a patient with known AGCT. Tumors can also be positive for CD99 and calretinin, although the latter is not as useful in pelvic washings because it also stains background mesothelial cells. Here, the cells have such distinct morphology from mesothelial cells that even a calretinin stain could be useful. These markers can also be positive in other sex cord-stromal tumors.

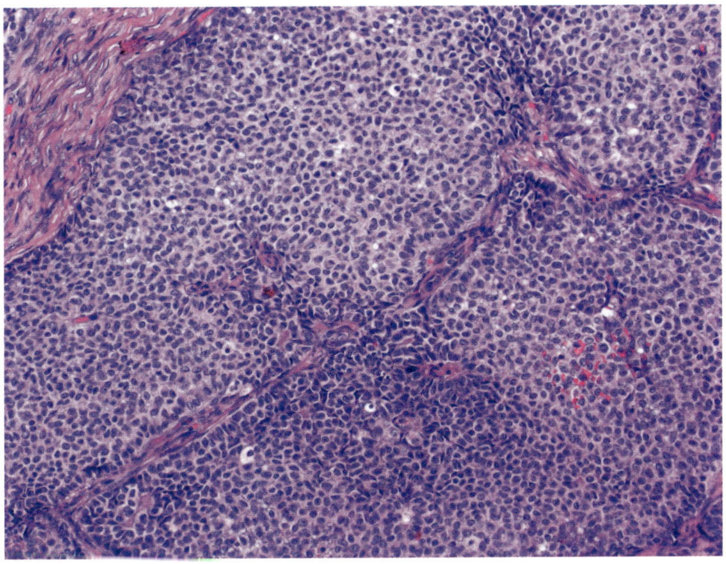

Figure 2.15a — Adult Granulosa Cell Tumor (AGCT), Oophorectomy [H&E stain; low-intermediate power]. The ovary in this oophorectomy specimen is largely replaced by solid nests and sheets of small, bland, and monotonous cells with a high nuclear-to-cytoplasmic ratio and minimal cytoplasm. The differential diagnosis from low-power examination includes Brenner tumor, carcinoid tumor, metastatic lobular or gastric carcinoma, and small cell carcinoma. AGCT is a type of sex cord-stromal tumor resembling ovarian follicular granulosa cells, and the majority of AGCTs are benign. However, a small subset recurs or metastasizes, and no good histologic features predict behavior.

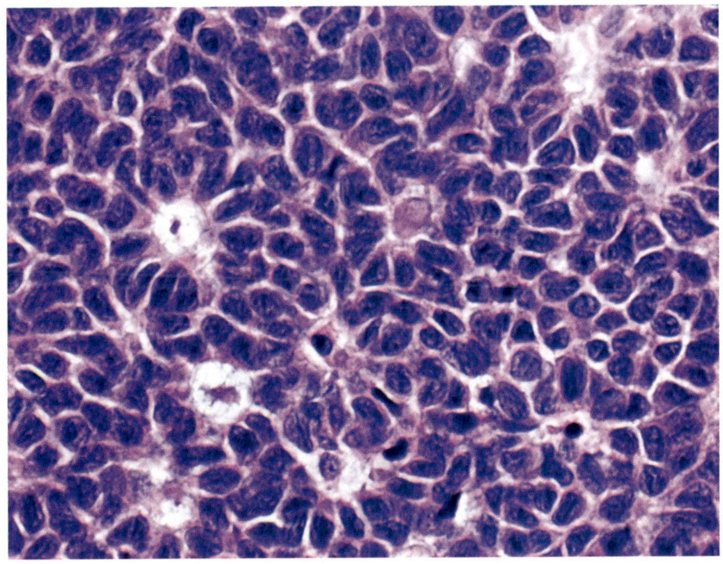

Figure 2.15b — Adult Granulosa Cell Tumor (AGCT), Oophorectomy [H&E stain; high power]. AGCTs are composed of cells with minimal cytoplasm and oval nuclei with prominent grooves, resembling coffee beans. This field also demonstrates the characteristic feature of Call-Exner bodies, which are small, round, follicle-like structures filled with pink material. AGCTs are immunoreactive for inhibin, calretinin, actin, and CD99.

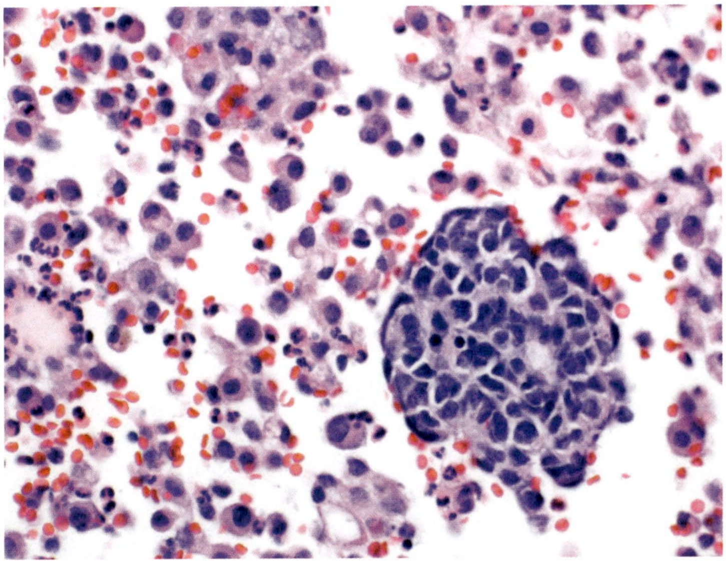

Figure 2.16a — Malignant Ovarian Teratoma, Pelvic Washing [Diff-Quik stain; high power]. Although most ovarian teratomas are benign, this unfortunate pediatric patient had a rare malignant ovarian teratoma. On resection, the tumor contained components of primitive neuroectodermal tissue and yolk sac tumor. This malignant fragment contains small cells with little cytoplasm that mold together like that seen in small cell carcinoma. This tissue fragment likely represents primitive neuroectodermal tumor (PNET).

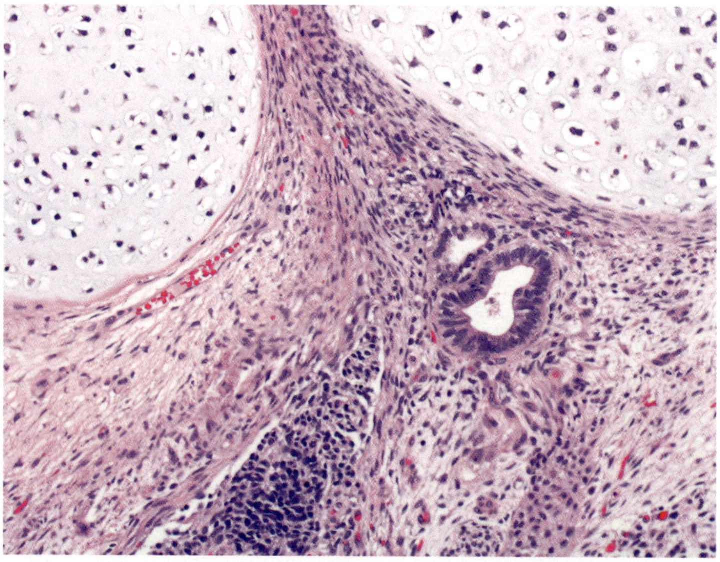

Figure 2.16b — Malignant Mixed Germ Cell Tumor, Teratoma Component, Retroperitoneal Lymph Node Dissection [H&E stain; intermediate power]. Several lymph nodes in this retroperitoneal lymph node dissection contain regions of normal-appearing tissue, including the bland hyaline cartilage and glands seen here. These are the teratomatous components of the metastatic mixed germ cell tumor. Identifying and removing the teratoma is pertinent because as typically bland cell types with low proliferation rates, they are relatively resistant to the systemic chemotherapy administered for germ cell tumors. Thus, surgical resection is the treatment of choice. It is also important to thoroughly evaluate the teratomatous component for any malignant transformation to a somatic malignancy (e.g., adenocarcinoma arising in a teratoma).

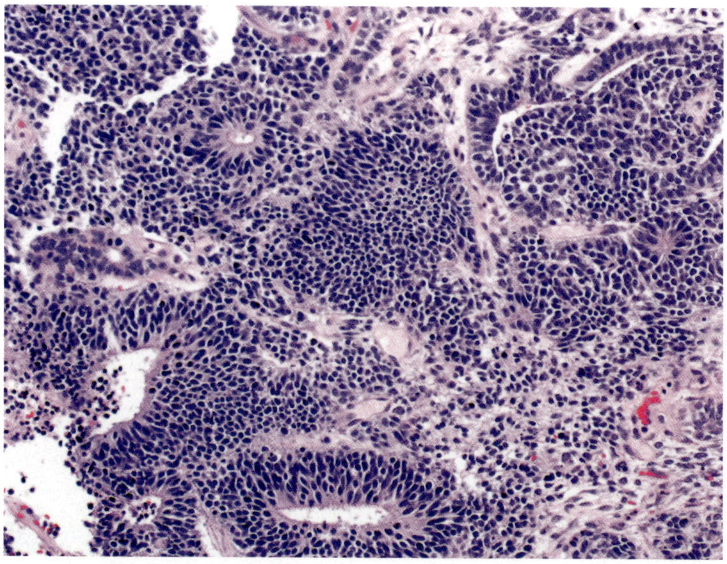

Figure 2.16c — Malignant Mixed Germ Cell Tumor, Teratoma Component, Retroperitoneal Lymph Node Dissection [H&E stain; intermediate power]. This field of the teratomatous component of the mixed germ cell tumor consists entirely of immature neural elements. This is an intermediate magnification, and the process is only focal. However, if the primitive neural elements become expansile and compose greater than a 4× low-power field, malignant transformation to a primitive neuroectodermal tumor (PNET) should be considered.

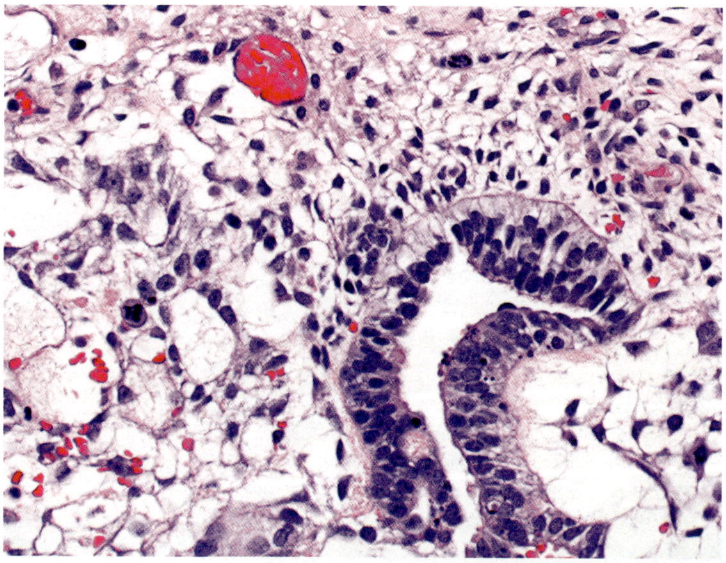

Figure 2.17a — Malignant Mixed Germ Cell Tumor, Yolk Sac Component, Retroperitoneal Lymph Node Dissection [H&E stain; intermediate power]. Yolk sac tumors display numerous variant histologic patterns, even within one tumor. The predominant pattern seen here is the reticular or microcystic pattern, consisting of a honeycomb or lace-like pattern of bland vacuolated cells, which is the most common histologic pattern. In addition, illustrated here is also an admixed glandular pattern, with cuboidal to columnar glandular-type cells with subnuclear vacuoles. Other variant histologic patterns include the endodermal sinus pattern, hepatoid pattern, papillary pattern, macrocystic pattern, myxomatous pattern, and solid pattern, among several others. Yolk sac tumors label for alpha-fetoprotein (AFP), cytokeratin, and glypican 3 by immunohistochemistry.

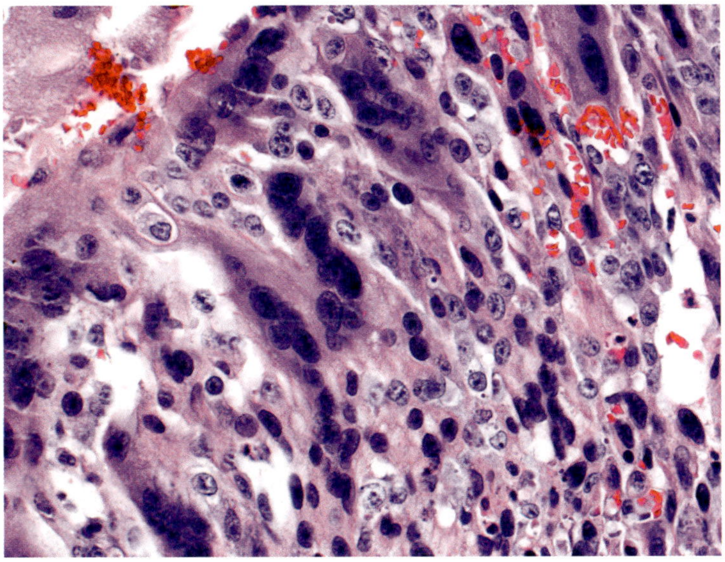

Figure 2.17b — Malignant Mixed Germ Cell Tumor, Choriocarcinoma Component, Retroperitoneal Lymph Node Dissection [H&E stain; intermediate power]. Choriocarcinoma rarely presents as a pure germ cell tumor but is more often seen as a component of a mixed malignant germ cell tumor. As this field illustrates, choriocarcinomas are composed of two cells types: cytotrophoblasts, which are smaller, round cells with single bland nuclei and distinct cell borders, as well as syncytiotrophoblasts, which are larger, multinucleated cells with eosinophilic and vacuolated cytoplasm. The presence of syncytiotrophoblasts alone does not classify as a choriocarcinoma, as syncytiotrophoblasts can be seen even in otherwise pure seminomas. Choriocarcinomas are also frequently accompanied by extensive hemorrhage and necrosis. They label for human chorionic gonadotropin (HCG), human placental lactogen (HPL), placental alkaline phosphatase (PLAP), and GATA-binding protein 3 (GATA3) by immunohistochemistry.

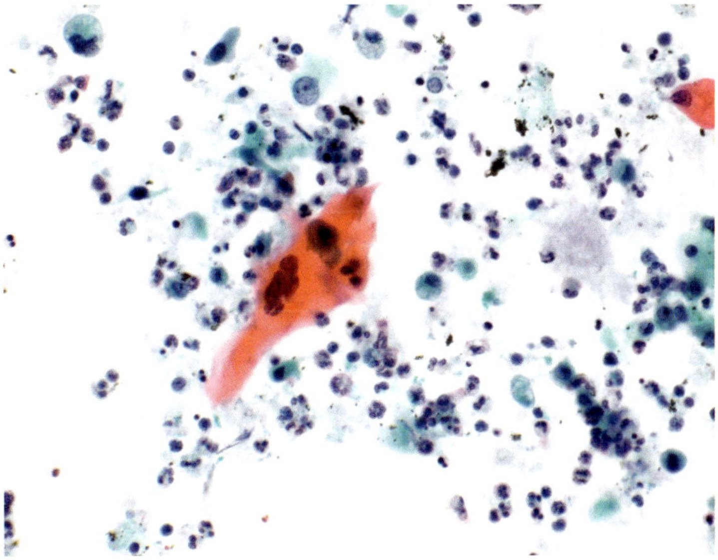

Figure 2.18a — Metastatic Cervical Squamous Cell Carcinoma, Ascites Fluid [Pap stain; high power]. The field contains a background of granular debris and acute inflammation; both are common findings in squamous cell carcinoma due to areas of necrosis. On a Pap-stained preparation, keratinizing squamous cell carcinoma is easy to spot because of the pink-to-orange staining of the cytoplasm. Keratinization seen within a serous cavity is a highly unusual finding and not typically associated with a benign process. Most metastatic carcinomas to the serous cavities are adenocarcinomas, with squamous cell carcinoma being rare.

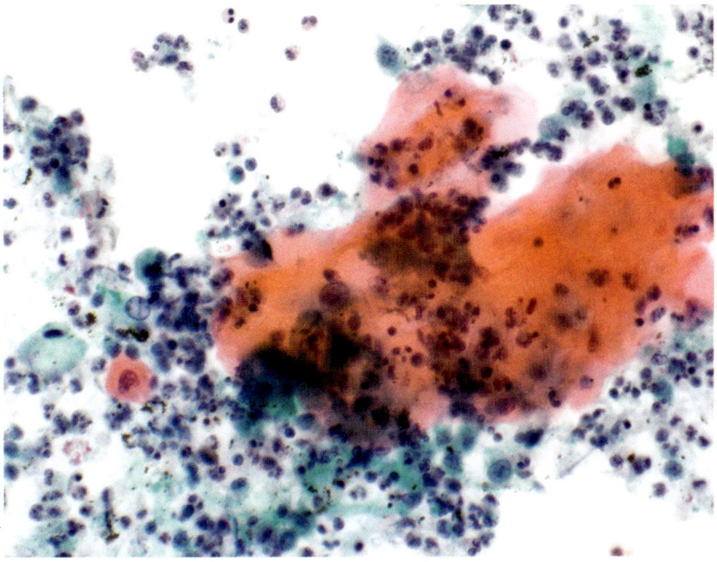

Figure 2.18b — Metastatic Cervical Squamous Cell Carcinoma, Ascites Fluid [Pap stain; high power]. Keratinizing squamous cell carcinoma often contains dense fragments of irregularly shaped keratin, and nuclei may appear small and pyknotic and stain poorly. Note the smaller keratinized cell in the bottom left corner of the field. Although the cell has abundant cytoplasm and a pale-staining nucleus, it is virtually diagnostic when found in a serous cavity in a patient with a history of squamous cell carcinoma.

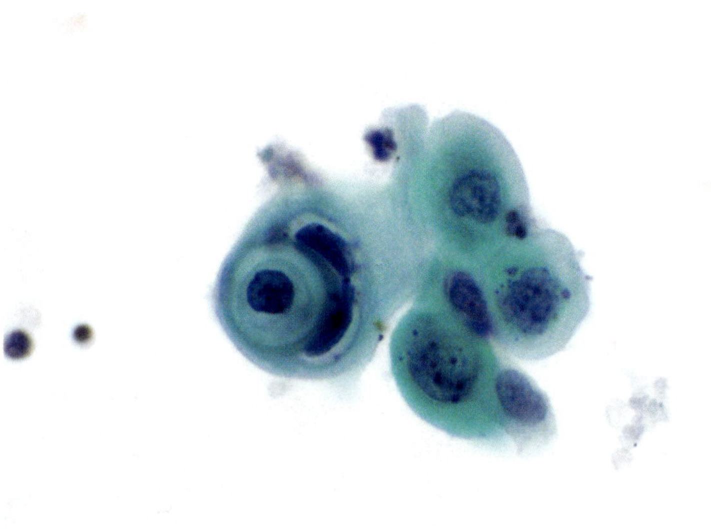

Figure 2.18c — Metastatic Cervical Squamous Cell Carcinoma, Ascites Fluid [Pap stain; high power]. A malignant squamous pearl is seen here, with one malignant cell surrounded by another, giving an overall lamellated appearance. The cytoplasm is dense and has a glassy appearance. Some nuclei are dark and pyknotic, whereas others are larger with coarse chromatin. Squamous cell carcinoma rarely metastasizes to serous cavities; in the pleural cavity, lung carcinoma is most common, and in the peritoneal cavity, cervical carcinoma is a consideration. One should always consider an adenocarcinoma with squamous differentiation, such as squamous differentiation within an endometrial carcinoma. Head and neck squamous cell carcinomas rarely metastasize to serous cavities.

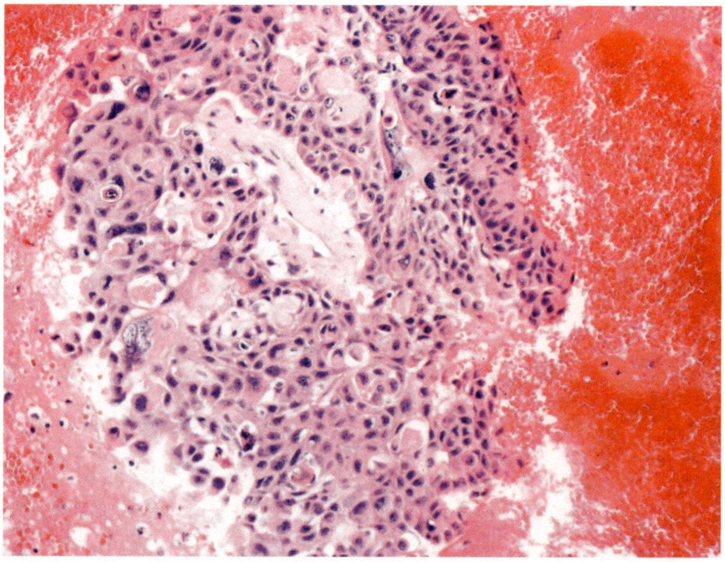

Figure 2.18d — Metastatic Cervical Squamous Cell Carcinoma, Ascites Fluid Cell Block Section [H&E stain; medium power]. Typically, patient history is known by the time a malignancy metastasizes to a serous cavity. Because cervical carcinomas are caused by human papillomavirus (HPV), diffuse staining with p16 by immunohistochemistry or HPV positivity by in situ hybridization can distinguish cervical squamous cell carcinoma from lung squamous cell carcinoma. The cells here have cytomorphologic features consistent with squamous cell carcinoma—pink, glassy cytoplasm, squamous pearls, and dark pyknotic nuclei. Some cells have large, pleomorphic nuclei. Expression of squamous markers (e.g., p40, p63) indicates a squamous carcinoma, but adenocarcinomas with squamous differentiation may express squamous markers as well.

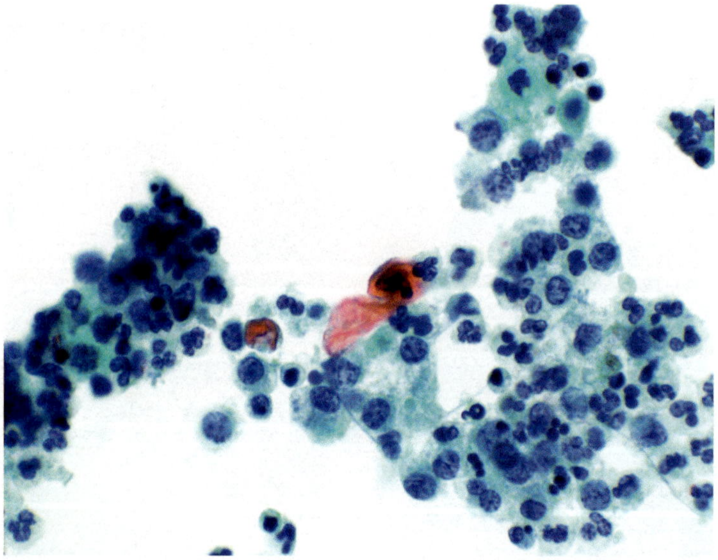

Figure 2.19a — Endometrioid Carcinoma With Squamous Differentiation, International Federation of Gynecology and Obstetrics (FIGO) Grade 1, Pelvic Washing [Pap stain; high power]. Cells with dense, bright orangeophilic cytoplasm and hyperchromatic nuclei with irregular nuclear contours are characteristic of malignant keratinizing squamous carcinoma. Parakeratotic cells and anucleate keratin fragments (hyperkeratosis) may also be present. In peritoneal fluid cytology, squamous cells are most commonly associated with endometrioid adenocarcinoma with extensive squamous differentiation but may also be seen in other metastatic carcinomas from nongynecologic sites.

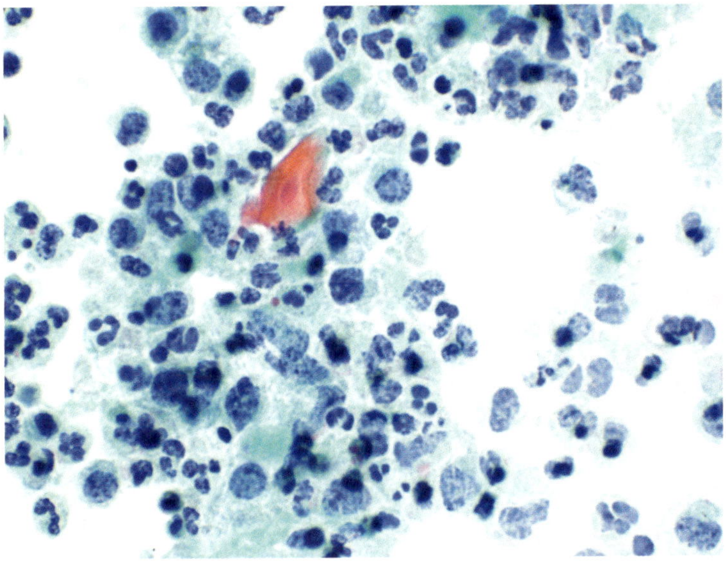

Figure 2.19b — Endometrioid Carcinoma With Squamous Differentiation, International Federation of Gynecology and Obstetrics (FIGO) Grade 1. Pelvic Washing [Pap stain; high power]. This field shows rare atypical keratinized cells with orangeophilic cytoplasm in a background of acute inflammation, reactive mesothelial cells, histiocytes, and lymphocytes. The presence of keratinized cells and keratinaceous material is likely related to the presence of the patient's endometrioid carcinoma with squamous differentiation; however, these elements alone, without an accompanying glandular component, are not considered diagnostic of involvement by the endometrioid carcinoma for staging purposes.

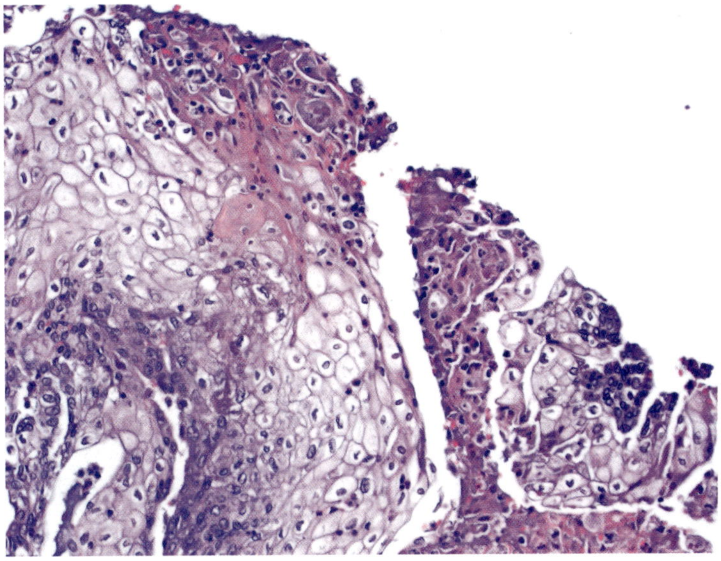

Figure 2.19c — Endometrioid Carcinoma With Squamous Differentiation, International Federation of Gynecology and Obstetrics (FIGO) Grade 1, Total Abdominal Hysterectomy and Bilateral Salpingo-oophorectomy [H&E stain; high power]. In this patient, endometrioid carcinoma (FIGO grade 1) involved the uterus with no definite evidence of myoinvasion; it also involved both ovaries. Morphologically similar synchronous endometrial and ovarian endometrioid carcinomas may represent two independent primary neoplasms or one primary tumor and metastases (either primary endometrial with ovarian metastasis or vice versa). In the setting of a noninvasive or minimally/superficially invasive primary endometrial endometrioid carcinoma without cervical involvement, an isolated morphologically similar ovarian tumor is traditionally interpreted as most likely representing an independent primary ovarian neoplasm. When either endometriosis or an underlying atypical proliferative (borderline) endometrioid tumor is identified in the ovary, interpretation as an independent primary ovarian neoplasm is supported. The presence of bilateral ovarian tumors can suggest metastases unless both tumors have evidence of either endometriosis or an underlying atypical proliferative (borderline) endometrioid tumor. However, it is known that metastases can demonstrate growth patterns simulating atypical proliferative/borderline tumors; thus such growth patterns do not always provide unequivocal evidence of a primary ovarian neoplasm.

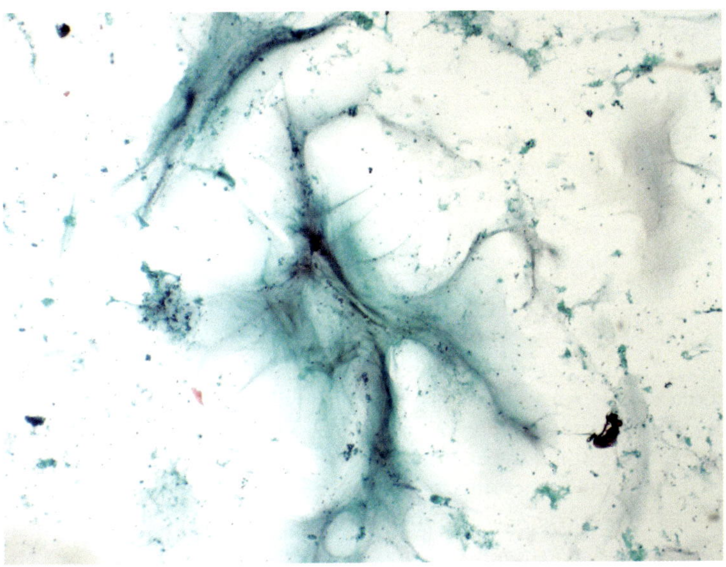

Figure 2.20a — Pseudomyxoma Peritonei, Ascites Fluid [Pap stain; low power]. At low power, the specimen appears paucicellular, and thick mucinous strands can be seen. On occasion, mucin may be the only finding but is sufficient to confirm the clinical diagnosis. The differential diagnosis includes the presence of physiologic mucus leaked from ruptured organs, ruptured ovarian or appendiceal mucinous cystadenomas, and myxoid matrix (mistaken for mucin) produced by myxoid neoplasms.

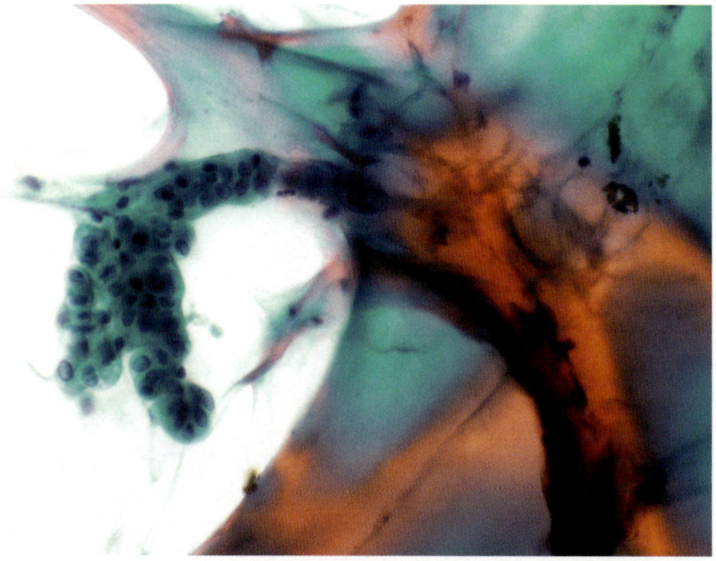

Figure 2.20b — Pseudomyxoma Peritonei, Ascites Fluid [Pap stain; high power]. Pseudomyxoma peritonei is a clinical syndrome associated with the extravasation of mucin into the abdomen that can involve the peritoneal cavity. The mucin originates from mucinous gastrointestinal tumors, most commonly the appendix. In this case, the mucin stains a brilliant mixture of orange, green, and blue. Relatively bland-appearing cuboidal cells are associated with the mucin. The amount of cytologic atypia and number of malignant cells seen is usually concordant with the grade of the primary neoplasm; that is, higher-grade tumors present with higher cellularity within the mucin, and these cells demonstrate more severe cytologic atypia.

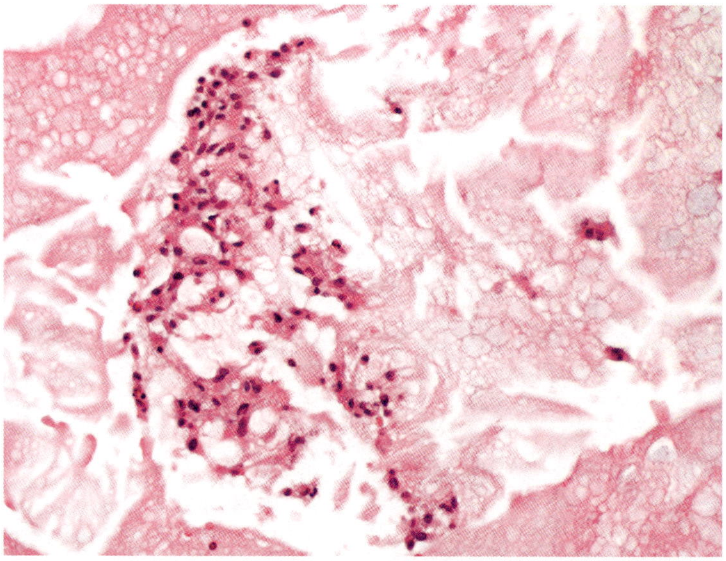

Figure 2.20c — Pseudomyxoma Peritonei, Ascites Fluid [H&E stain; medium power]. This patient had pseudomyxoma peritonei in conjunction with gallbladder adenocarcinoma and gallbladder rupture. Bubbly mucinous matrix surrounds bland-appearing glandular cells. The different materials used to create cell block preparations (e.g., gels, fibrin) can make it difficult to decipher extracellular material when present; in this case, the presence of abundant mucin is an important clue to the diagnosis. A special stain for mucin could be ordered on this block to confirm its presence, if one is unsure.

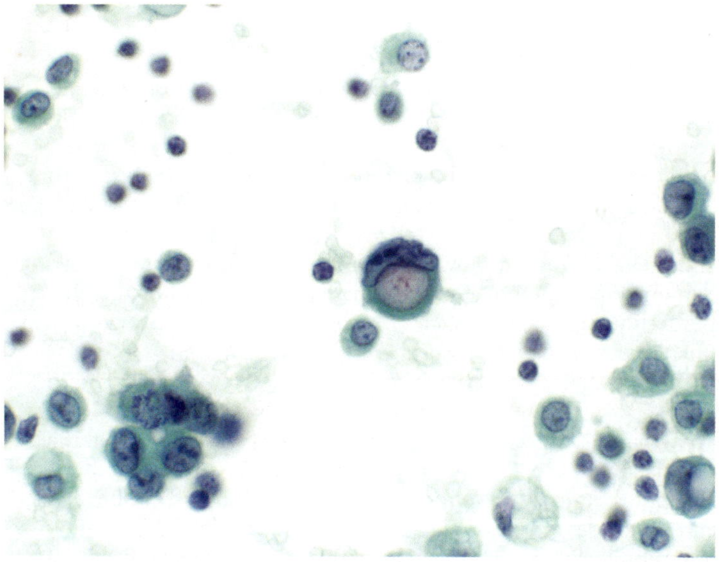

Figure 2.21a — Metastatic Colorectal Adenocarcinoma, Ascites Fluid Cell Block Section [Pap stain; high power]. This malignant cell has a signet-ring appearance. Although mucin vacuoles may simply appear white and be difficult to distinguish from benign vacuoles within mesothelial cells, occasionally the Pap stain will stain mucinous vacuoles pink. The nucleus is compressed by the vacuole; close examination reveals its malignant features: hyperchromasia, irregular nuclear borders, and coarse chromatin. Signet ring cells in serous fluids can be seen in many adenocarcinomas, especially those arising from the gastrointestinal tract.

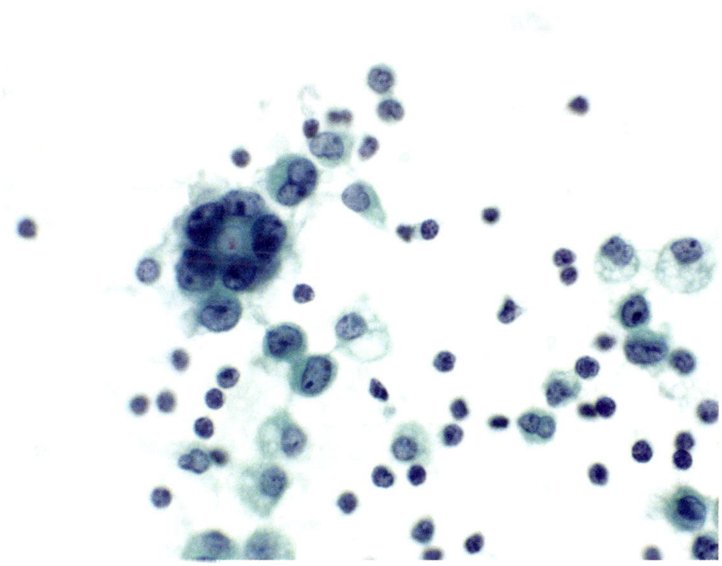

Figure 2.21b — Metastatic Colorectal Adenocarcinoma, Ascites Fluid [Pap stain; high power]. A glandular structure seen here forms a readily identifiable glandular space containing pink-staining mucin. In this case, the patient had a history of colon cancer, but this finding is nonspecific beyond being from an adenocarcinoma with mucinous features. If a history is not known, immunoreactivity for cytokeratin 20 (CK20) is useful because benign components within serous cavities are not immunoreactive for CK20. However, not all adenocarcinomas arising from the colon are immunoreactive for CK20.

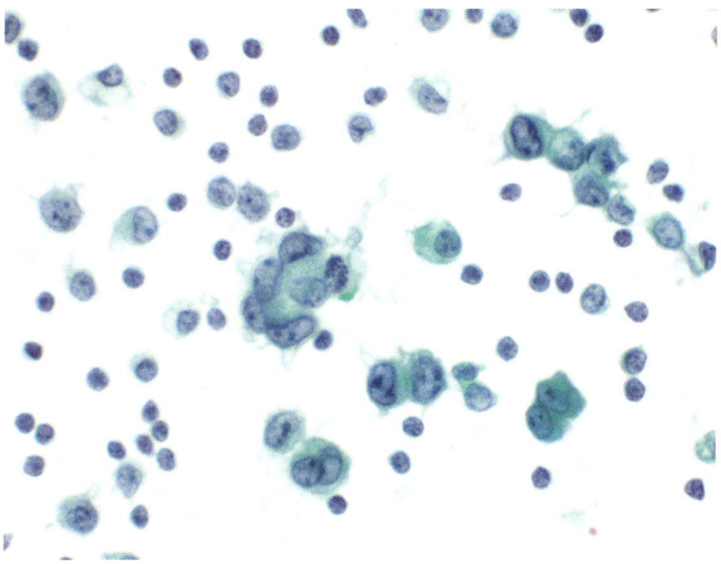

Figure 2.21c — Metastatic Colorectal Adenocarcinoma, Ascites Fluid [Pap stain; high power]. Numerous malignant cells can be seen in this field, intermixed with smaller lymphocytes. The malignant cells have irregular nuclear borders and coarse chromatin. In the center of the field are malignant cells arranged in a glandular formation, although no glandular contents can be identified. One of the cells in the glandular structure is forming an aneuploid mitotic figure; these may be seen more prominently in patients undergoing chemotherapy.

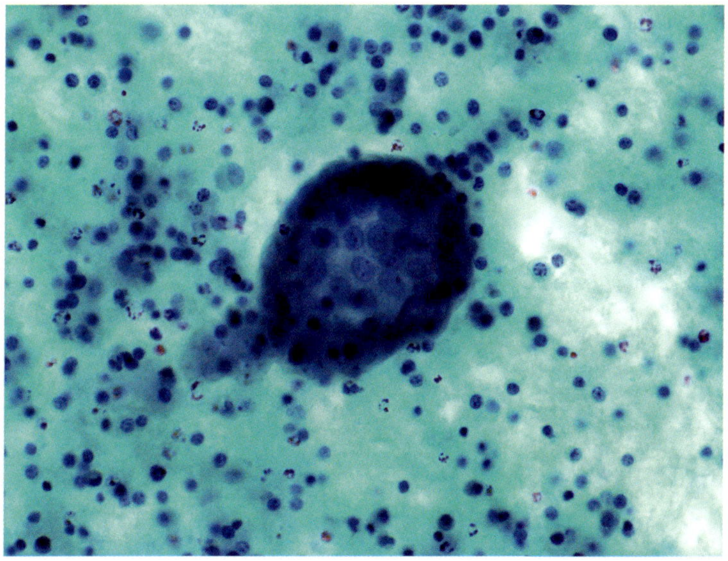

Figure 2.21d — Metastatic Colorectal Adenocarcinoma, Pericardial Fluid [Pap stain; high power]. One cytomorphologic feature of many colorectal adenocarcinomas is necrosis. In this case, the necrotic material fills the field as granular debris and obscures the cells. Numerous acute inflammatory cells can also be seen, associated with the necrosis. The differential diagnosis includes fibrinous pericarditis, as well as other inflammatory conditions. Pericardial specimens often demonstrate dramatic reactive atypia with numerous mesothelial cells, so a high threshold for a diagnosis of malignancy is required. For instance, the cells composing a ball-like structure in the center of the field have very bland cytomorphology and may represent reactive mesothelial cells rather than malignant cells.

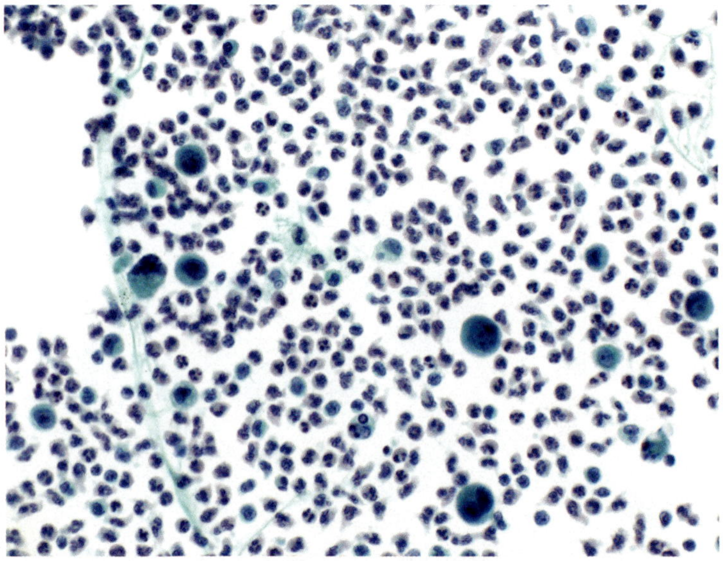

Figure 2.22a — Metastatic Colonic Adenocarcinoma, Pleural Fluid [Pap stain; medium power]. This is a case of metastatic poorly differentiated adenocarcinoma from a patient with a history of a right-sided colon tumor, peritoneal carcinomatosis, and multiple lymph node metastases. Malignant pleural effusions in this setting typically arise when lung metastases are present. Here, the malignant cells are discohesive but easily identifiable by their high nuclear-to-cytoplasmic ratio and hyperchromatic nuclei with chromatin clumping and macronucleoli. There is nothing characteristic of a colonic metastasis and the final diagnosis will depend on the known history with confirmatory immunostaining. A background of associated dirty necrosis with debris (not seen here) and inflammation is a common finding in colonic metastases particularly in fine-needle aspiration (FNA) specimens.

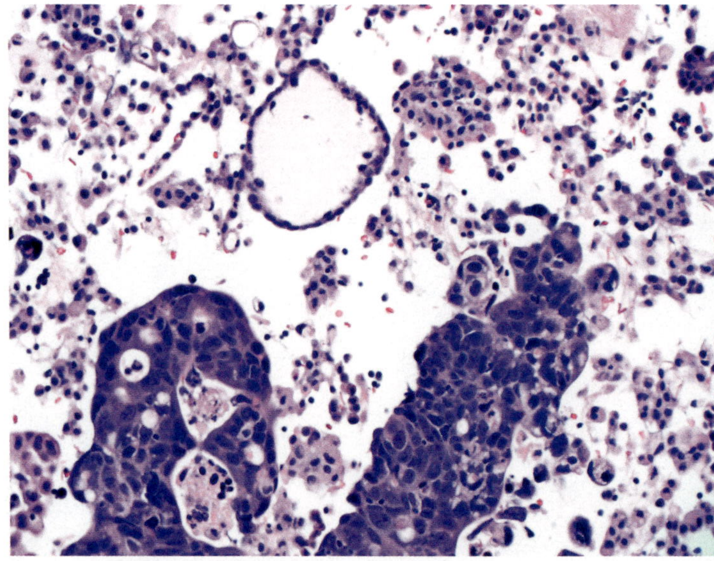

Figure 2.22b — Metastatic Colorectal Adenocarcinoma, Ascites Fluid Cell Block Section [H&E stain; medium power]. In this case, cell block provides better diagnostic material than the cytospin preparation. The top center field contains a ball of bland mesothelial cells, whereas the fragment of cells at the bottom of the field are undoubtedly malignant. The cell block preparation allows cytomorphology to be visualized without obstruction by necrotic material. Even without a history, the presence of necrosis and punched-out microcystic spaces within the fragment are strongly suggestive of a gastrointestinal primary. Although not entirely sensitive for colorectal primaries, positive for CDX2 and cytokeratin 20 (CK20) are relatively specific markers in this situation.

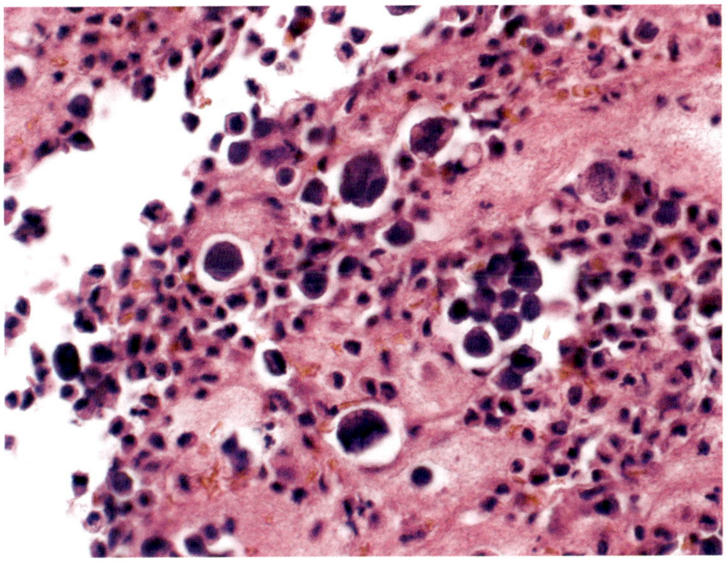

Figure 2.22c — Metastatic Colonic Adenocarcinoma, Pleural Fluid Cell Block Section [H&E stain; high power]. Metastatic adenocarcinoma can resemble mesothelioma, especially when poorly differentiated, as in this case. Poorly differentiated tumors must be evaluated with a broad differential diagnosis in mind, including carcinoma, mesothelioma, hematologic malignancies, and melanoma. A full panel of immunostains should be performed and directly influenced by the likelihood of any given tumor. Markers expressed by tumors of gastrointestinal origin include CDX2, villin, carcinoembryonic antigen (CEA), and mucins, although none are entirely specific. Cytokeratin 7 (CK7) and cytokeratin 20 (CK20) immunostaining may aid in localizing the primary tumor to the colon (CK7 negative, CK20 positive); otherwise, gastric, pancreatic, biliary, and esophageal carcinomas may have similar profiles (CK7 positive).

Figure 2.23a — Metastatic Colonic Adenocarcinoma, Pleural Fluid Cell Block Section [Ber-EP4 immunohistochemical stain; high power]. Ber-EP4 membranous positivity correlates with epithelial cell adhesion molecule (EpCAM) protein expression in most carcinomas. Deletions involving the polyadenylation site in last two exons of the *EPCAM* gene have recently been identified as a novel cause of Lynch syndrome. Such deletions disrupt the 3′ end of the *EPCAM* gene, leading to transcriptional read-through of the mutated *EPCAM* allele and allele-specific epigenetic inactivation of its neighboring gene *MSH2* via hypermethylation of the *MSH2* promoter. Lynch syndrome, also known as hereditary nonpolyposis colorectal cancer (HNPCC), is an autosomal dominantly inherited cancer predisposition syndrome caused by inactivating germline mutations in DNA mismatch repair genes (*MSH2*, *MLH1*, *MSH6*, and *PMS2*). The lack of EpCAM protein expression requires biallelic *EPCAM* deletion and is often associated with concomitant loss of *MSH2* expression.

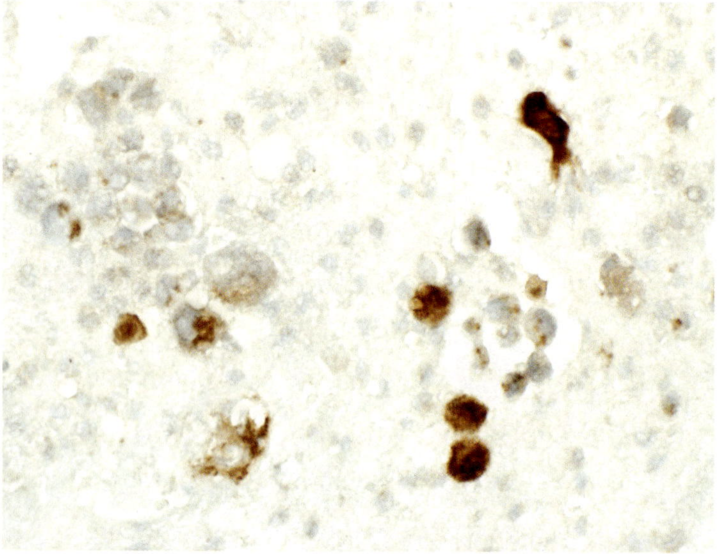

Figure 2.23b — Metastatic Colonic Adenocarcinoma, Pleural Fluid Cell Block Section [Calretinin immunohistochemical stain; high power]. Only nuclear and cytoplasmic Calretinin staining should be interpreted as positive. In this field, the carcinoma cells demonstrate spotty, granular cytoplasmic staining, whereas strong nuclear and cytoplasmic staining is seen in a few benign mesothelial cells. Colon carcinomas with solid growth patterns, microsatellite instability (MSI), and prominent lymphocytic infiltration within and around the tumor are a WHO-recognized subtype of colon carcinoma often referred to as *medullary carcinoma*. Unlike poorly differentiated colonic adenocarcinoma, medullary carcinoma is typically positive for calretinin.

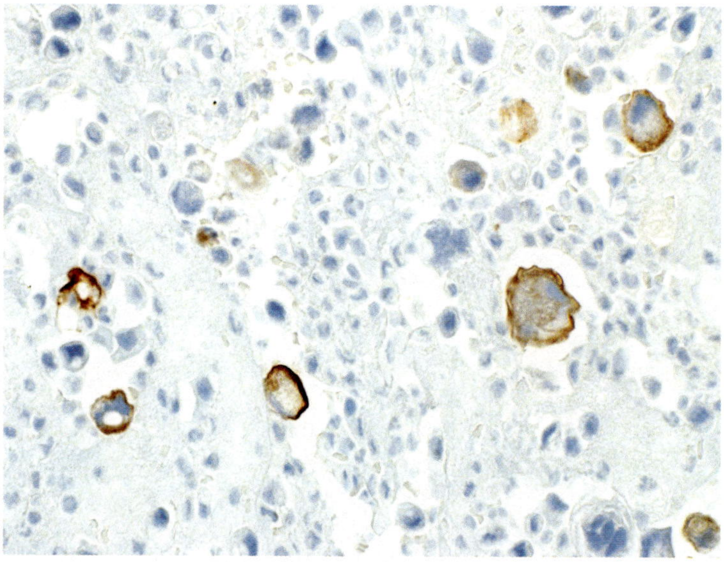

Figure 2.23c — Metastatic Colonic Adenocarcinoma, Pleural Fluid Cell Block Section [Cytokeratin 20 (CK20) immunohistochemical stain; high power]. CK20 is a low-molecular-weight epithelial marker commonly used in combination with cytokeratin 7 (CK7) and others to determine tumor site of origin for metastatic lesions. Expression of CK7 is notably absent in lower gastrointestinal tract tumors. Positive staining for both CK7 and CK20 would favor an upper gastrointestinal (e.g., stomach) or urothelial cell carcinoma; however, any intestinal-type or mucinous tumor may show a similar pattern of staining. Expression of CK20 is mostly absent in nonenteric type lung, breast, and pancreatobiliary tumors, and positive in gastrointestinal (especially colon) and urothelial carcinomas. Negative staining for both CK7 and CK20 would favor prostatic, renal, or hepatocellular carcinoma.

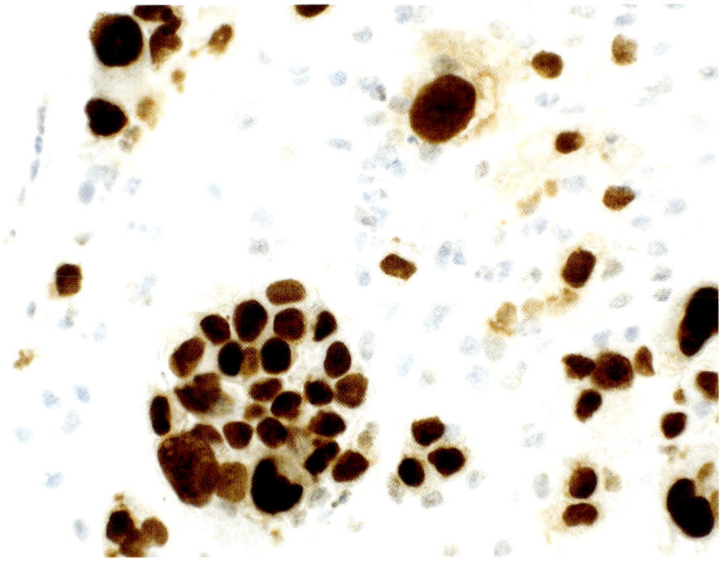

Figure 2.23d — Metastatic Colonic Adenocarcinoma, Pleural Fluid Cell Block Section [CDX2 immunohistochemical stain; high power]. *CDX2* is a homeobox gene that encodes a nuclear transcription factor critical for intestinal embryonic development. It is relatively specific for enteric differentiation and is typically positive in colorectal adenocarcinomas (~85%) but can be negative in poorly differentiated tumors. Positive staining is also seen in bladder/urachal adenocarcinoma, small intestinal carcinoma, gastric adenocarcinoma, gastrointestinal carcinoid tumors, and mucinous adenocarcinomas of the lung, ovary, and cervix.

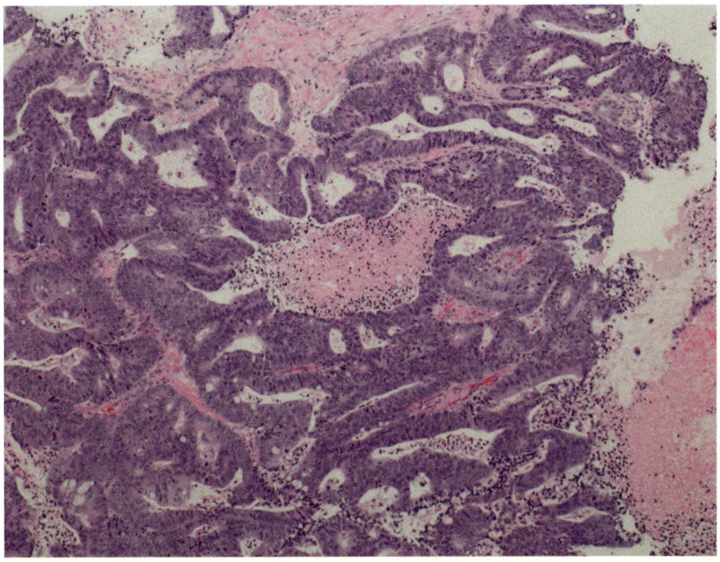

Figure 2.24a — Metastatic Colorectal Adenocarcinoma, Excision [H&E stain; low power]. This excisional biopsy reveals complex glandular structures composed of columnar type cells with zones of necrosis. The histologic features are in keeping with adenocarcinoma with intestinal differentiation. The presence of garland-pattern necrosis consisting of glandular epithelium with incomplete lumina lining necrotic lakes is a characteristic feature of metastatic colorectal carcinoma. Although the leading differential is metastatic colorectal adenocarcinoma, primary lung and ovarian carcinomas can also show intestinal differentiation and be histologically and immunophenotypically indistinguishable from colorectal adenocarcinomas.

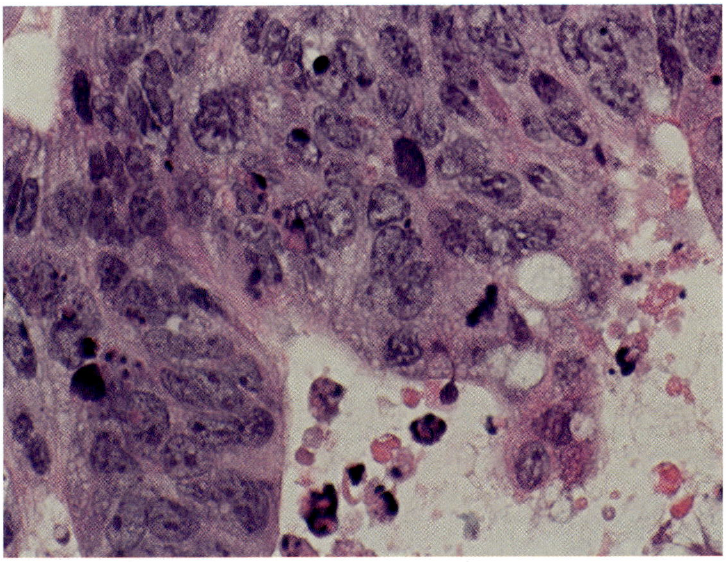

Figure 2.24b — Metastatic Colorectal Adenocarcinoma, Excision [H&E stain; high power]. High-power examination of the metastatic colorectal adenocarcinoma reveals atypical glandular cells with high nuclear-to-cytoplasmic ratios, frequent mitotic figures, and associated cellular necrosis. By immunohistochemistry, colorectal adenocarcinomas are typically immunoreactive for cytokeratin 20 (CK20) and CDX2 and negative for cytokeratin 7 (CK7) and thyroid transcription factor-1 (TTF-1). In contrast, conventional primary pulmonary adenocarcinomas are typically immunoreactive for CK7 and TTF-1, and negative for CK20 and CDX2. However, primary pulmonary adenocarcinomas that display intestinal-type differentiation may lose TTF-1 and CK7 immunoreactivity and instead show CDX2 and CK20 reactivity. In the setting of an intestinal-type adenocarcinoma in the lung, clinical history is key in supporting a diagnosis of metastatic colorectal adenocarcinoma (history of colorectal carcinoma) versus primary pulmonary adenocarcinoma (new solitary mass).

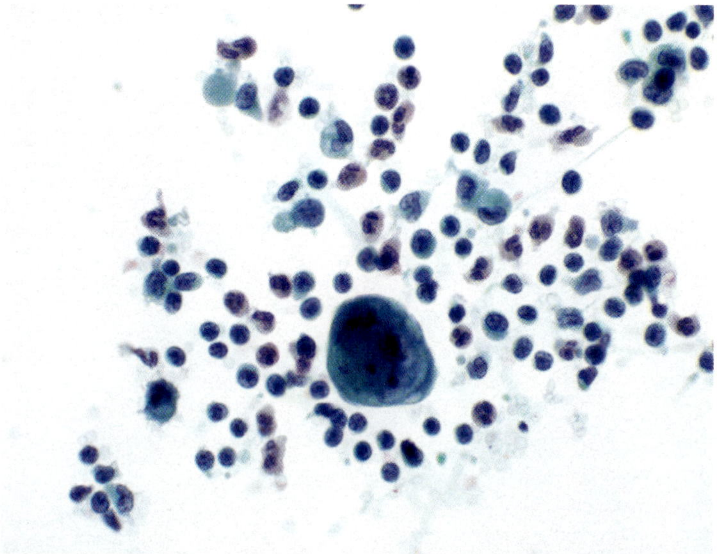

Figure 2.25 — Metastatic Pancreatic Adenocarcinoma, Ascites Fluid [Pap stain; high power]. A single giant malignant cell is present in the middle of the field and surrounded by inflammatory cells, histiocytes, and mesothelial cells. The cell is hyperchromatic and has irregular nuclear borders and coarse chromatin. As is typical for most adenocarcinomas involving fluid specimens, the cytomorphologic findings are nonspecific for even carcinoma.

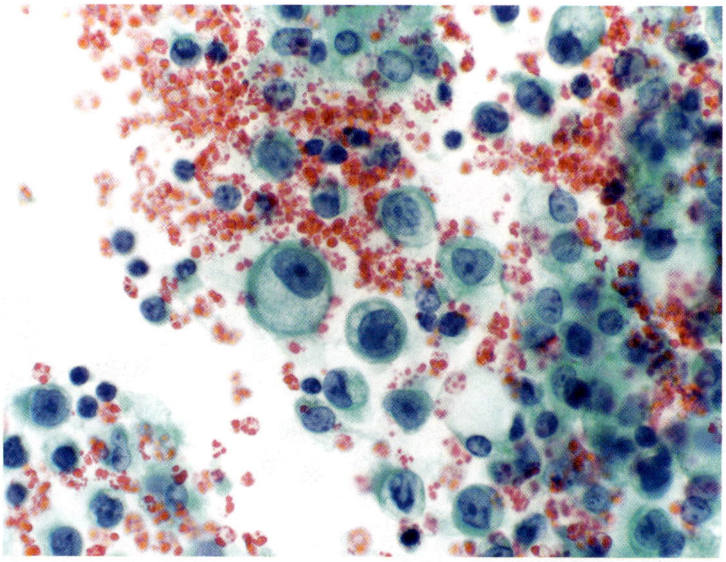

Figure 2.26a — Metastatic Poorly Differentiated Adenocarcinoma, Gastric Primary, Peritoneal Fluid [Pap stain; high power]. This field shows pleomorphic cells with vacuolated cytoplasm, enlarged nuclei, prominent nucleoli, and irregular nuclear borders. The patient has nodular studding of the omentum, small bowel mesentery, and surface of the appendix along with gastric thickening and cystic lesions in both ovaries. Biopsy of the stomach showed a poorly differentiated adenocarcinoma in a background of oxyntic mucosa with active chronic *Helicobacter pylori* gastritis.

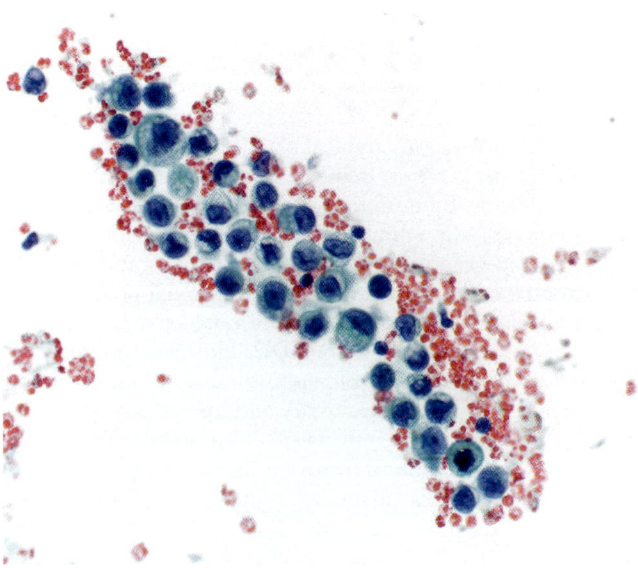

Figure 2.26b — Metastatic Poorly Differentiated Adenocarcinoma, Gastric Primary, Peritoneal Fluid [Pap stain; high power]. Some of the cells show intracytoplasmic mucin vacuoles displacing the nucleus to one side without intraluminal gland formation, whereas others have folded hyperchromatic nuclei with increased nuclear-to-cytoplasmic ratios. Although immunohistochemical stains are helpful in distinguishing metastatic carcinoma from benign processes with signet-ring-cell change (e.g., degenerating histiocytes and mesothelial cells), there are no gastric specific markers to date.

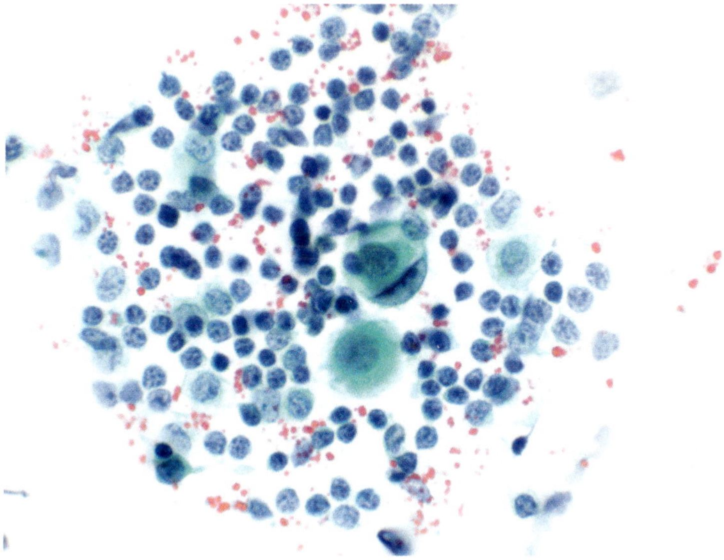

Figure 2.27a — Metastatic Gastric Adenocarcinoma, Ascites Fluid [Pap stain; high power]. Gastric adenocarcinoma frequently displays signet ring morphology. In this case, the tumor cells are discohesive, and blend in with scattered benign mesothelial cells in a serous cavity specimen. A large signet ring cell can be seen in the center of the field; the nucleus is compressed against one side of the cell. Because this cell is so large, it is tempting to imagine that the adjacent epithelioid cells are lymphocytes or mesothelial cells. However, compared with the red blood cells in the field, these cells are much larger. They also have coarse chromatin and irregular nuclear borders; in fact, most of the cells in this field are malignant.

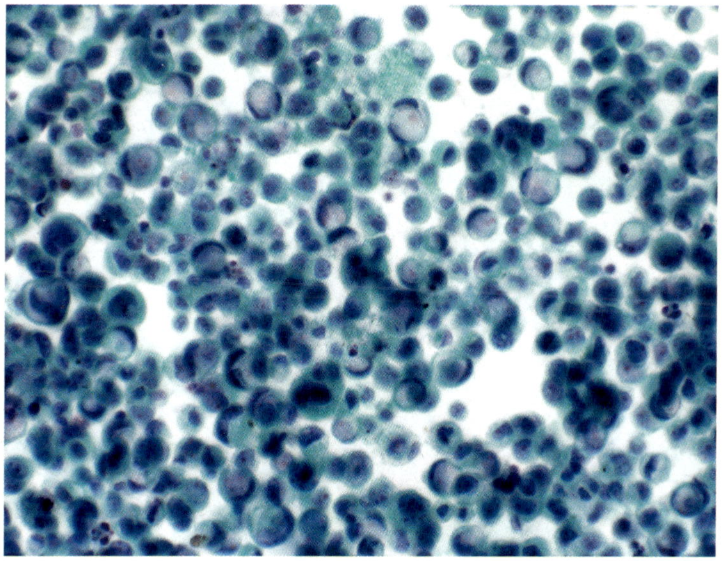

Figure 2.27b — Metastatic Gastric Adenocarcinoma, Ascites Fluid [Pap stain; high power]. This specimen is hypercellular with numerous malignant cells taking on a signet-ring appearance. In this case, the mucin vacuoles sometimes contain pink or cyan bodies that give them a targetoid appearance; this distinguishes them from nonmucinous vacuoles that sometimes form within benign mesothelial cells. In some instances, the malignant cells have engulfed neighboring cells, giving a cell-in-cell appearance. Although this phenomenon is often associated with neoplasia, it can also occur in benign mesothelial cells.

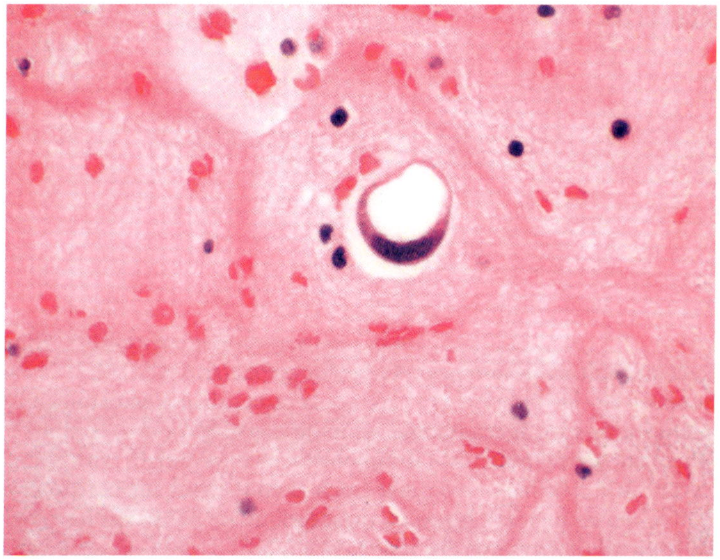

Figure 2.27c — Metastatic Gastric Adenocarcinoma, Ascites Fluid Cell Block Section [H&E stain; high power]. This high-power view shows a large malignant cell with a signet-ring configuration. Notice the retraction artifact of the cell block's gel material away from the cell; this artifact is not seen as prominently in association with benign background cells.

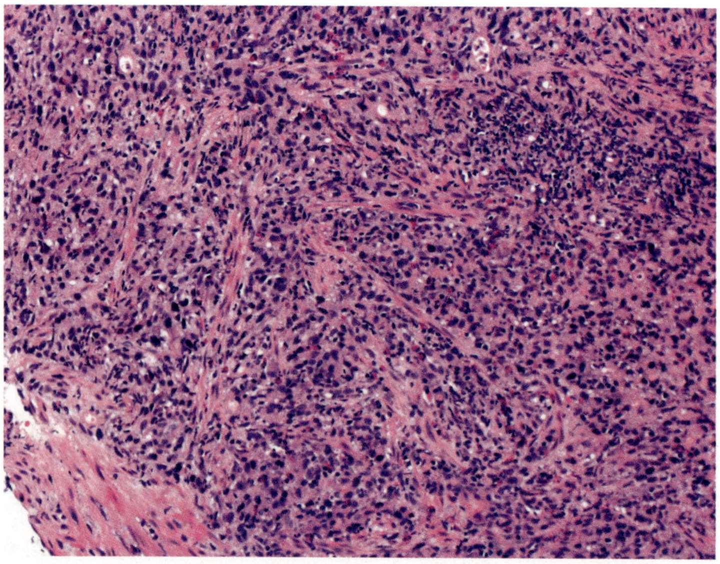

Figure 2.28a — Gastric Signet-Ring Adenocarcinoma, Biopsy [H&E stain; low-intermediate power]. This biopsy contains a cellular infiltrate involving connective tissue and smooth muscle bundles. From low to intermediate power, the exact nature of the cellular infiltrate cannot be definitively characterized. There are abundant lymphocytes, as well as eosinophils, and the differential diagnosis includes benign inflammatory processes, malignant hematologic processes, and signet-ring adenocarcinoma. Signet-ring adenocarcinoma can be particularly treacherous to identify and diagnose on frozen-section evaluation of peritoneal biopsies because the material can be limited, the frozen-section processing creates artifact, and the cells may be rare and histologically mimic histiocytes.

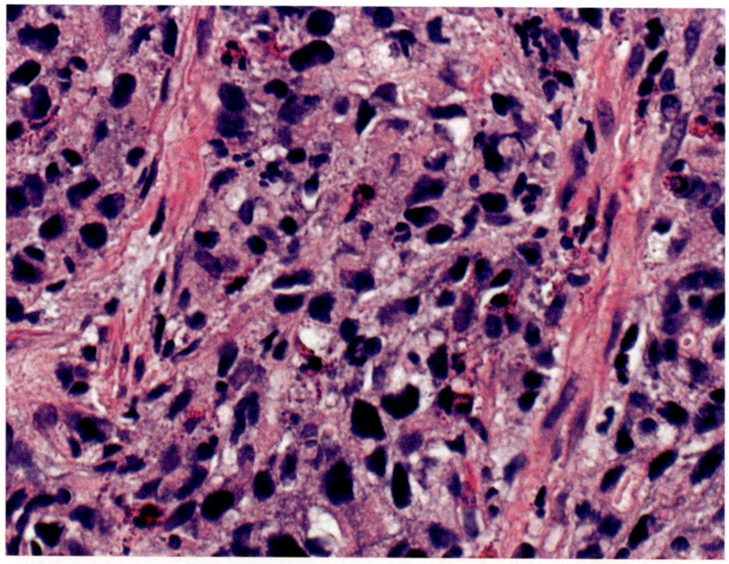

Figure 2.28b — Gastric Signet-Ring Adenocarcinoma, Biopsy [H&E stain; high power]. Higher-power evaluation of this biopsy reveals an atypical cellular infiltrate composed of cells with hyperchromatic and irregular nuclei admixed with plasma cells, lymphocytes, and eosinophils. There is a suggestion of cytoplasmic vacuoles containing mucin, but they are not well developed. The differential diagnosis includes a hematologic malignancy and signet-ring adenocarcinoma, including those of gastrointestinal, lung, gynecologic, and breast origin. A targeted immunopanel is necessary to reach the correct diagnosis.

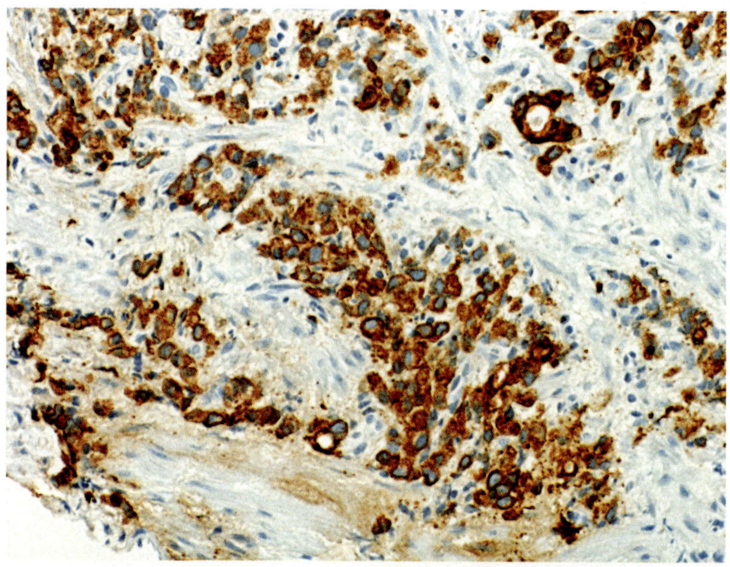

Figure 2.28c — Gastric Signet-Ring Adenocarcinoma, Biopsy [AE1/AE3 immunohistochemical stain; intermediate power]. This cellular infiltrate is immunoreactive for cytokeratin AE1/AE3, supporting classification as a carcinoma rather than a hematologic malignancy such as lymphoma, which is negative for cytokeratins. There are several potential immunohistochemical pitfalls with aberrant cytokeratin labeling of sarcomas (such as epithelioid angiosarcoma and leiomyosarcoma); however, those lesions are not in the histologic differential diagnosis of this lesion.

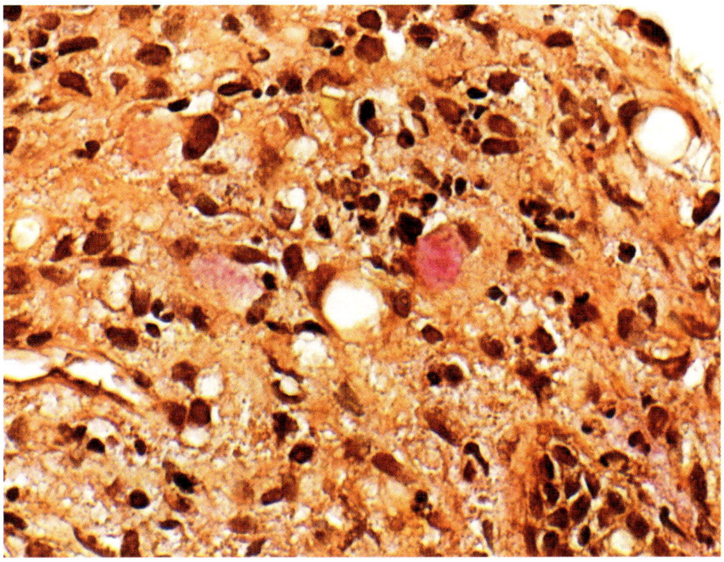

Figure 2.29a — Gastric Signet-Ring Adenocarcinoma, Biopsy [Mucicarmine special stain; high power]. Seen here at high magnification, a special stain for mucicarmine highlights focal intracellular mucin (pink) within the malignant cells, supporting classification as a signet-ring adenocarcinoma. The mucicarmine labeling in signet-ring adenocarcinomas can be diffuse or focal, necessitating complete examination of the whole slide to identify any possible positive areas. The mucicarmine labeling should be intracellular to support classification as an adenocarcinoma; extracellular labeling is not diagnostic and can also be seen in other lesions such as mesothelioma.

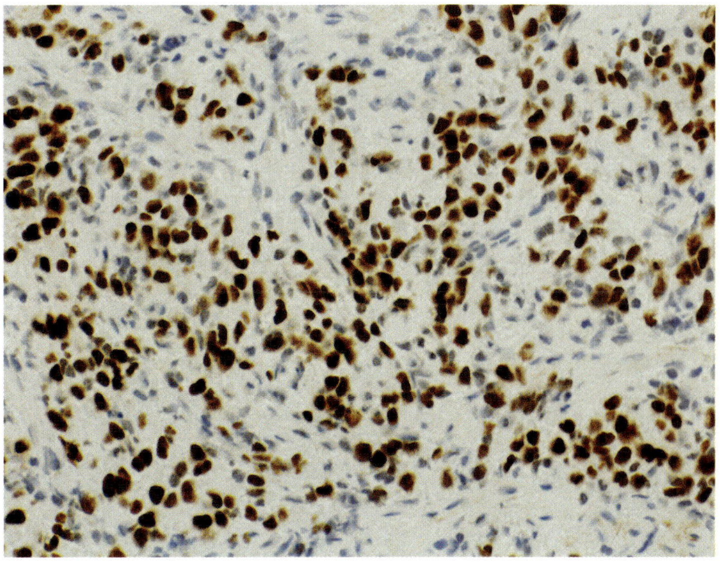

Figure 2.29b — Gastric Signet-Ring Adenocarcinoma, Biopsy [CDX2 immunohistochemical stain; intermediate power]. This signet-ring adenocarcinoma displays strong and diffuse nuclear CDX2 labeling, supporting classification as a primary gastric signet-ring adenocarcinoma in this setting. CDX2 is a nuclear transcription factor with a role in development of the intestinal tract, and immunohistochemical reactivity for CDX2 is used to support intestinal-type differentiation in an adenocarcinoma. However, CDX2 labeling can be seen in an adenocarcinoma of *any origin* that shows intestinal-type differentiation, such as in a primary lung or ovarian adenocarcinoma. In the setting of a signet-ring cell adenocarcinoma in a female patient, the primary differential includes gastric origin and lobular breast carcinoma. CDX2 is useful here because breast lobular carcinomas do not label for CDX2.

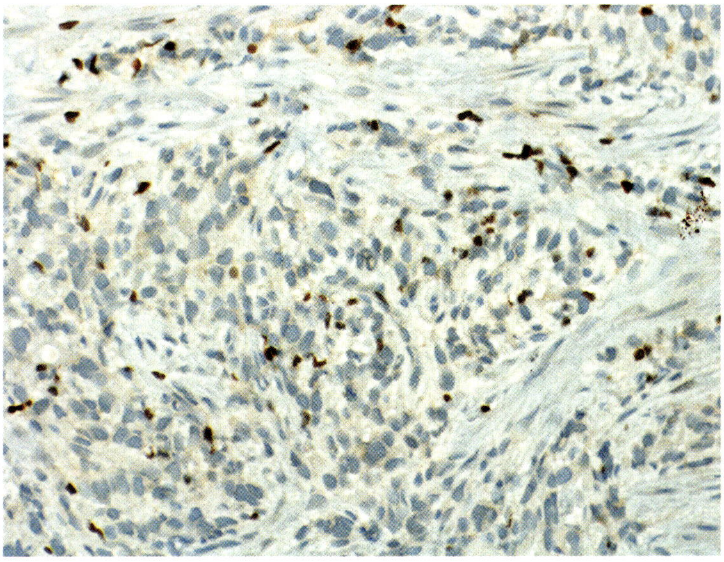

Figure 2.29c — Gastric Signet-Ring Adenocarcinoma, Biopsy [GATA-binding protein 3 (GATA3) immunohistochemical stain; intermediate power]. Gastric signet-ring adenocarcinomas are negative for GATA3 (lymphocytes serve as a positive internal control). The differential diagnosis of a signet-ring adenocarcinoma in a female patient includes metastatic breast lobular carcinoma, the majority of which label for estrogen receptor (ER) and GATA3 but are negative for CDX2. An immunopanel including AE1/AE3, CDX2, ER, and GATA3 are useful in making the distinction between gastric signet-ring adenocarcinoma and breast lobular carcinoma. These two lesions mimic each other not only on the basis of morphology, but also with the pattern of spread, and lobular breast carcinomas can metastasize to the stomach, ovary, and peritoneum. In addition, both carcinomas occur in patients with germline *CDH1* mutations, resulting in the hereditary diffuse gastric and lobular breast cancer syndrome.

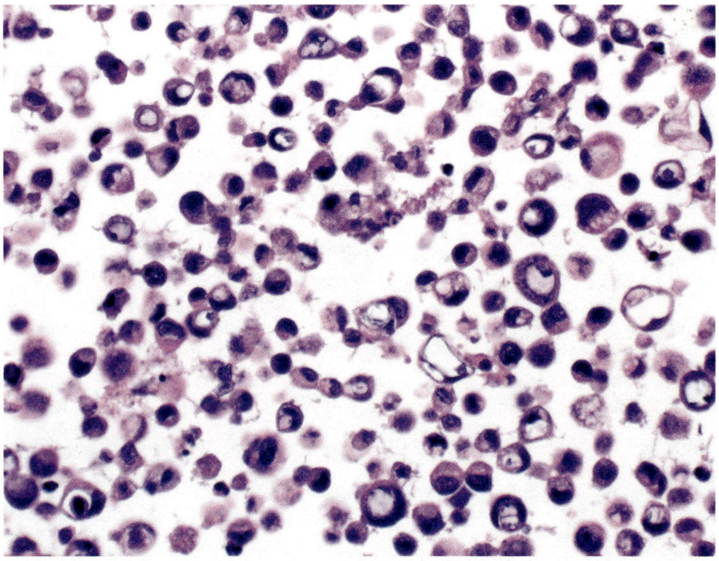

Figure 2.30a — Metastatic Esophageal Adenocarcinoma, Pleural Effusion Cell Block Section [H&E stain; medium power]. On H&E, the cells maintain a signet-ring cell appearance, with the mucinous vacuoles deforming the nuclei to varying degrees, depending on the cell and the angle of sectioning. Significant hyperchromasia can be seen. A simple, special stain for mucin (such as mucicarmine) would confirm the diagnosis if there were any question and the patient's history was known. Otherwise, even immunohistochemical studies may not distinguish between different primary sites within the upper gastrointestinal tract.

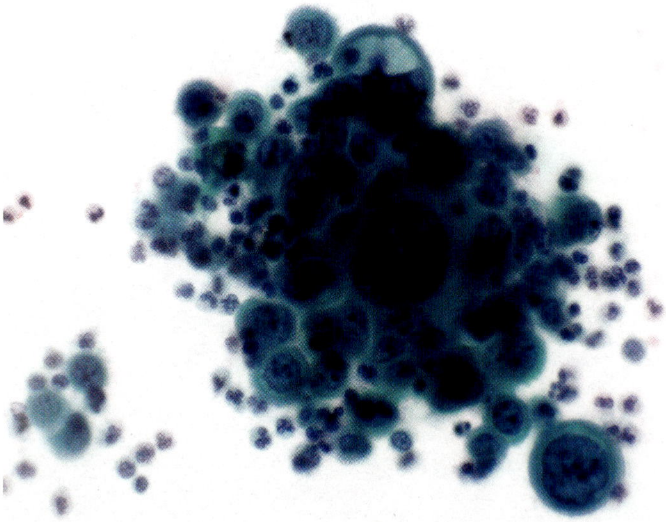

Figure 2.30b — Metastatic Esophageal Squamous Cell Carcinoma, Pleural Effusion [Pap stain; high power]. These cells have all the features of malignancy, with great variation in nuclear size and coarse chromatin. The cells form a fragment, so this is likely a carcinoma. It is poorly differentiated but maintains some cytomorphologic features that suggest a squamous cell carcinoma—the nuclei are mostly centrally located, and the cytoplasm appears dense. The large cell at the top of the field has a vacuolated appearance that may at first suggest this to be an adenocarcinoma. In this case, the patient had an esophageal squamous cell carcinoma.

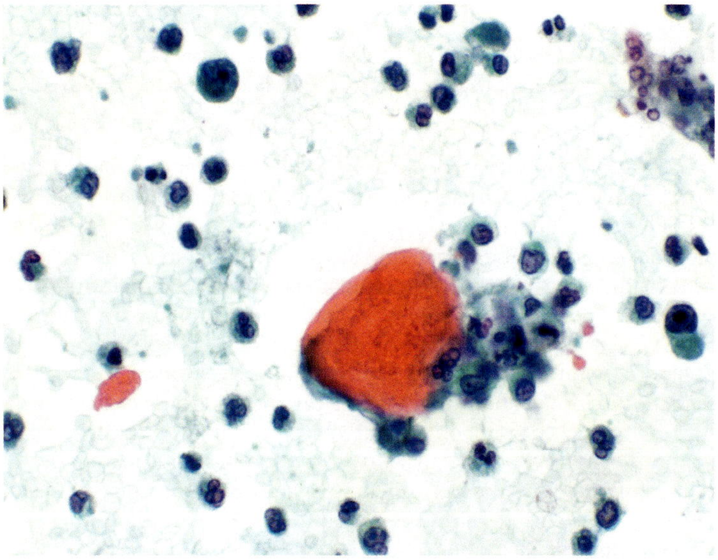

Figure 2.31a — Metastatic Pulmonary Squamous Cell Carcinoma, Pleural Fluid [Pap stain; high power]. This field demonstrates a large fragment of acellular material with orangeophilic staining. Taken alone, it is difficult to determine the origin of acellular material; however, in this case, a cell on the right side of the field has an enlarged nucleus and a gigantic nucleolus. Although nucleoli can be seen in reactive conditions, a nucleolus this large is almost certainly malignant. Keratinaceous material is not typically seen in serous effusions because squamous cell carcinoma only rarely involves serous cavities. In patients with a history of squamous cell carcinoma, the presence of keratin alone should cause one to look more closely for malignant cells.

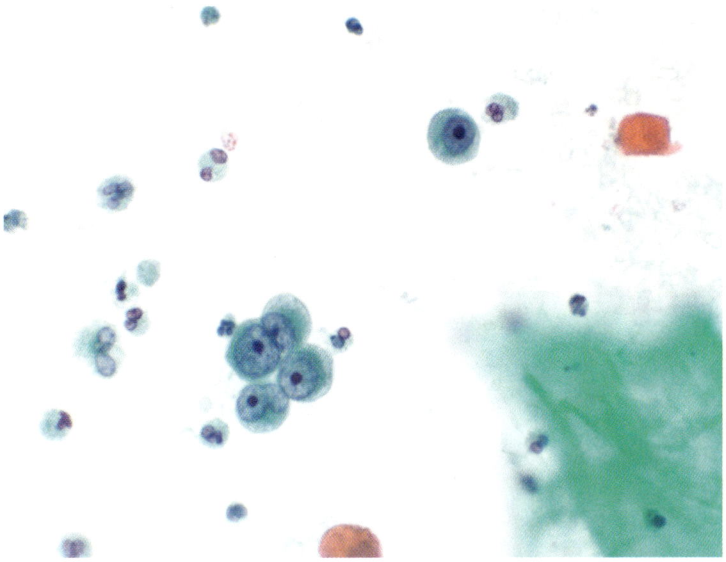

Figure 2.31b — Metastatic Pulmonary Squamous Cell Carcinoma, Pleural Fluid [Pap stain; high power]. Rare specks of keratin, along with several malignant cells, can be seen in the field. The malignant cells are relatively uniform and only slightly enlarged, and they could be mistaken for mesothelial cells if it were not for their prominent nucleoli. In this case, the cytomorphologic features seen could also be found in an adenocarcinoma, and thus the fragments of keratin in the background are the largest clue to a squamous cell carcinoma. Although cytokeratin 5/6 (CK5/6) expression is shared between cells of squamous and mesothelial lineage, squamous cells tend to express p40 and p63, whereas calretinin is specific for mesothelial cells in this differential.

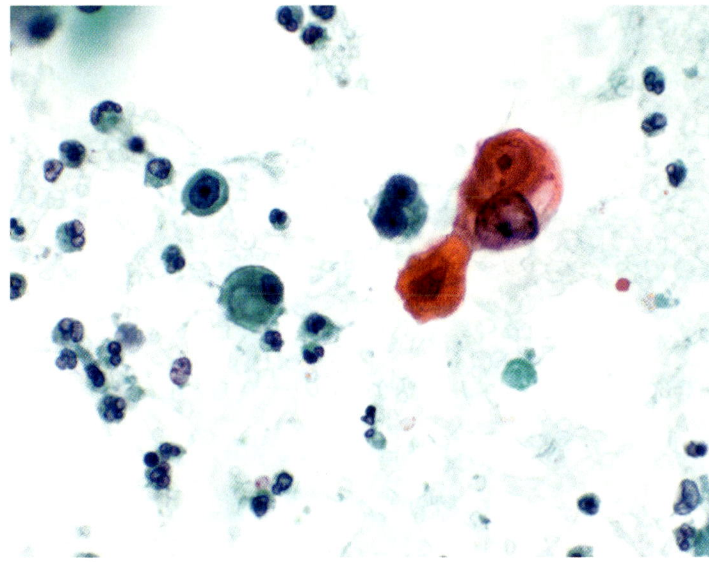

Figure 2.31c — Metastatic Pulmonary Squamous Cell Carcinoma, Pleural Fluid [Pap stain; high power]. Although all the malignant cells in this field contain keratin because they are carcinoma cells, the group of cells in the right upper corner demonstrates the orangeophilic keratin staining seen in keratinizing squamous cell carcinoma with the Pap stain. The dense keratin results in a much fainter staining, and thus hyperchromasia is less apparent. Note how the keratinized cells also have irregular cellular borders and irregular shapes compared with the other malignant cells in the field.

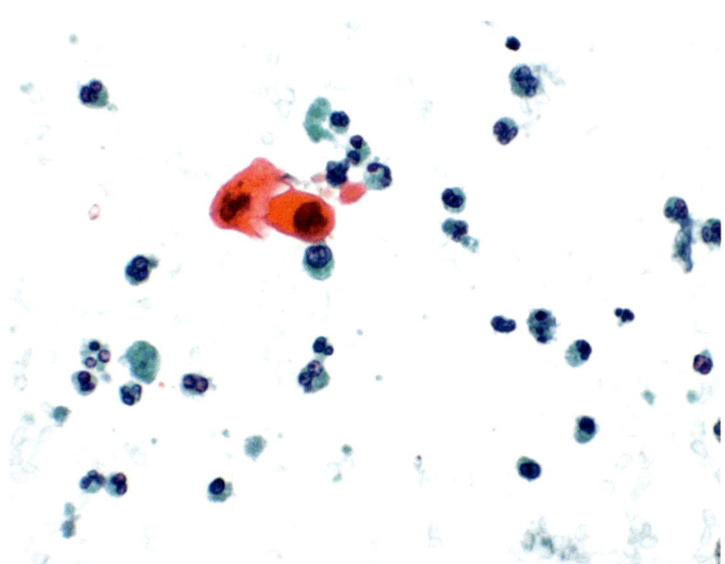

Figure 2.32a — Metastatic Pulmonary Squamous Cell Carcinoma, Pleural Fluid [Pap stain; high power]. The two keratinized cells seen here have pyknotic nuclei and irregular shapes. They are greatly enlarged compared with the background inflammatory cells. Because of the pyknotic nature of the nuclei, the nuclear-to-cytoplasmic ratios are not increased, as is often seen in other carcinomas.

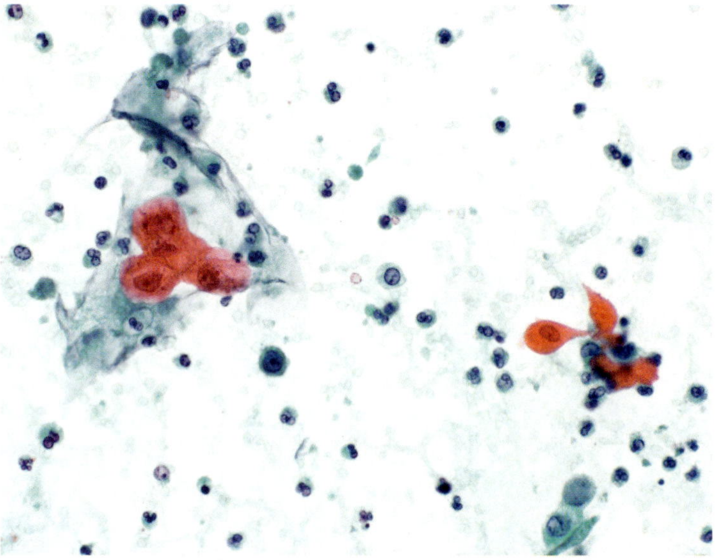

Figure 2.32b — Metastatic Pulmonary Squamous Cell Carcinoma, Pleural Fluid [Pap stain; medium power]. The field contains malignant keratinized cells, inflammation, and acellular debris, which clings to the malignant cells like the "clinging" tumor diathesis seen in liquid-based Pap tests. Because squamous cell carcinomas are often necrotic, this material could represent necrosis in a metastatic tumor or indicate invasion of the carcinoma into the serous cavity.

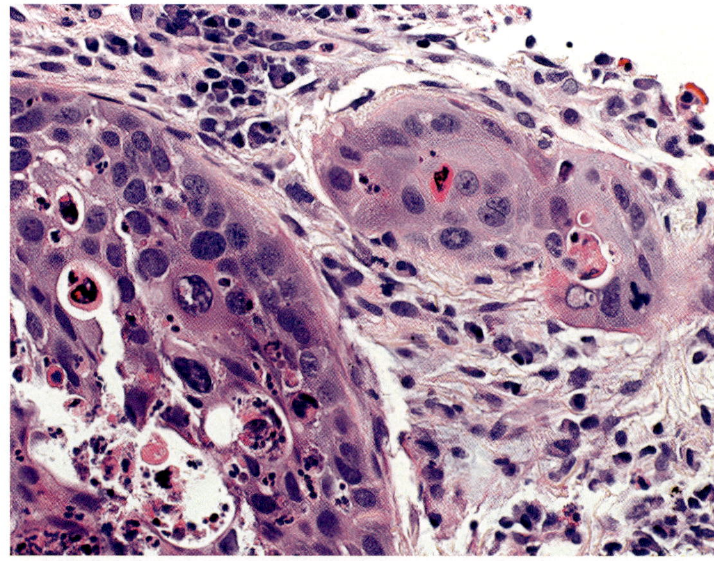

Figure 2.33a — Primary Lung Squamous Cell Carcinoma, Biopsy [H&E stain; intermediate power]. Lung carcinoma is the most common cancer type worldwide, and the incidence of lung adenocarcinoma is higher than that of lung squamous cell carcinoma. Although both are considered non–small cell lung carcinomas, distinguishing between lung adenocarcinoma and squamous cell carcinoma is essential because some treatments do not work in or have excessive toxicity in patients with squamous cell carcinomas. At times, the histologic features alone may be sufficient to distinguish between squamous and glandular differentiation. However, at other times, immunostains may be necessary to differentiate the two. This field shows a non–small cell lung carcinoma with focal dyskeratotic cells and atypical mitotic figures, suggestive of a squamous cell carcinoma.

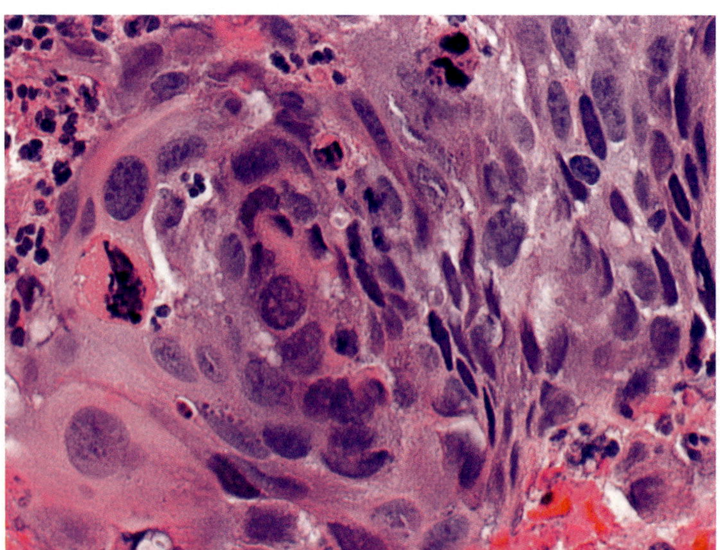

Figure 2.33b — Primary Lung Squamous Cell Carcinoma, Biopsy [H&E stain; high power]. Higher-power examination of this non–small cell lung carcinoma confirms the presence of intracellular keratin, dyskeratotic cells, and intercellular bridges, supporting classification as a squamous cell carcinoma. In poorly differentiated or nonkeratinizing carcinomas, an immunopanel containing thyroid transcription factor-1 (TTF-1), Napsin A, and p40 may be necessary to confirm the diagnosis of squamous cell carcinoma versus adenocarcinoma. Adenocarcinomas should be evaluated via molecular sequencing for the presence of targetable mutations or rearrangements. All non–small cell carcinomas should now be evaluated for programmed death-ligand 1 (PD-L1) status via approved assays because anti–PDL-1 immunotherapy is now approved as first- and second-line therapy.

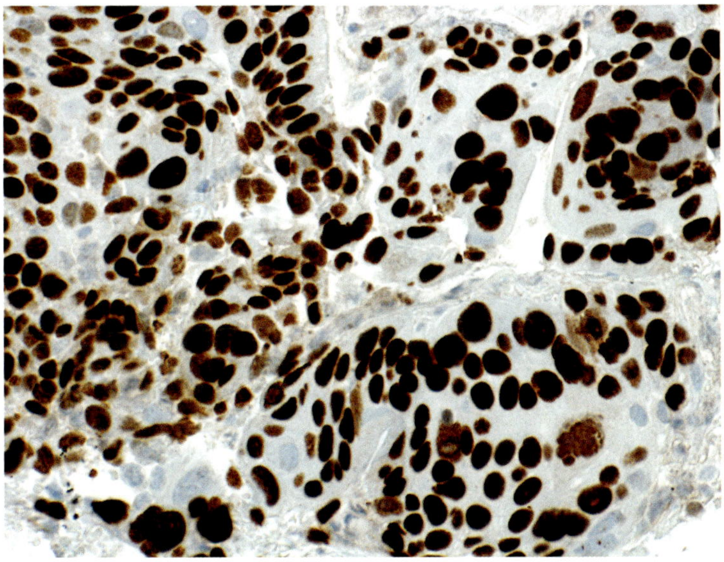

Figure 2.33c — Primary Lung Squamous Cell Carcinoma, Biopsy [p40 immunohistochemical stain; intermediate power]. Squamous cell carcinomas are immunoreactive for p40, which is an isoform of p63. p63 also labels squamous cell carcinomas, but p63 is less specific for squamous differentiation than p40, and a greater degree of p63 labeling is seen in pulmonary adenocarcinomas. Squamous cell carcinomas of any primary site label for p40, such that clinical history and radiographic findings are often necessary to distinguish between a primary lung squamous cell carcinoma and metastasis from another site. The one useful feature may be the human papillomavirus (HPV) status of a carcinoma, as primary lung carcinomas are rarely if ever HPV-related.

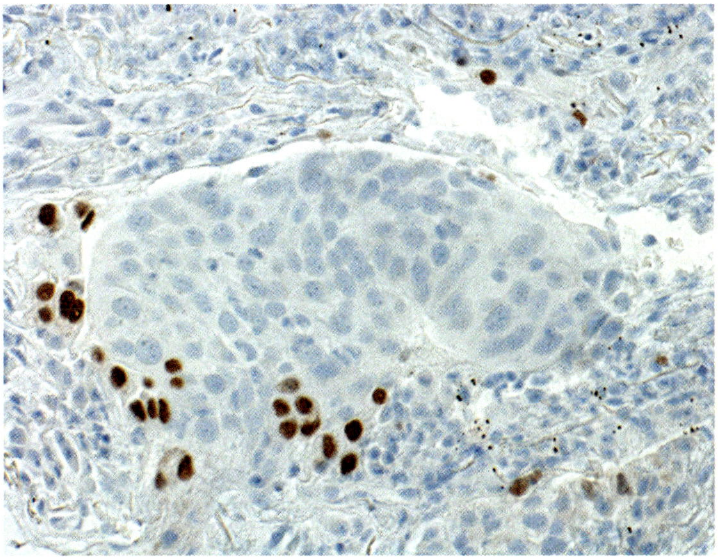

Figure 2.34a — Primary Lung Squamous Cell Carcinoma, Biopsy [Thyroid transcription factor-1 (TTF-1) immunohistochemical stain; intermediate power]. This non–small cell carcinoma is negative for TTF-1, arguing against a diagnosis of adenocarcinoma and supporting the morphologic impression of squamous cell carcinoma. Note that TTF-1 labeling is seen in the benign, reactive pneumocytes adjacent to the carcinoma cells; this labeling should not be overinterpreted as TTF-1 labeling within the carcinoma cells. Positive TTF-1 labeling in a tumor trumps the p63 or p40 status and would favor classification as an adenocarcinoma, even if there is focal p63 or p40 labeling in the tumor.

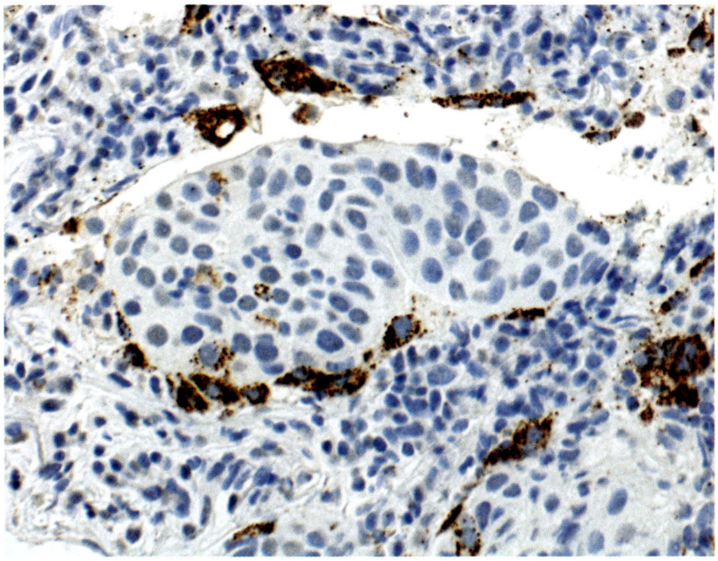

Figure 2.34b — Primary Lung Squamous Cell Carcinoma, Biopsy [Napsin A immunohistochemical stain; intermediate power]. This non–small cell carcinoma is Napsin A-negative, arguing against a diagnosis of pulmonary adenocarcinoma and instead supporting the morphologic impression of squamous cell carcinoma. Note again that the benign pneumocytes, as well as admixed alveolar macrophages, show the characteristic cytoplasmic, nuclear labeling for Napsin A, and this should not be overinterpreted as positive Napsin A labeling in the carcinoma.

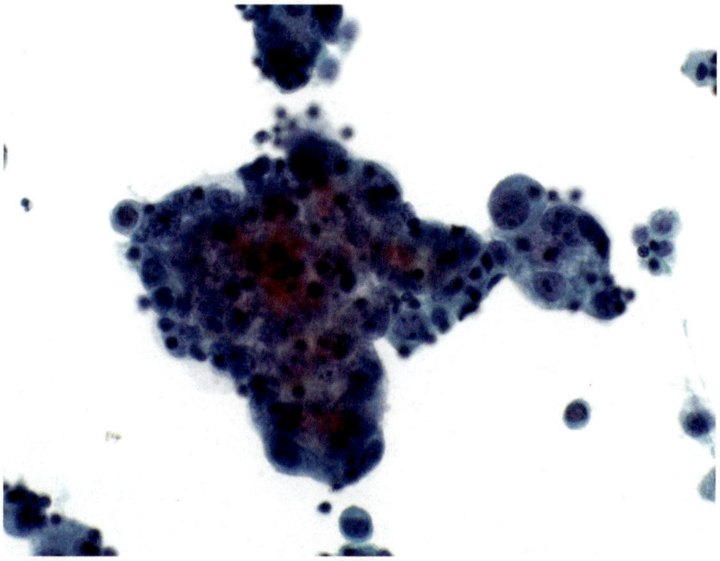

Figure 2.35a — Metastatic Pulmonary Adenocarcinoma, Pleural Fluid [Pap stain; high power]. Three-dimensional clusters, papillary fragments, and acinar structures with smooth community borders are commonly encountered in lung adenocarcinomas. The cells tend to be pleomorphic with hyperchromatic nuclei, coarsely granular chromatin, prominent nucleoli, and delicately vacuolated cytoplasm. Note the scattered inflammatory cells and reactive mesothelial cells in the background. Molecular profiling is feasible on cytology specimens and clinically helpful in identifying patients who may benefit from targeted therapies. Approximately 10% of patients with lung adenocarcinoma in the United States have tumor-associated *EGFR* mutations, which are potentially targetable with epidermal growth factor receptor-tyrosine kinase inhibitors (EGFR-TKIs). The location and type of mutation are important, however, because certain variants confer increased sensitivity to TKIs (e.g., L858R, exon 19 deletions), whereas others confer acquired resistance (e.g., threonine-790 to methionine [T790M]) or decreased sensitivity (e.g., exon 20 insertions). Never smokers (defined as <100 cigarettes in a lifetime) with adenocarcinoma have the highest incidence of *EGFR, HER2, ALK, RET,* and *ROS1* mutations. *KRAS* mutations are relatively rare in never smokers and are less common in East Asian versus U.S./European patients.

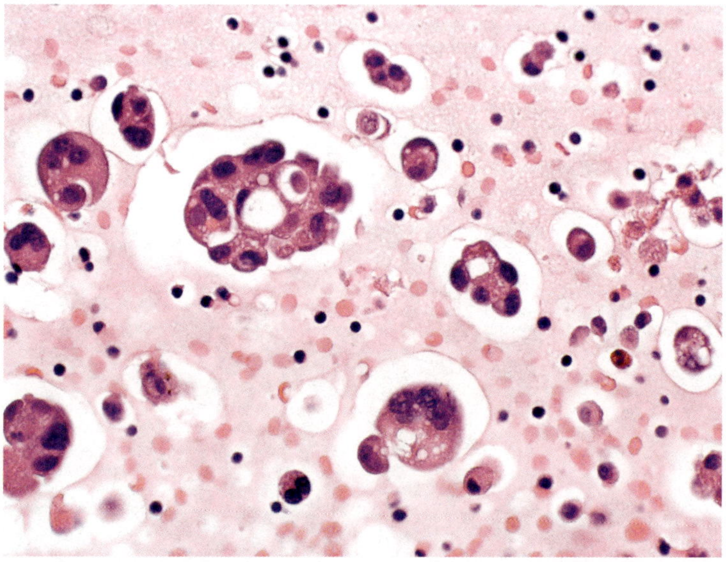

Figure 2.35b — Metastatic Pulmonary Adenocarcinoma, Pleural Fluid Cell Block Section [H&E stain; high power]. Cytologically, adenocarcinoma differentiation can be expressed in several architectural patterns, including flat cohesive sheets, papillae, and micropapillary tufts, as shown here. Although pulmonary adenocarcinoma with a micropapillary component is considered a poor prognostic factor for overall survival and for recurrence in patients with limited resections, the presence of micropapillary tufts in cytology material is not a good indicator of micropapillary pattern in surgically resected specimens (positive predictive value ~64%).

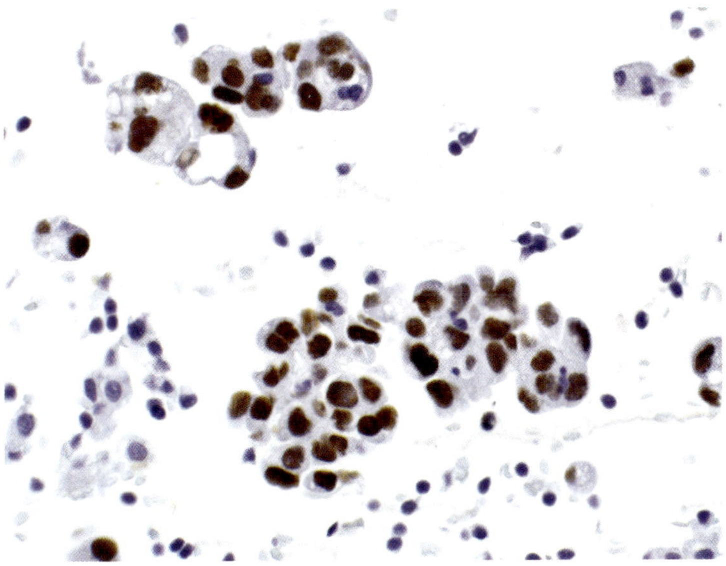

Figure 2.36a — Metastatic Pulmonary Adenocarcinoma, Pleural Fluid Cell Block Section [Thyroid transcription factor-1 (TTF-1) immunohistochemical stain; high power]. TTF-1 is a nuclear protein that regulates transcription activity in the lung and thyroid. It is normally expressed in type II pneumocytes and Clara cells, thyroid follicular and parafollicular C cells, and diencephalon. Overall, it is expressed in 75% of nonmucinous lung adenocarcinomas, 75% to 90% of small cell carcinomas of the lung, 10% of mucinous lung adenocarcinomas, and 40% of large cell carcinomas of the lung. Other tumor subsets that are known to stain with TTF-1 include small cell carcinoma of nonpulmonary sites, pulmonary sclerosing hemangioma, cholangiocarcinoma, endometrial and endocervical adenocarcinoma, and esophageal adenocarcinoma. Neoplastic thyroid tissue is also TTF-1 positive, but immunoreactivity tends to be less common or at lower levels in undifferentiated thyroid carcinomas.

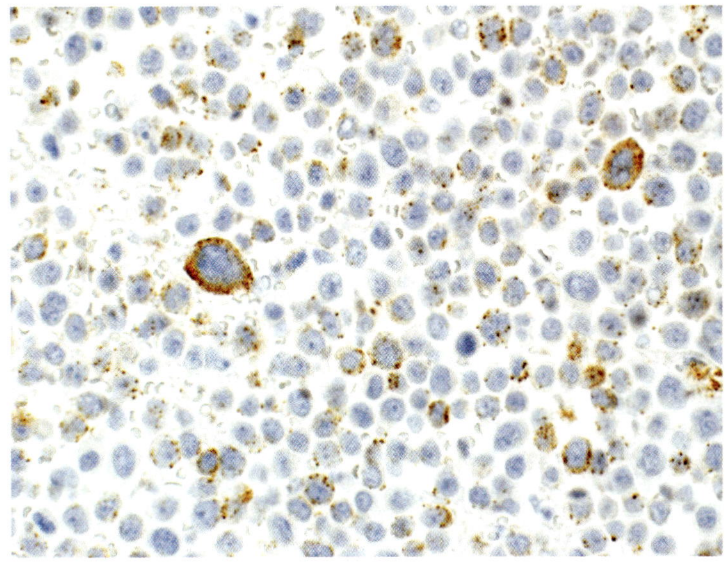

Figure 2.36b — Metastatic Pulmonary Adenocarcinoma, Pleural Fluid Cell Block Section [Napsin A immunohistochemical stain; high power]. Napsin A is an aspartic proteinase that is abundantly expressed in normal lung (type 2 pneumocytes and alveolar macrophages) and kidney tissue (both fetal and adult). It is involved in processing pulmonary surfactant protein B produced by alveolar type 2 pneumocytes and protein catabolism in the proximal tubules of the adult kidney. In neoplasms other than primary lung adenocarcinoma, Napsin A expression is most frequently observed in ovarian clear cell adenocarcinoma and papillary Renal Cell Carcinoma (RCC), followed by other subtypes of RCC, thyroid carcinoma, and adrenal cortical neoplasms.

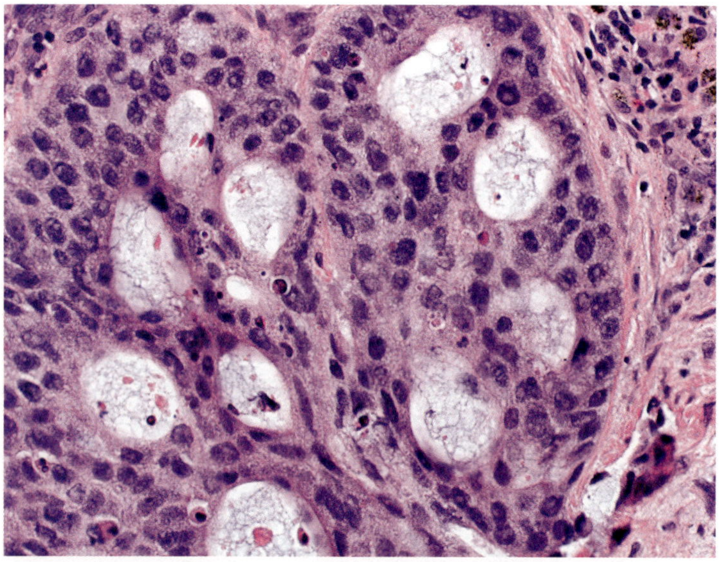

Figure 2.37a — Pulmonary Adenocarcinoma Involving the Pleura, Biopsy [H&E stain; intermediate power]. Pulmonary adenocarcinoma can display a range of morphologic features. In this field, the adenocarcinoma displays a cribriform architecture with gland formation and intraluminal mucin, atypical cells with nuclear hyperchromasia, and readily apparent mitoses and apoptotic figures. Other variant morphologies include acinar (gland-forming) pattern, solid pattern, micropapillary pattern, papillary pattern, and mucinous morphology, among others. Pulmonary adenocarcinoma is the most common type of lung carcinoma to arise in nonsmokers.

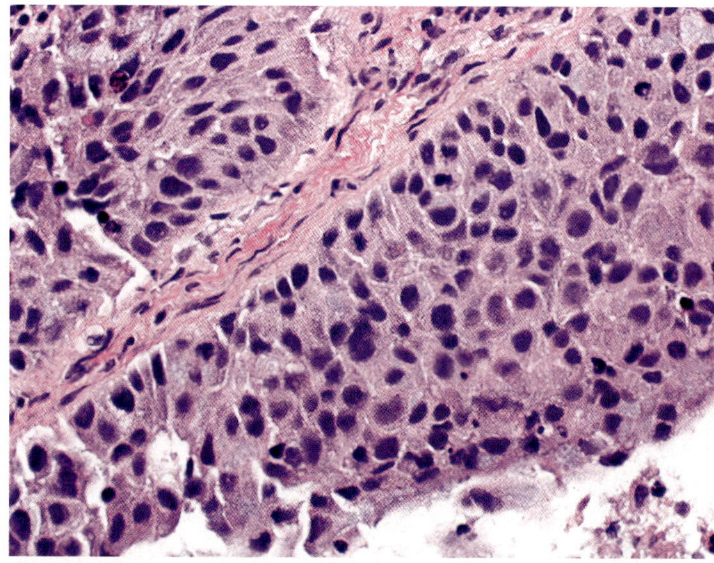

Figure 2.37b — Pulmonary Adenocarcinoma Involving the Pleura, Biopsy [H&E stain; intermediate power]. This field of a pulmonary adenocarcinoma involving the pleura is taken from the same case previously illustrated, which showed a cribriform glandular architecture. In this field, the adenocarcinoma displays solid pattern growth with scattered cells containing intracellular mucin vacuoles, as well as regions of necrosis. The differential diagnosis of a solid pattern carcinoma involving the pleura should include spread from a pulmonary adenocarcinoma, involvement by malignant mesothelioma, and metastatic spread from another site. Immunostains can aid in this differential.

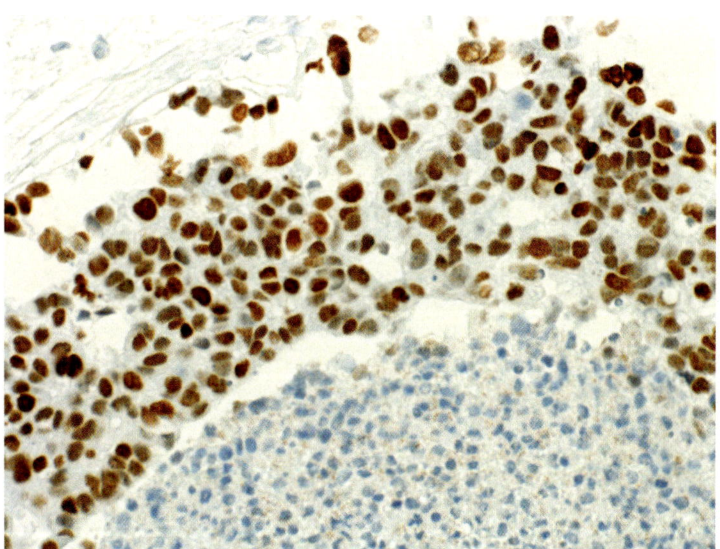

Figure 2.37c — Pulmonary Adenocarcinoma Involving the Pleura, Biopsy [Thyroid transcription factor-1 (TTF-1) immunohistochemical stain; intermediate power]. The majority of pulmonary adenocarcinomas are positive for TTF-1, which is used to support pulmonary or thyroid origin in a carcinoma. Because TTF-1 is a transcription factor, TTF-1 labeling should be nuclear in distribution. TTF-1 also labels small cell carcinomas of any site. TTF-1 labeling is rare in carcinomas of other origin, as well as pulmonary squamous cell carcinomas. In addition to TTF-1 and Napsin A, pulmonary adenocarcinomas also label for cytokeratin 7 (CK7) but are typically negative for cytokeratin 20 (CK20).

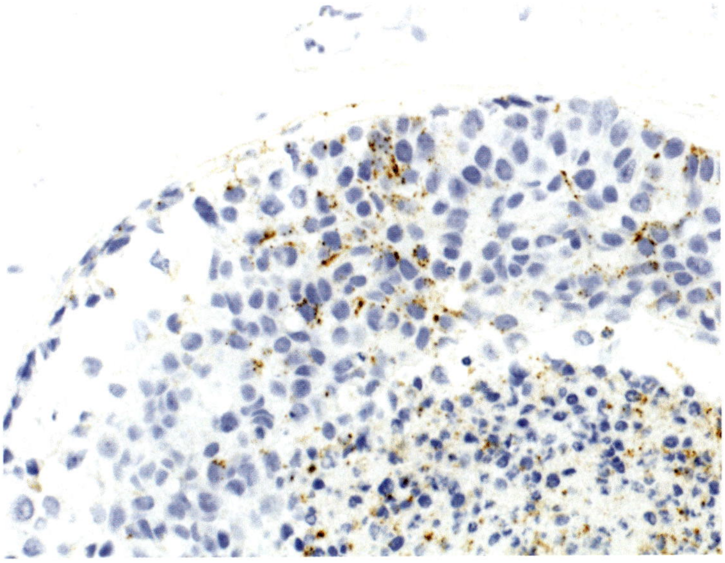

Figure 2.37d — Pulmonary Adenocarcinoma Involving the Pleura, Biopsy [Napsin A immunohistochemical stain; intermediate power]. The majority of pulmonary adenocarcinomas are positive for Napsin A, which is a protein involved in the production of surfactant. Labeling for Napsin A should be cytoplasmic, is typically granular in appearance, and may be focal. A potential diagnostic pitfall is Napsin A labeling in histiocytes, which should not be overinterpreted as labeling in neoplastic cells. In addition, Napsin A labeling is seen in papillary Renal Cell Carcinomas (RCCs).

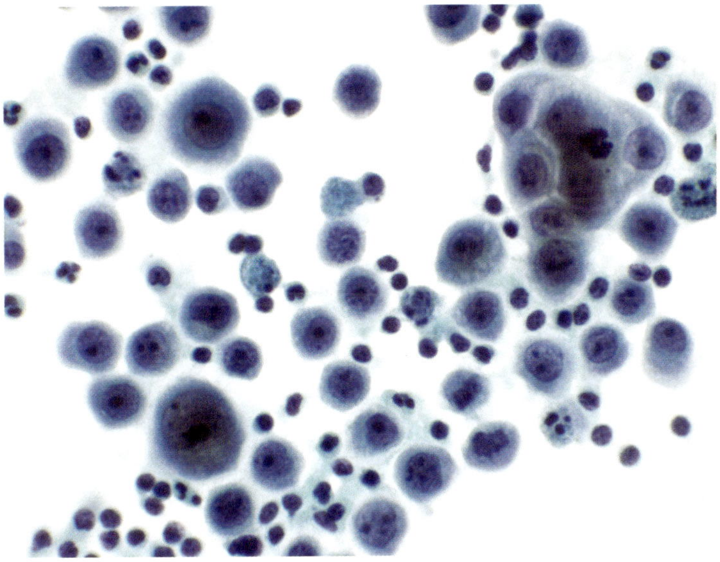

Figure 2.38a — Metastatic Adenosquamous Carcinoma, Pleural Fluid [Pap stain; high power]. This smear shows tumor cells with dense cytoplasm, distinct cell borders, hyperchromatic nuclei, prominent nucleoli, and rare mitotic figures (upper right). There is subtle irregularity of the nuclear membranes but no overt keratinization, necrosis, or spindle/caudate shapes. In the absence of unequivocal keratinization, immunohistochemistry with positive squamous markers such as p40 or p63 is required to diagnose nonkeratinizing squamous cell carcinoma.

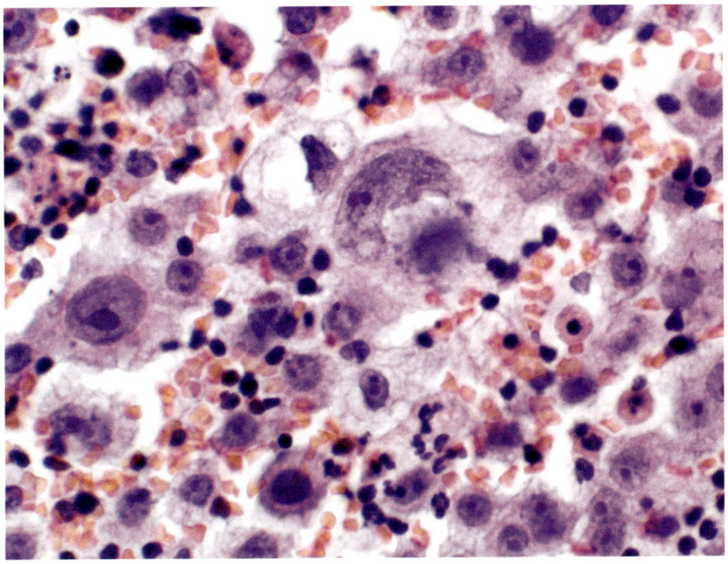

Figure 2.38b — Metastatic Adenosquamous Carcinoma, Pleural Fluid Cell Block Section [H&E stain; high power]. A cell block of the same specimen demonstrates pleomorphic cells with vacuolated cytoplasm, prominent nucleoli, and rare cells with intracytoplasmic blue mucin (center). Tumors are considered adenosquamous carcinoma when components of adenocarcinoma and squamous cell carcinoma each compose at least 10% of the tumor. Although this diagnosis can be made with certainty only on surgical specimens, the identification of glandular differentiation in cytology specimens should prompt molecular testing for potentially actionable alterations. Tumors with even minor components of adenocarcinoma have been shown to harbor *ALK, KRAS,* or *EGFR* alterations.

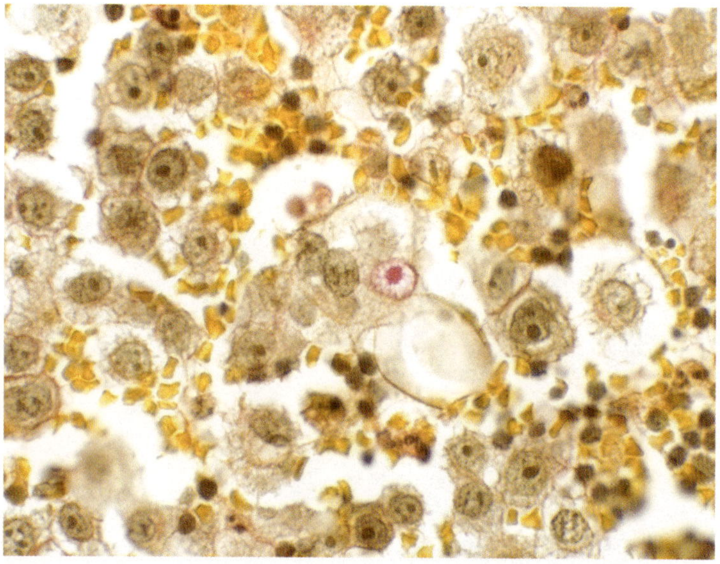

Figure 2.38c — Metastatic Adenosquamous Carcinoma, Pleural Fluid Cell Block Section [Mucicarmine special stain; high power]. Mucicarmine helps differentiate adenocarcinoma from squamous cell carcinoma with pseudoglandular features. True intracellular mucin (center) should stain pink with mucicarmine, as seen here. These areas should also be positive for thyroid transcription factor-1 (TTF-1) and Napsin A.

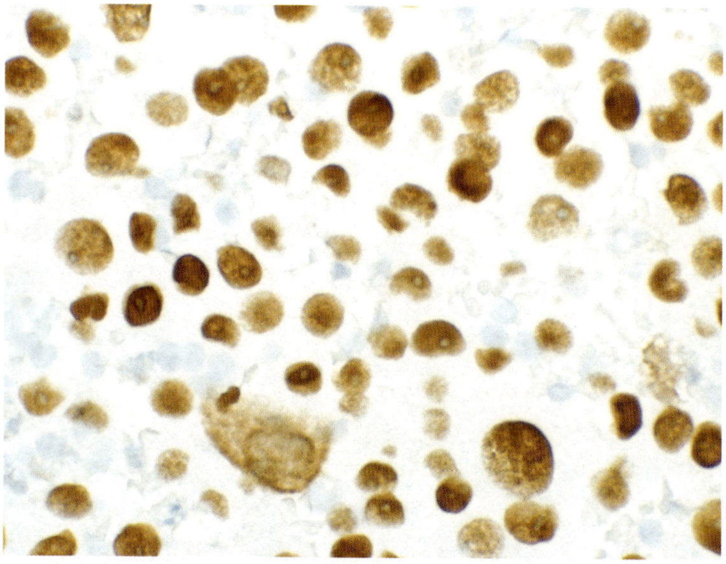

Figure 2.38d — Metastatic Adenosquamous Carcinoma, Pleural Fluid Cell Block Section [p40 immunohistochemical stain; high power]. P40 is a truncated, nontransactivating isoform of p63 that is highly specific for squamous cells and basal cells. Adenosquamous carcinoma typically demonstrates strong p40 immunostaining in the squamous cell carcinomatous component, which should also be negative for thyroid transcription factor-1 (TTF-1).

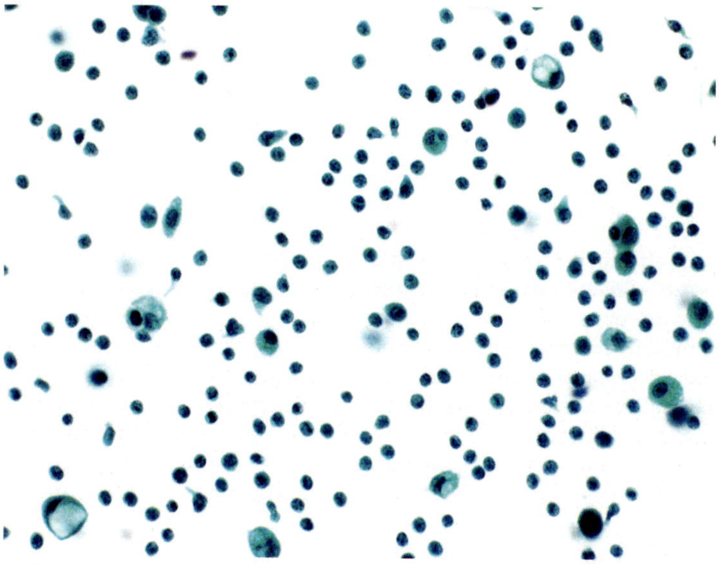

Figure 2.39a — Metastatic Small Cell Lung Carcinoma, Pleural Fluid [Pap stain; medium power]. Small cell carcinoma of the lung is characterized by small cells with scant cytoplasm, round to oval nuclei, finely granular chromatin, and absent or inconspicuous nucleoli. In this field, the cells are dispersed as single cells, mimicking mature lymphocytes. Occasionally, tear-dropped cells can be seen with cytoplasmic tails and elongation of nuclear material, which is not typical of mature lymphocytes. Larger mesothelial cells and histiocytes are scattered in the background.

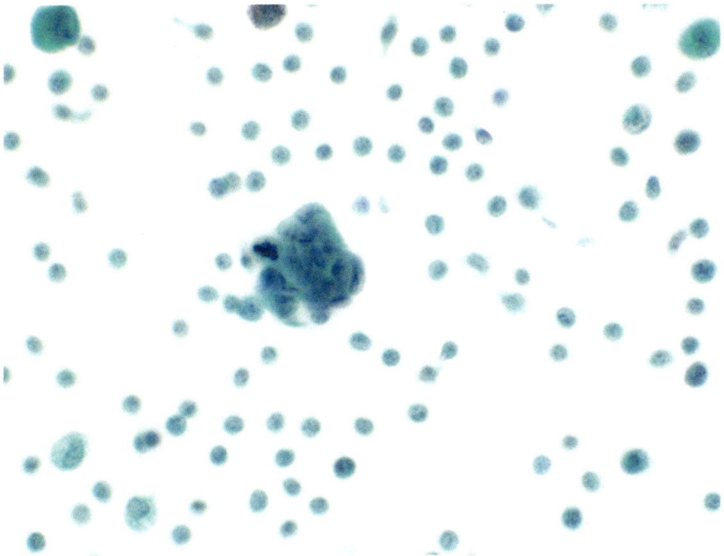

Figure 2.39b — Metastatic Small Cell Lung Carcinoma, Pleural Fluid [Pap stain; medium power]. Small tight clusters of cells (center) with nuclear molding and ill-defined cell borders are common. Nuclei should maintain a finely distributed granular chromatin texture. Mitotic figures are usually numerous, and tumors frequently have a short-term response to cytotoxic chemotherapy and radiotherapy. After chemotherapy, however, the proliferation rate of small cell lung cancer may be markedly diminished to levels more fitting for carcinoid tumors.

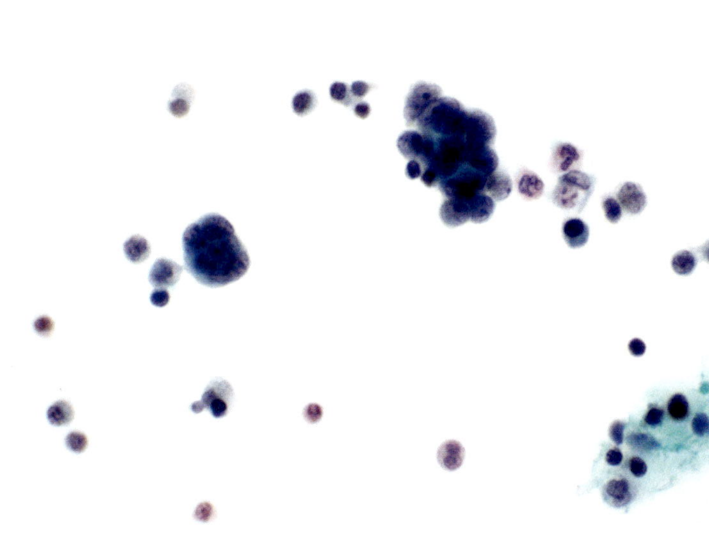

Figure 2.39c — Metastatic Small Cell Lung Carcinoma, Pleural Fluid [Pap stain; high power]. Features such as individual cell necrosis characterized by pyknotic nuclei and apoptotic bodies (lower right), nucleoli, and tumor cell cannibalism can be seen in some cases of serous effusions with metastatic (upper right) small cell carcinoma. Although ipsilateral pleural effusion (benign or malignant) may still be considered as limited-stage, the majority of patients show evidence of widespread dissemination at presentation. No genetic factors investigated thus far are predictive of prognosis.

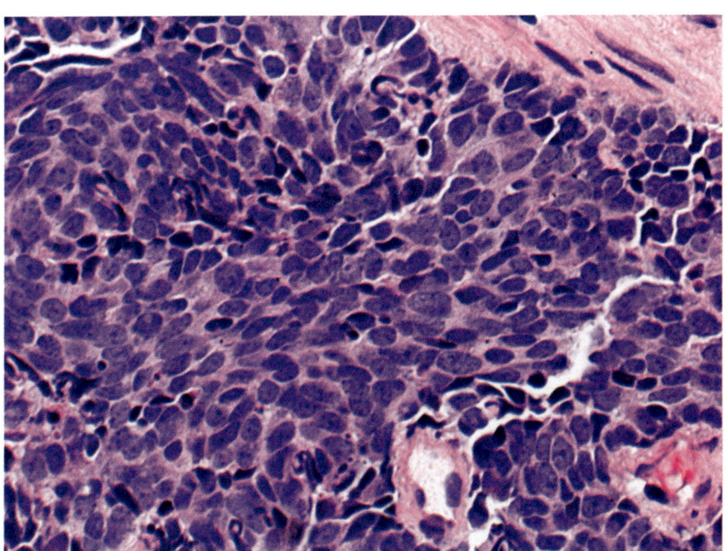

Figure 2.40a — Small Cell Lung Carcinoma, Transbronchial Biopsy [H&E stain; high power]. This transbronchial biopsy reveals sheets of highly atypical, small, round blue cells characterized by high nuclear-to-cytoplasmic ratios, nuclear hyperchromasia, nuclear molding, inconspicuous nucleoli, and numerous apoptotic bodies. The histologic features are characteristic of small cell carcinoma, which is highly aggressive. The histologic differential diagnosis includes diffuse large B-cell lymphoma, primitive neuroectodermal tumor (PNET), and basaloid squamous cell carcinoma.

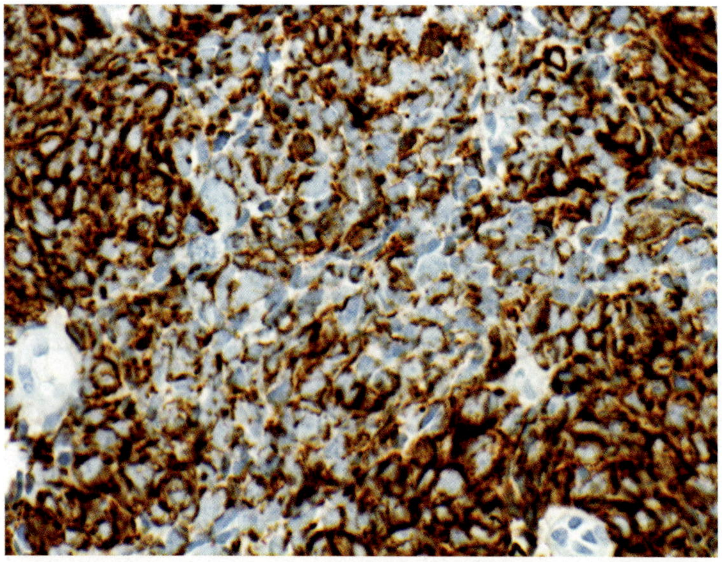

Figure 2.40b — Small Cell Lung Carcinoma, Transbronchial Biopsy [CAM5.2 immunohistochemical stain; high power]. Small cell carcinomas are typically immunoreactive for cytokeratin and characteristically display perinuclear dot-like labeling, as illustrated by this CAM5.2 immunostain. Small cell carcinomas may display only focal cytokeratin labeling, and it may be necessary to include several cytokeratin antibodies on an immunopanel to demonstrate epithelial differentiation. The cytokeratin immunoreactivity distinguishes small cell carcinoma from lymphoma and primitive neuroectodermal tumor (PNET).

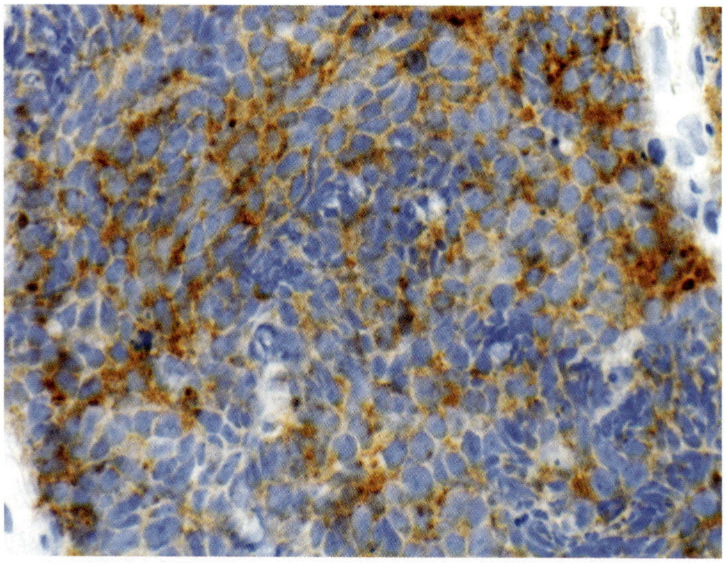

Figure 2.40c — Small Cell Lung Carcinoma, Transbronchial Biopsy [synaptophysin immunohistochemical stain; high power]. Small cell carcinomas are immunoreactive for neuroendocrine markers, such as synaptophysin, as seen here, as well as chromogranin and CD56. CD56 is the most sensitive but least specific of these three neuroendocrine markers, such that non-neuroendocrine neoplasms may display CD56 labeling as well. The reactivity for neuroendocrine markers may be focal in small cell carcinoma, as well as absent on any one individual marker.

Figure 2.41a — Small Cell Lung Carcinoma, Transbronchial Biopsy [Thyroid transcription factor-1 (TTF-1) immunohistochemical stain; high power]. Small cell carcinoma displays immunoreactivity for TTF-1. Although TTF-1 labeling in an adenocarcinoma is used to support a pulmonary origin, TTF-1 labeling in small cell carcinoma can be seen in small cell carcinomas *of any primary site*. TTF-1 labeling in a small cell carcinoma does not indicate a pulmonary origin. The TTF-1 labeling does help distinguish small cell carcinoma from basaloid squamous cell carcinoma and Merkel cell carcinoma, which are negative for TTF-1.

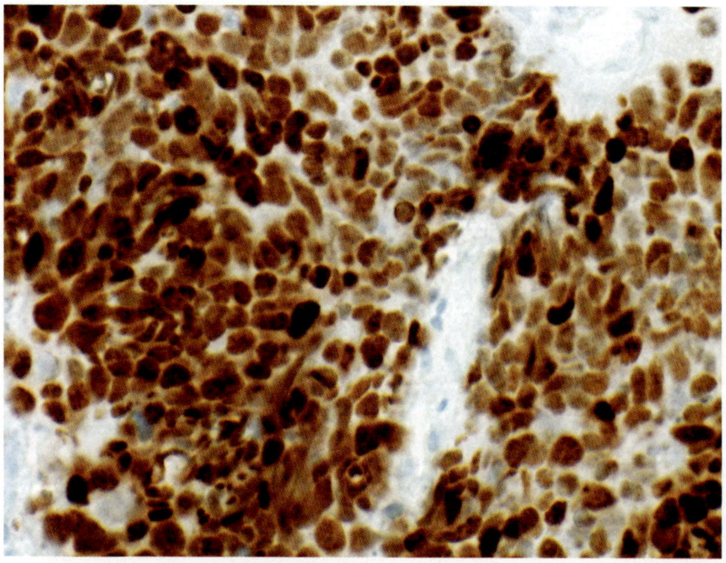

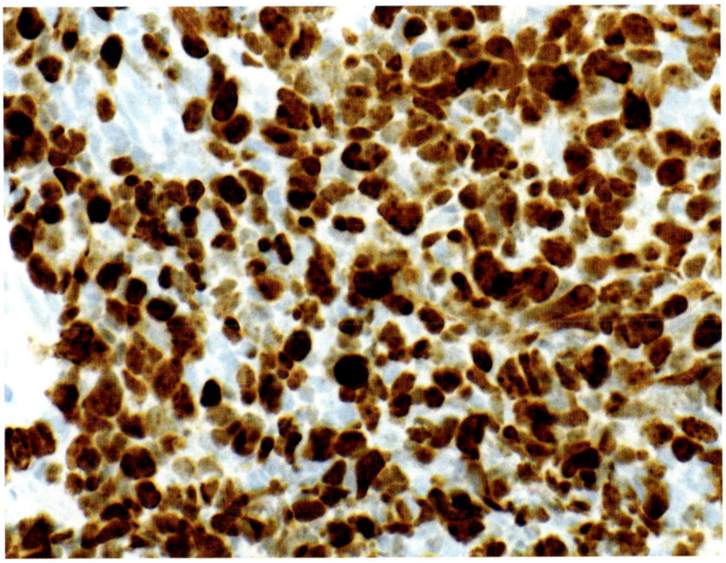

Figure 2.41b — Small Cell Lung Carcinoma, Transbronchial Biopsy [Ki-67 immunohistochemical stain; high power]. Small cell carcinomas display high Ki-67 (MIB1) proliferation indices, reflecting the high proportion of cells undergoing active cell cycle progression. This is also paralleled by the high mitotic rate and abundant apoptotic bodies present on the H&E. Diffuse large B-cell lymphomas, which are in the differential diagnosis for small cell carcinoma, also have high Ki-67 proliferation indices of greater than 90%; however, diffuse large B-cell lymphomas label for B-cell markers such as CD20 or CD19 and are negative for cytokeratins.

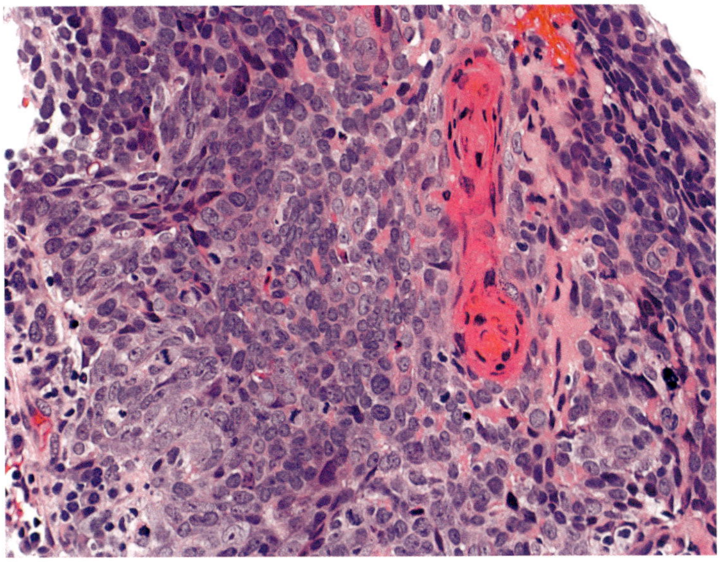

Figure 2.42a — Metastatic human papillomavirus–related (HPV-related) squamous cell carcinoma, Biopsy [H&E stain; intermediate power]. This transbronchial biopsy of a patient with a known history of head and neck squamous cell carcinoma reveals a basaloid neoplasm with abrupt keratinization. The presence of keratinization supports classification as a squamous cell carcinoma, and the differential diagnosis otherwise includes poorly differentiated adenocarcinoma, small cell carcinoma, lymphoma, and melanoma. The histologic features seen here do not distinguish between a primary or a metastatic squamous cell carcinoma. The presence of an abrupt transition between basaloid cells and keratinizing cells is also a feature seen in nuclear protein in testis (NUT) carcinomas, which occur in the midline of the body and are associated with rearrangements in the *NUT* gene; these carcinomas are immunoreactive for *NUT1.*

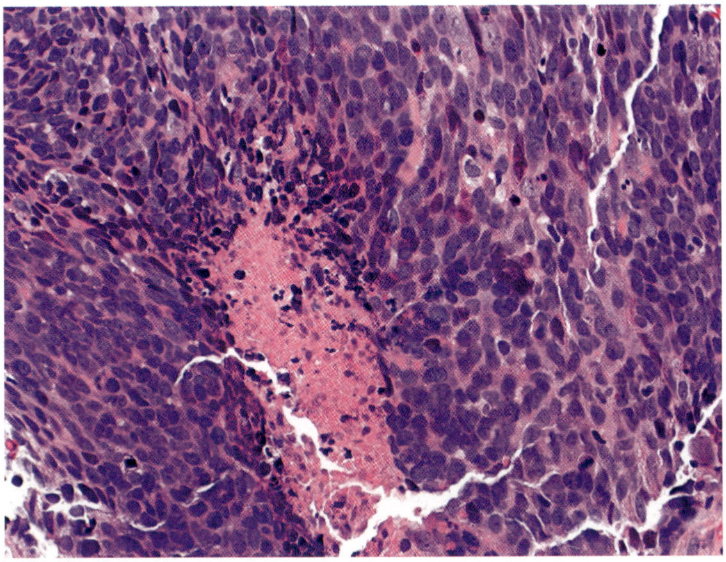

Figure 2.42b — Metastatic human papillomavirus–related (HPV-related) squamous cell carcinoma, Biopsy [H&E stain; intermediate power]. This field of the same transbronchial biopsy shows hyperchromatic cells, nuclear molding, a high nuclear-to-cytoplasmic ratio, and associated cellular necrosis. Although keratinization is seen elsewhere in the biopsy, the morphologic features seen here can be seen in squamous cell carcinomas with basaloid features and in small cell carcinomas. The presence of keratinization does not exclude an admixed small cell carcinoma component, as both can occur in the same tumor. A targeted immunopanel can aid in this differential.

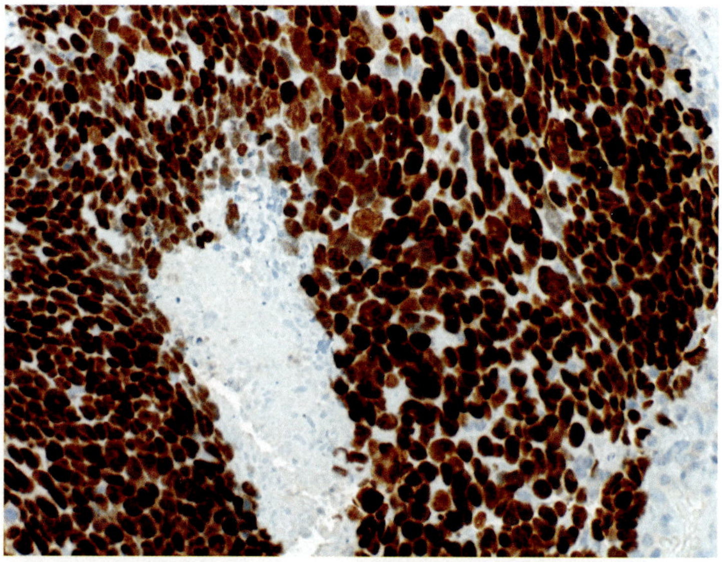

Figure 2.43a — Metastatic human papillomavirus–related (HPV-related) Squamous Cell Carcinoma, Biopsy [p40 immunohistochemical stain; intermediate power]. This basaloid carcinoma is strongly and diffusely immunoreactive for nuclear p40 labeling, and is negative for the neuroendocrine marker synaptophysin, the adenocarcinoma/small cell carcinoma marker thyroid transcription factor-1 (TTF-1), and the nuclear protein in testis (NUT) carcinoma marker NUT1. The morphology and immunoprofile support classification as a squamous cell carcinoma. The histologic features seen here do not distinguish between a primary or metastatic squamous cell carcinoma. The patient's history of a head and neck squamous cell carcinoma may favor a metastasis; however, smoking is a risk favor for both head and neck squamous cell carcinomas and pulmonary squamous cell carcinomas, such that the clinical history does not exclude the possibility of a second primary tumor in the lung.

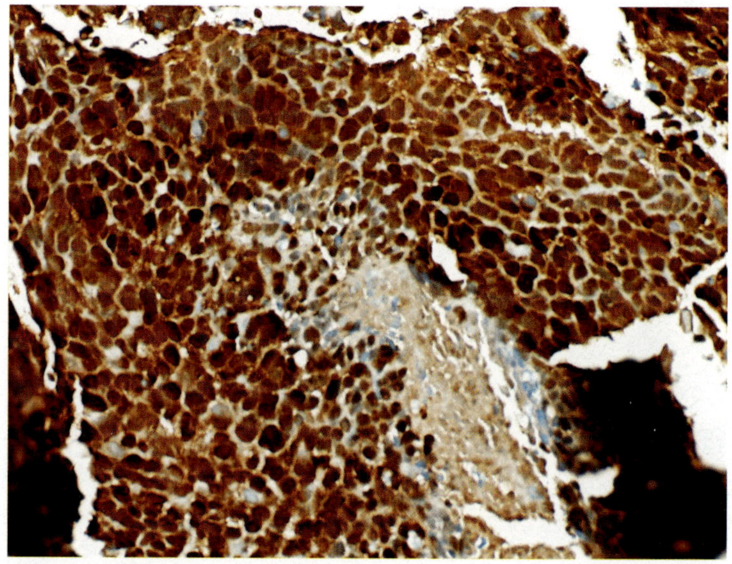

Figure 2.43b — Metastatic human papillomavirus–related (HPV-related) Squamous Cell Carcinoma, Biopsy [p16 immunohistochemical stain; intermediate power]. The patient's known head and neck squamous cell carcinoma was an HPV-related squamous cell carcinoma of the tonsil. This is useful information, because primary lung squamous cells are not HPV related, and special studies to assess the HPV status of the current lesion in the lung can address this question. Indeed, this carcinoma is strongly and diffusely immunoreactive for p16, which is a surrogate for HPV infection. The p16 labeling should indeed be strong and diffuse, and not focal or patchy, to consider it "positive" in this setting. The viral proteins inactivate Rb, leading to increased p16 protein levels and p16 labeling by immunohistochemistry. p16 labeling is not entirely specific to HPV infection, however, since any inactivation of Rb will lead to p16 positivity. For instance, Rb inactivation is a common features of small cell carcinomas, and p16 positivity does not imply HPV integration in those tumors. In situ hybridization studies for the actual viral nucleic acids are needed for confirmation.

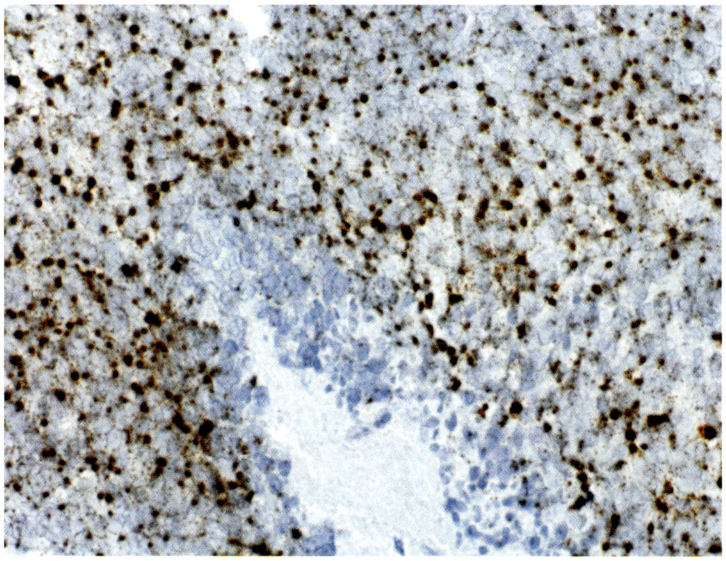

Figure 2.43c — Metastatic human papillomavirus–related (HPV-related) Squamous Cell Carcinoma, Biopsy [HPV RNA in situ hybridization; intermediate power]. The squamous cell carcinoma shows diffuse positivity for HPV by RNA in situ hybridization, as evidenced by the dark brown punctate staining throughout the carcinoma cell nuclei. Positive labeling should be interpreted as signal within the carcinoma cells, and any potential background signal in the stromal cells or extracellular space should be excluded. This finding supports classification as a metastatic HPV-related squamous cell carcinoma.

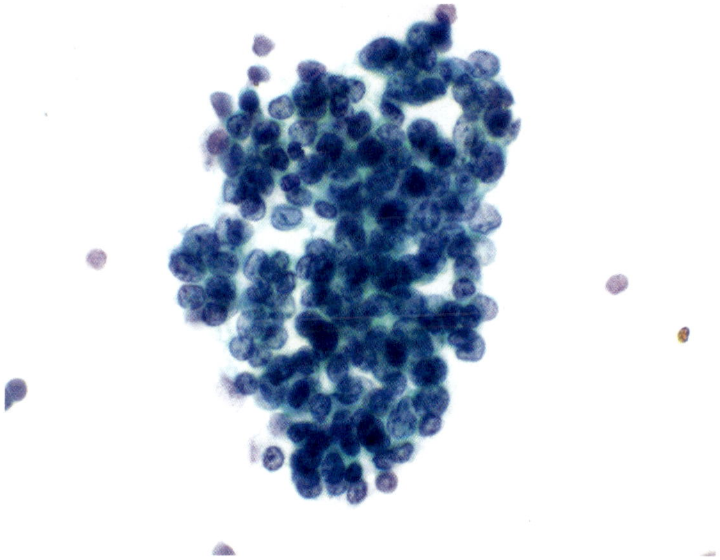

Figure 2.44a — Metastatic Merkel Cell Carcinoma, Pleural Fluid [Pap stain; high power]. Merkel cell carcinoma falls into the category of tumors with small, round blue cells, in which the differential diagnosis includes other neuroendocrine tumors, melanoma, lymphoma, and Ewing's sarcoma (EWS), among others. Immunosuppression increases the odds of developing Merkel cell carcinoma, particularly in individuals with chronic lymphocytic leukemia, advanced HIV, or solid organ transplants, compared with the general population.

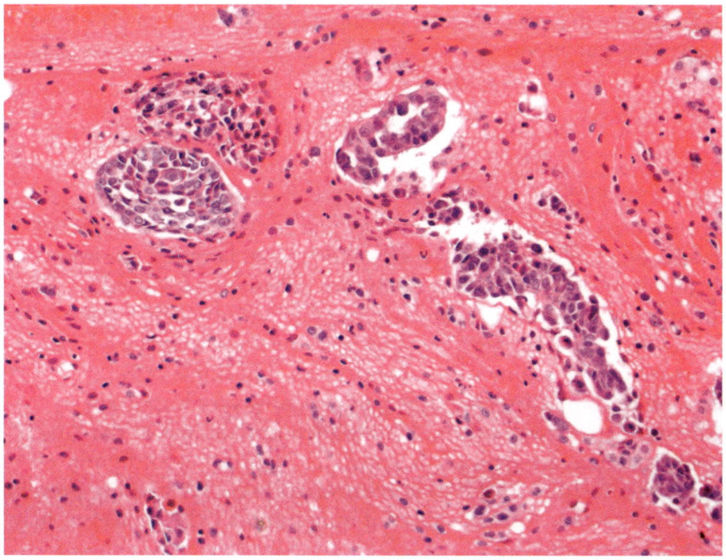

Figure 2.44b — Metastatic Merkel Cell Carcinoma, Pleural Fluid Cell Block Section [H&E stain; medium power]. Here, the malignant cells are predominantly arranged in cohesive groups. Apoptotic bodies and nuclear molding are often present. Metastatic small cell carcinoma from both lung and nonpulmonary sites is usually in the differential diagnosis. Most Merkel cell carcinomas are positive for neurofilament, neuron-specific enolase, CD56, and cytokeratin 20 (CK20) (>80% of cases). CK20 characteristically shows punctate perinuclear dot-like staining. Rare cases with thyroid transcription factor-1 (TTF-1) or cytokeratin 7 (CK7) protein expression have been reported.

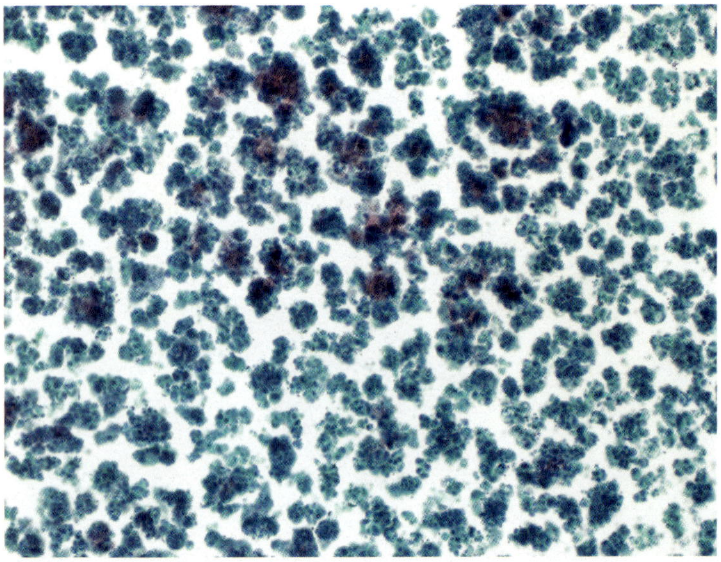

Figure 2.45a — Metastatic Ductal Carcinoma of the Breast, Pleural Fluid [Pap stain; low power]. Metastatic breast cancer accounts for about 25% of all malignant pleural effusions and is the second most common cause of malignant pleural effusions after lung cancer. The cells of metastatic ductal carcinoma are usually monomorphic with irregular nuclei, multiple nucleoli, and a high nuclear-to-cytoplasmic ratio. They are typically seen in abundant, optically dense proliferation spheres or cannonballs, morulae, and papillae, as shown here.

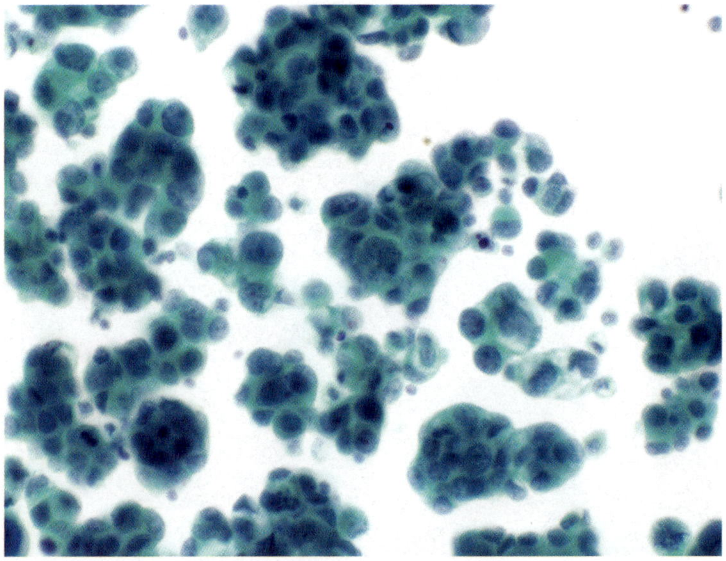

Figure 2.45b — Metastatic Ductal Carcinoma of the Breast, Pleural Fluid [Pap stain; high power]. This field shows large, malignant glandular cells from a metastatic ductal carcinoma. The cells within the clusters show loss of polarity, nuclear molding, and a rare mitotic figure (bottom left corner). Clear spaces between adjacent tumor cells are also present, mimicking mesothelial cells. Occasionally, cytoplasmic vacuolization or intracellular lumina can be seen. Metastatic ductal carcinoma should be suspected in patients with prior history, even if remote. Mammaglobin and gross cystic disease fluid protein 15 (GCDFP-15) are useful immunostains if positive, but up to 50% of cases may be negative. GATA-binding protein 3 (GATA3) and epithelial membrane antigen (EMA), on the other hand, are positive in a majority of breast carcinomas and are only rarely seen in lung adenocarcinoma.

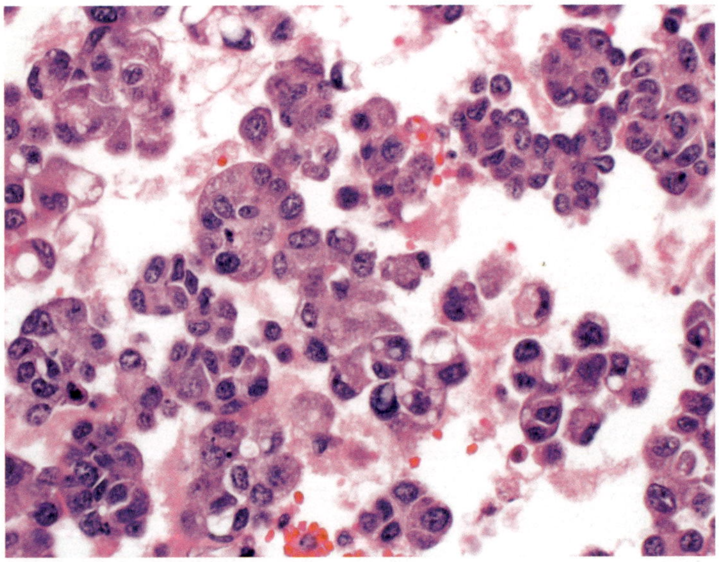

Figure 2.45c — Metastatic Ductal Carcinoma of the Breast, Pleural Fluid Cell Block Section [H&E stain; high power]. In patients with a history of breast cancer and smoking, the differential diagnosis of a malignant pleural effusion is often between a primary lung tumor or metastasis from the breast. Effusions associated with lung adenocarcinoma can have variable morphologic features that overlap with metastatic breast cancer, including papillary clusters, acini, three-dimensional aggregates, and single cells. Therefore, immunohistochemical stains become very important in distinguishing the two entities. Unlike adenocarcinomas of the lung, breast carcinomas are typically negative for Napsin A and thyroid transcription factor-1 (TTF-1) and positive for GATA-binding protein 3 (GATA3) (80%). A small subset of lung tumors (3%), however, may show GATA3 positivity.

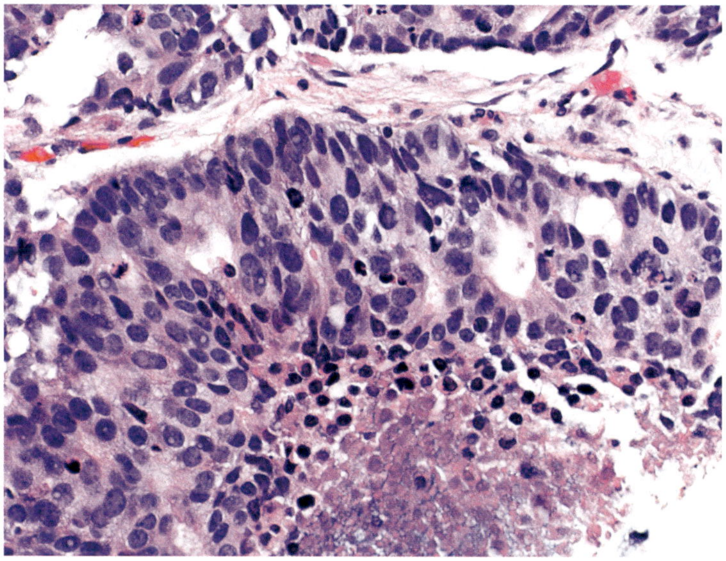

Figure 2.46a — Metastatic Breast Ductal Carcinoma to the Lung and Pleura, Biopsy [H&E stain; intermediate power]. This core needle biopsy of a pleural-based lung mass in a woman with a history of breast carcinoma displays an adenocarcinoma with cribriform gland formation, cellular necrosis, nuclear hyperchromasia, and frequent mitotic figures. Breast carcinoma is the most common cancer type in women in the United States, and in the setting of a history of breast carcinoma, metastatic breast carcinoma is high on the differential. However, lung carcinoma is the second most common type of cancer in women in the United States, and thus a new primary lung carcinoma must also be excluded. A targeted immunopanel can aid in this differential.

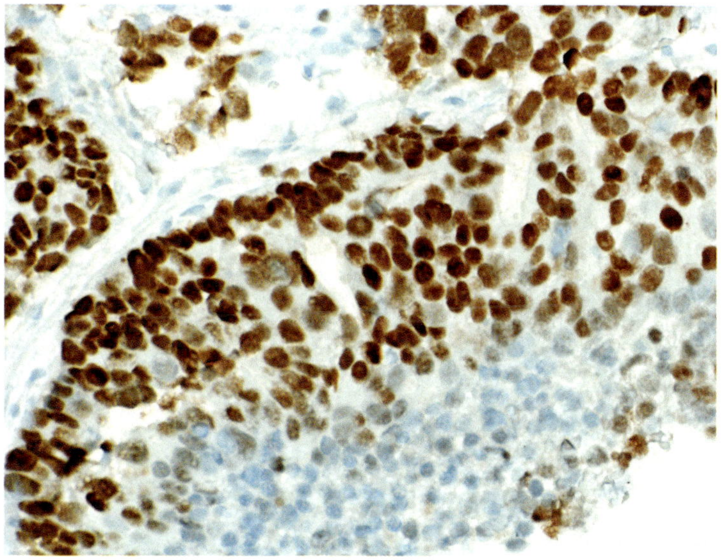

Figure 2.46b — Metastatic Breast Ductal Carcinoma to the Lung and Pleura, Biopsy [Estrogen receptor (ER) immunohistochemical stain; intermediate power]. This adenocarcinoma in the lung/pleura is strongly and diffusely immunoreactive for ER, which is in keeping with spread from this patient's known ER-positive ductal carcinoma of the breast. There are several potential pitfalls with ER labeling in adenocarcinomas in the lung. The majority of breast carcinomas are ER positive, however; of course, a significant subset is ER negative such that ER negativity does not exclude the diagnosis of metastatic breast carcinoma. Furthermore, breast carcinomas can lose ER expression from the primary tumor to the metastasis. Finally, primary lung adenocarcinomas can also express ER; however, the expression is typically focal. The inclusion of a lung-specific marker such as thyroid transcription factor-1 (TTF-1) is useful in this setting.

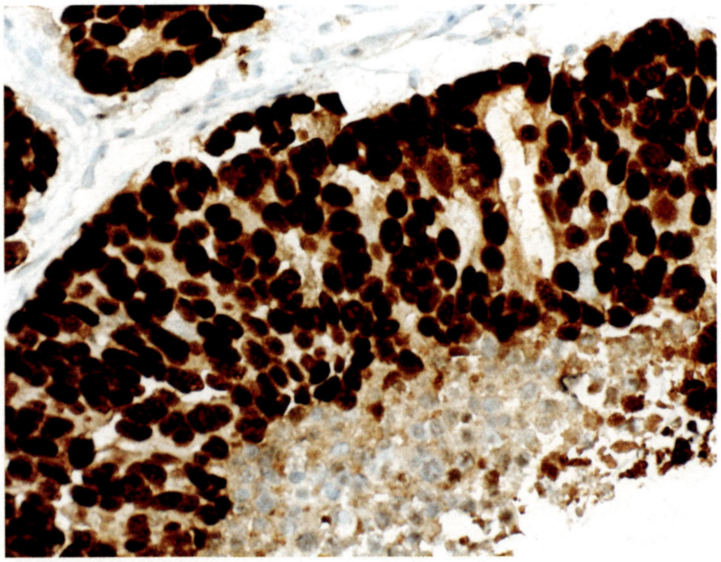

Figure 2.46c — Metastatic Breast Ductal Carcinoma to the Lung and Pleura, Biopsy [GATA-binding protein 3 (GATA3) immunohistochemical stain; intermediate power]. This adenocarcinoma in the lung/pleura is strongly and diffusely immunoreactive for GATA3, which supports classification as metastatic breast ductal carcinoma. GATA3, gross cystic disease fluid protein (GCDFP-15), and mammaglobin are immunostains that are used to support mammary origin in an adenocarcinoma. They are all highly sensitive for ER-positive breast carcinomas, and GATA3 has superior sensitivity than the other markers for ER-negative breast carcinomas. GATA3 is not entirely specific for breast carcinoma, however, and labeling can also be seen in urothelial carcinomas, squamous cell carcinomas, mesotheliomas, and others. An immunopanel approach is recommended.

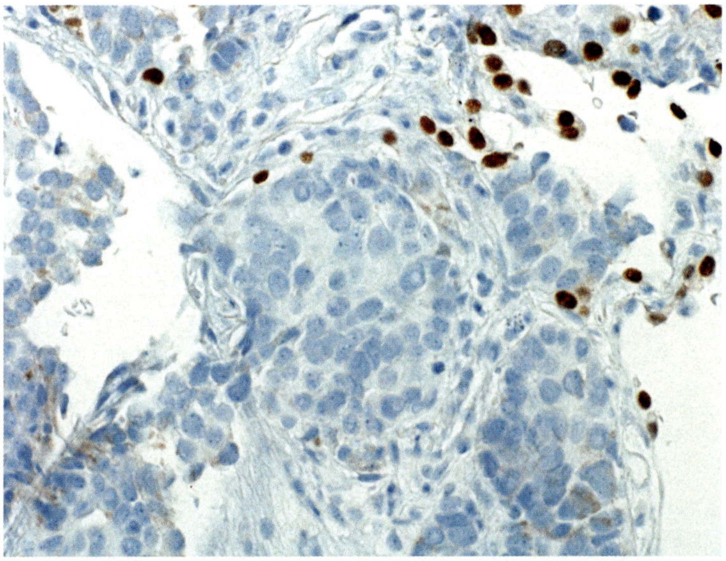

Figure 2.46d — Metastatic Breast Ductal Carcinoma to the Lung and Pleura, Biopsy [Thyroid transcription factor-1 (TTF-1) immunohistochemical stain; intermediate power]. An immunostain for TTF-1 is negative in the metastatic ductal carcinoma to the lung/pleura, and it highlights the background benign alveolar architecture. The TTF-1 labeling in benign pneumocytes should not be misinterpreted as labeling of the carcinoma cells. The benign pneumocytes are smaller than the carcinoma cells and are arranges in linear rows along the alveolar septa. TTF-1 labeling in a carcinoma is used to support pulmonary or thyroid origin. TTF-1 is usually negative in breast carcinoma, although very rarely, it can be focally positive.

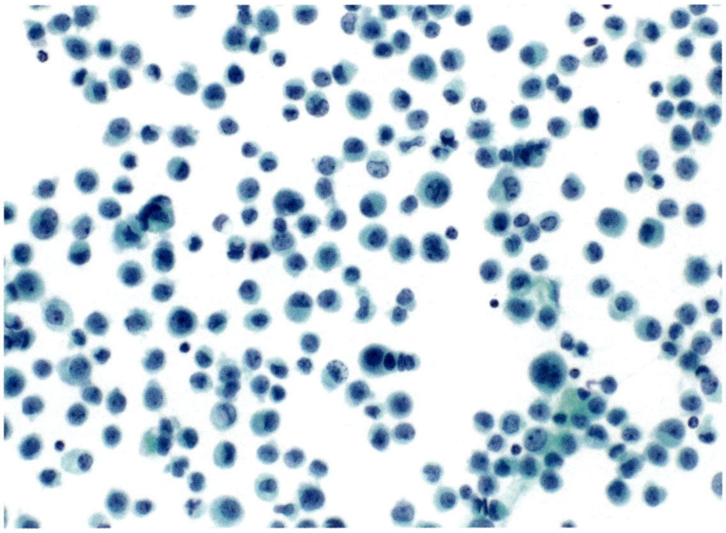

Figure 2.47a — Metastatic Lobular Carcinoma of the Breast, Pleural Fluid [Pap stain; medium power]. Lobular carcinoma of the breast is perhaps the subtlest of all adenocarcinomas and can be very difficult to distinguish from histiocytes and mesothelial cells. The cells are quite monotonous with plasmacytoid/histiocytoid features and fairly uniform, bland nuclei. Apart from the great number of fairly uniform isolated cells, small groups of distinctly cohesive cells in small, caterpillar-like chains (center) are often visible.

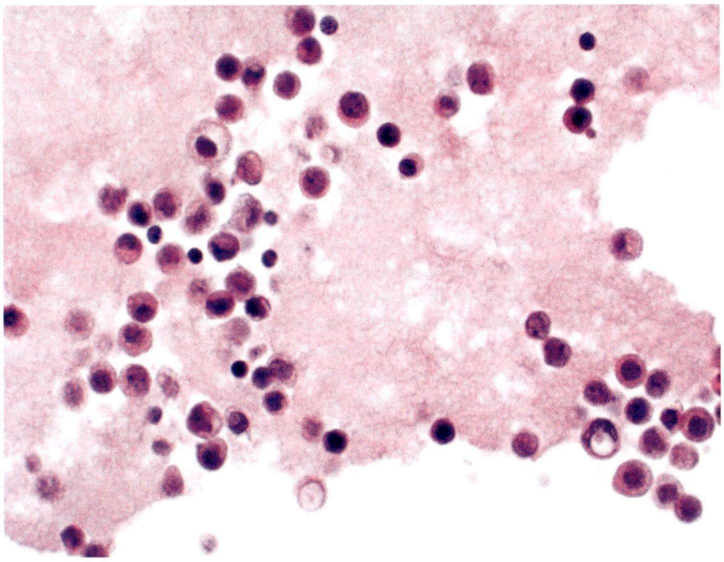

Figure 2.47b — Metastatic Lobular Carcinoma of the Breast, Pleural Fluid Cell Block Section [H&E stain; high power]. Lobular carcinoma consists of small uniform cells with eccentrically placed, variably hyperchromatic nuclei with mild to moderate irregularity. Intracytoplasmic lumens containing a central inclusion of mucin may be present and in some cells may be large enough to create a signet ring cell–like appearance (lower right corner). These cells are best demonstrated with mucicarmine stain.

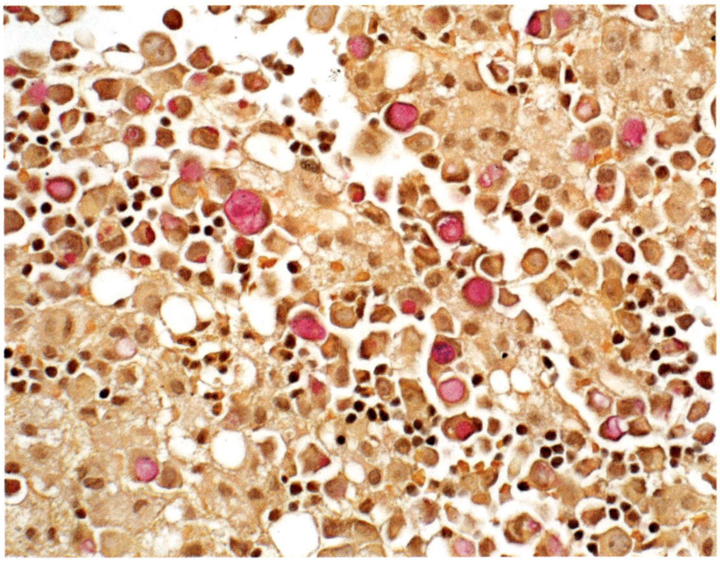

Figure 2.48a — Metastatic Lobular Carcinoma of the Breast, Pleural Fluid Cell Block Section [Mucicarmine special stain; high power]. A mucicarmine stain highlights many intracytoplasmic lumina, some showing a central eosinophilic mucin droplet within the cytoplasmic vacuole ("targetoid" or "magenta bodies"). The mucicarmine solution is composed of aluminum chloride, carmine (a red coloring material), and alcohol. The aluminum is believed to form a chelation complex with the carmine, changing the molecule to a positive charge, allowing it to bind with epithelial sulfated acid mucins. Although very specific for epithelial mucins, mucicarmine is not sensitive enough to be used on its own.

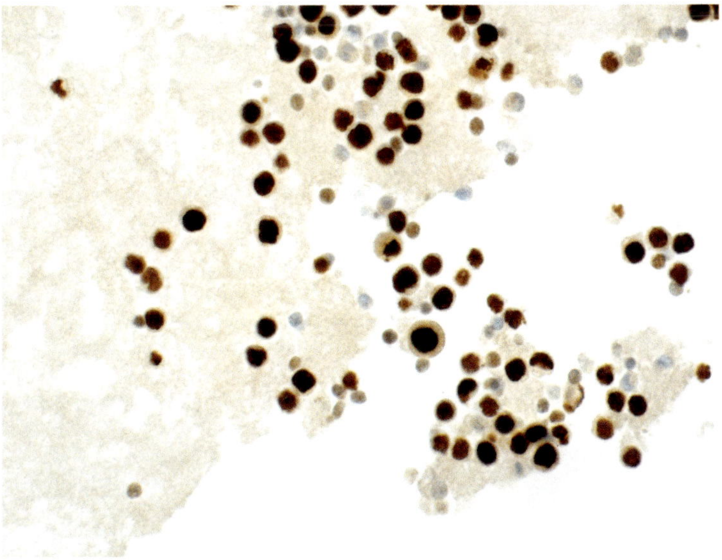

Figure 2.48b — Metastatic Lobular Carcinoma of the Breast, Pleural Fluid Cell Block Section [GATA-binding protein 3 (GATA3) immunohistochemical stain; high power]. GATA3 function is known to be important in the regulation of genes such as *MUC1*/epithelial membrane antigen (EMA) involved in the luminal differentiation of breast epithelium and genes related to T-cell development. Immunoexpression is positively correlated with estrogen receptor–positive (ER-positive; luminal subtypes) breast carcinomas and is more sensitive than gross cystic disease fluid protein 15 (GCDFP-15) and mammaglobin in luminal-like and human epidermal growth factor receptor type 2 (HER2) overexpression subtypes. Expression is less common in pleomorphic lobular, apocrine, papillary, and metaplastic subtypes.

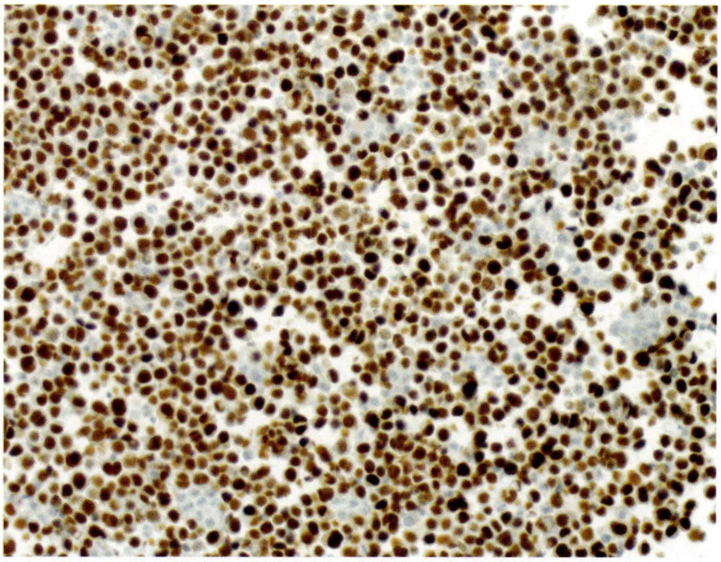

Figure 2.48c — Metastatic Lobular Carcinoma of the Breast, Pleural Fluid Cell Block Section [Estrogen receptor (ER) immunohistochemical stain; low power]. Metastatic lobular carcinoma cells in body fluids are easy to misdiagnose as reactive mesothelial cells because of their relatively bland cytomorphology. As with ductal breast cancer, metastases are most often to bone, lung, liver, and brain. Lobular carcinoma also has a predilection for metastasis to the abdomen, peritoneum, and leptomeninges. Several immunostains are helpful for differentiating metastatic lobular carcinoma from reactive/atypical mesothelial cells. Most lobular carcinomas are positive for ER, progesterone receptor (PR), and GATA-binding protein 3 (GATA3) and negative for E-cadherin, whereas mesothelial cells are typically positive for calretinin, cytokeratin 5/6 (CK5/6), and WT-1 and negative for ER, PR, and GATA3. In the peritoneum, distinction from metastatic gastric cancer with signet-ring-cell features is also problematic, but gastric carcinoma is typically negative for GATA3.

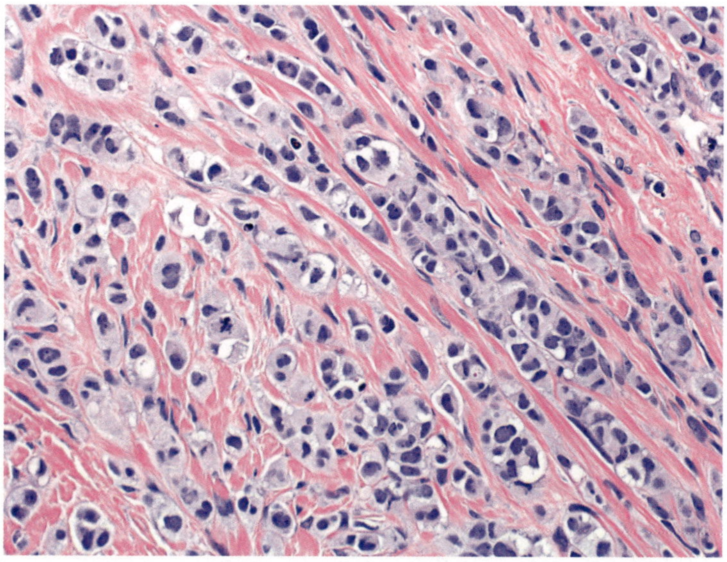

Figure 2.49a — Metastatic Breast Lobular Carcinoma to the Ovary, Oophorectomy [H&E stain; intermediate power]. This oophorectomy specimen in a middle-aged woman shows ovarian parenchyma largely replaced by signet ring cells. In this field, the cells are arranged as loosely cohesive, individual cells or as single-file rows of cells. The single-file pattern of infiltration should raise suspicion for involvement by metastatic lobular carcinoma of the breast. Invasive lobular carcinoma is the second most common type of breast carcinoma, second in frequency to invasive ductal carcinoma (i.e., invasive mammary carcinoma of no special type). Invasive lobular carcinoma has an unusual pattern of metastatic spread, which differs from that of ductal carcinomas; lobular carcinomas more frequently involve the peritoneal cavity, the luminal gastrointestinal tract, and the gynecologic tract.

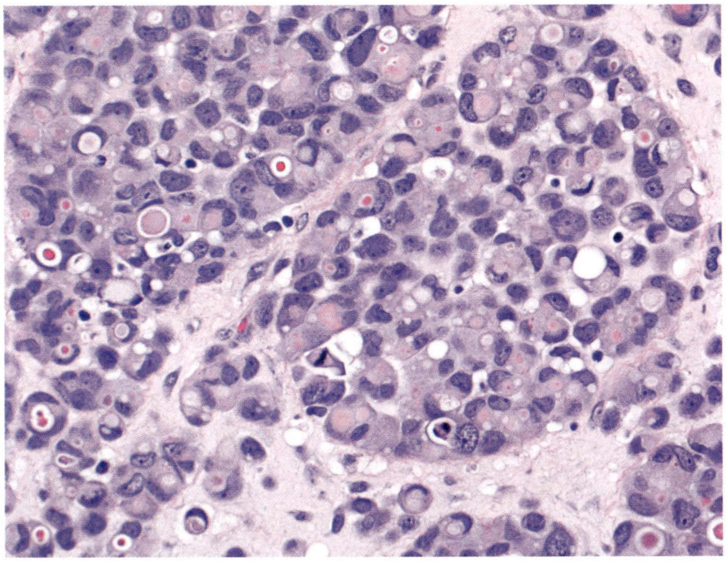

Figure 2.49b — Metastatic Breast Lobular Carcinoma to the Ovary, Oophorectomy [H&E stain; intermediate power]. Although metastatic lobular carcinoma typically displays a single-file pattern of infiltrative growth, variant patterns of growth can also be seen. This field taken from the same case of metastatic lobular carcinoma to the ovary demonstrates a solid pattern of growth, but the cells are still loosely cohesive and display signet vacuoles. Another variant pattern morphology seen in lobular carcinoma is the formation of small tubules or glands.

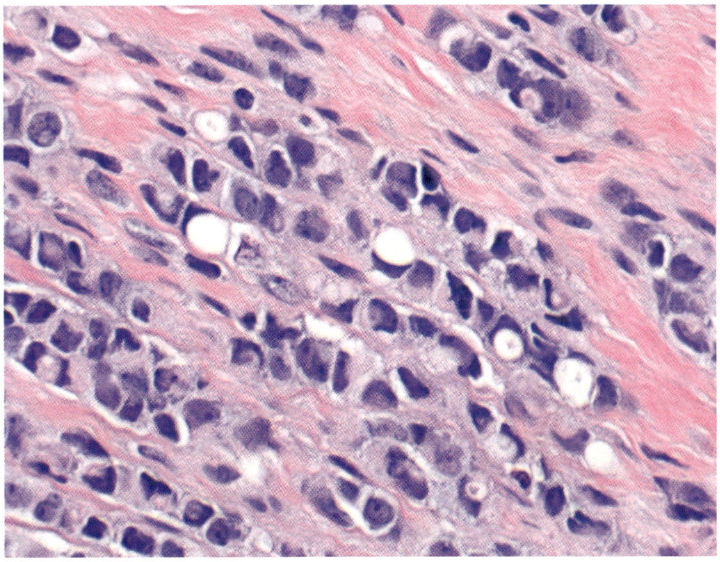

Figure 2.49c — Metastatic Breast Lobular Carcinoma to the Ovary, Oophorectomy [H&E stain; high power]. At high power, metastatic lobular carcinoma characteristically displays a loosely cohesive growth pattern with prominent cytoplasmic vacuoles. These cytoplasmic vacuoles create the signet-ring morphology in which the nucleus is peripherally compressed by the vacuole and resembles the gem of a class ring. The lack of cellular adhesion is due to mutations in *CDH1*, which encodes the E-cadherin cell adhesion protein. Loss of membranous labeling for E-cadherin by immunohistochemistry is a defining feature of breast lobular carcinoma. Of note, signet ring adenocarcinomas of the stomach can also display *CHD1* mutations and loss of membranous E-cadherin labeling; additional, site-specific immunohistochemistry is necessary to resolve this differential.

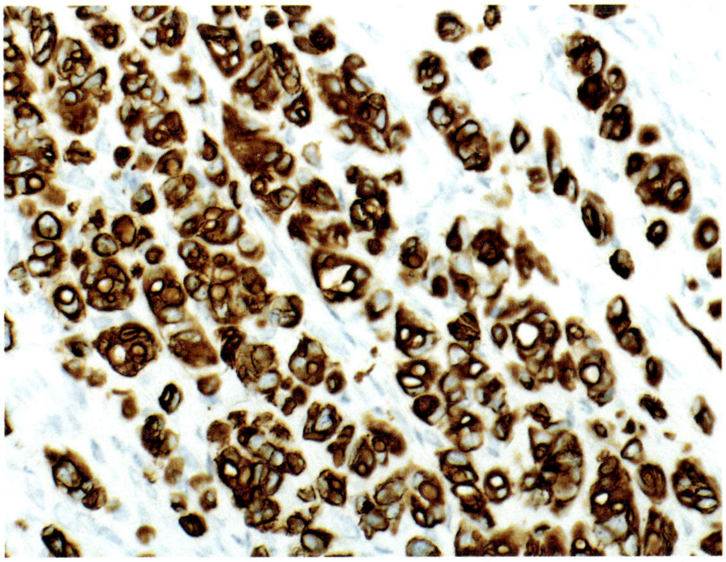

Figure 2.50a — Metastatic Breast Lobular Carcinoma to the Ovary, Oophorectomy [cytokeratin 7 (CK7) immunohistochemical stain; intermediate power]. Breast carcinomas, including both those of the lobular and ductal phenotypes, are immunoreactive for CK7 and are negative for cytokeratin 20 (CK20). In the setting of loosely cohesive cells with minimal to variable cytoplasm, cytokeratin immunoreactivity supports the diagnosis of a carcinoma rather than a lymphoma. However, the pattern of CK7 positivity/CK20 negativity is not specific to breast carcinoma and can also be seen in gastric adenocarcinomas and gynecologic adenocarcinomas, among others.

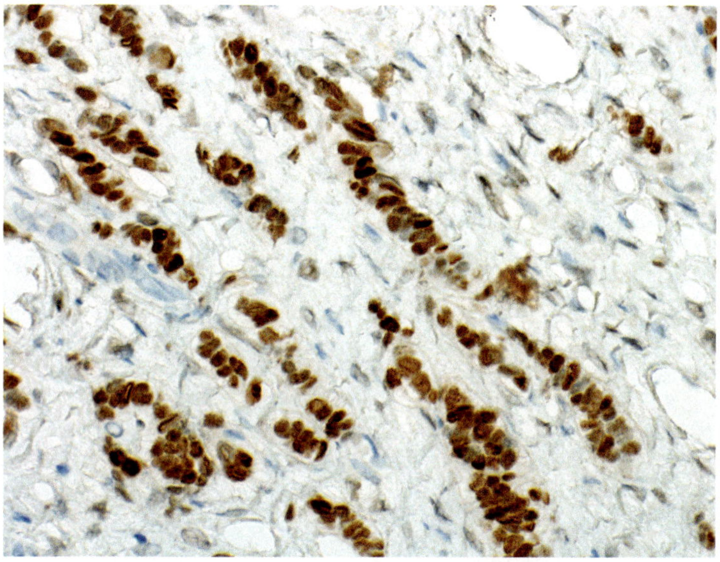

Figure 2.50b — Metastatic Breast Lobular Carcinoma to the Ovary, Oophorectomy [Estrogen receptor (ER) immunohistochemical stain; intermediate power]. This metastatic lobular carcinoma to the ovary displays nuclear positivity for ER. Although the majority of lobular carcinomas are ER positive, a subset can be ER negative. In addition, the hormone receptor status of breast carcinomas can change from the primary to the metastatic site, and the metastatic carcinoma can lose ER immunoreactivity. It is important to recheck both the ER and human epidermal growth factor receptor 2 (HER2) status of the metastatic carcinoma because changes in ER and HER2 status drives treatment management. ER immunoreactivity can be seen in the ovarian stromal cells, as well as in gynecologic tract malignancies.

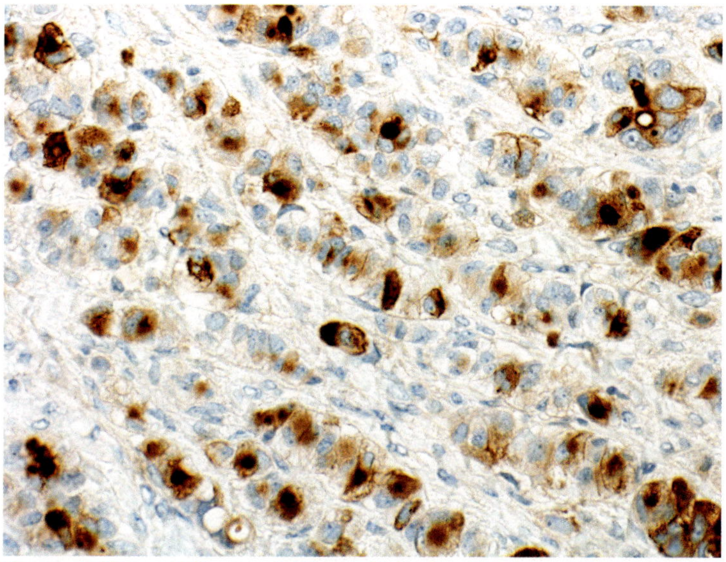

Figure 2.50c — Metastatic Breast Lobular Carcinoma to the Ovary, Oophorectomy [Gross cystic disease fluid protein (GCDFP-15) immunohistochemical stain; intermediate power]. GCDFP-15 is a glycoprotein that was originally isolated in breast cyst fluid, and immunohistochemistry for GCDFP-15 is used to support mammary origin in an adenocarcinoma. Labeling for GCDFP-15 in benign breast tissue, primary breast carcinomas, and metastatic breast carcinomas is typically focal, as seen here. GCDFP-15 labels any cell with apocrine-type differentiation, so adnexal carcinomas also label for GCDFP-15. Other immunostains used to support a mammary origin in a metastatic adenocarcinoma include mammaglobin and GATA-binding protein 3 (GATA3).

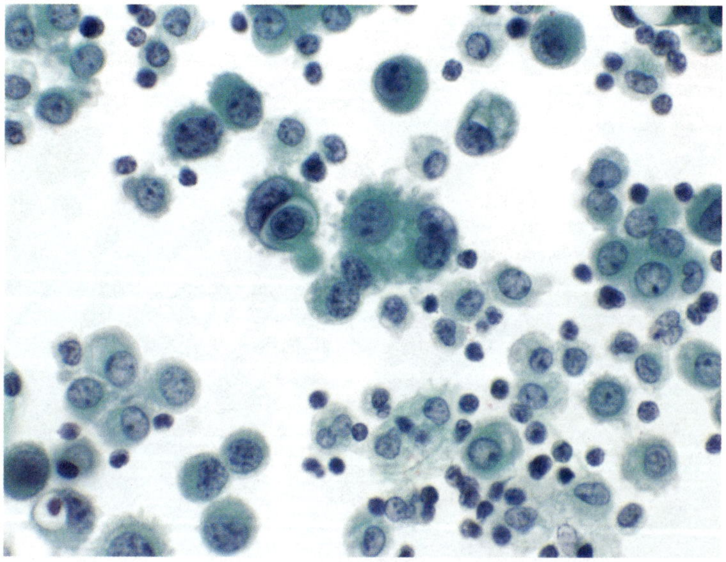

Figure 2.51a — Metastatic Urothelial Carcinoma, Pleural Fluid [Pap stain; high power]. This image shows singly dispersed malignant cells with deeply convoluted nuclear outlines, hyperchromatic nuclei, and rare cytoplasmic vacuoles (bottom left). Cells may occasionally exhibit window-like spaces between adjacent cells and two-tone cytoplasm, mimicking mesothelial cells.

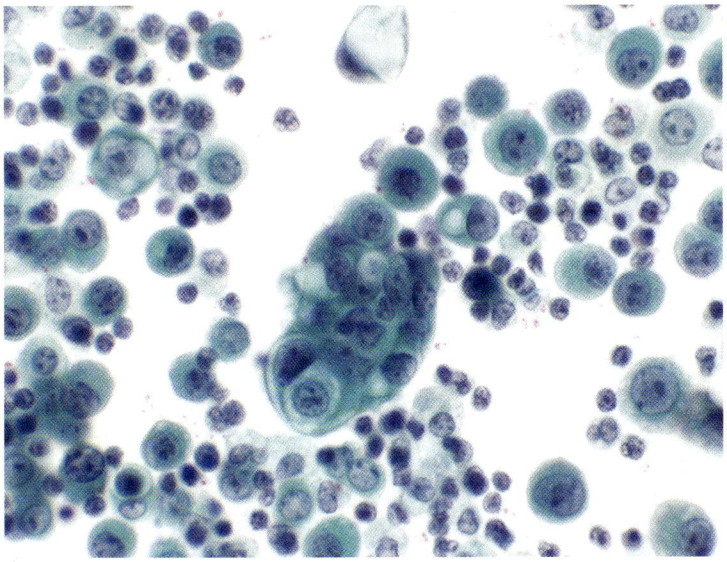

Figure 2.51b — Metastatic Urothelial Carcinoma, Pleural Fluid [Pap stain; high power]. Although the cytomorphologic features (pleomorphic epithelioid cells, high nuclear-to-cytoplasmic ratio, irregular nuclear contours, prominent nucleoli) shown here are diagnostic of malignancy, they are relatively nonspecific for tumor cell of origin. Note the malignant cells with cytoplasmic vacuoles present in the background, along with numerous small lymphocytes. Metastatic urothelial carcinoma is seldom accurately identified without knowledge of a patient's preexisting urothelial carcinoma or clinical context and should always be included in the differential diagnosis for patients with even remote histories of bladder cancer.

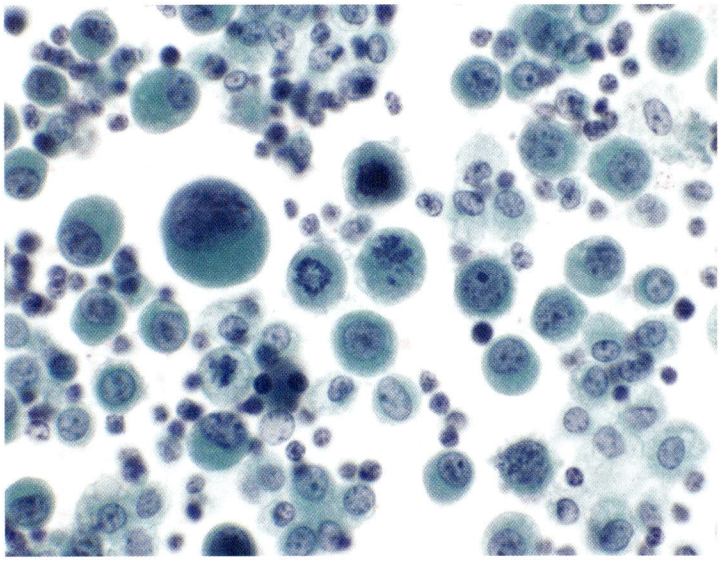

Figure 2.51c — Metastatic Urothelial Carcinoma, Pleural Fluid [Pap stain; high power]. This field shows numerous single pleomorphic epithelioid cells with dense granular cytoplasm and coarse nuclear chromatin. Atypical mitotic figures (center) are easily identified. Most metastatic urothelial carcinomas are high grade, such as this, but metastases from predominantly low-grade papillary carcinomas have also been reported.

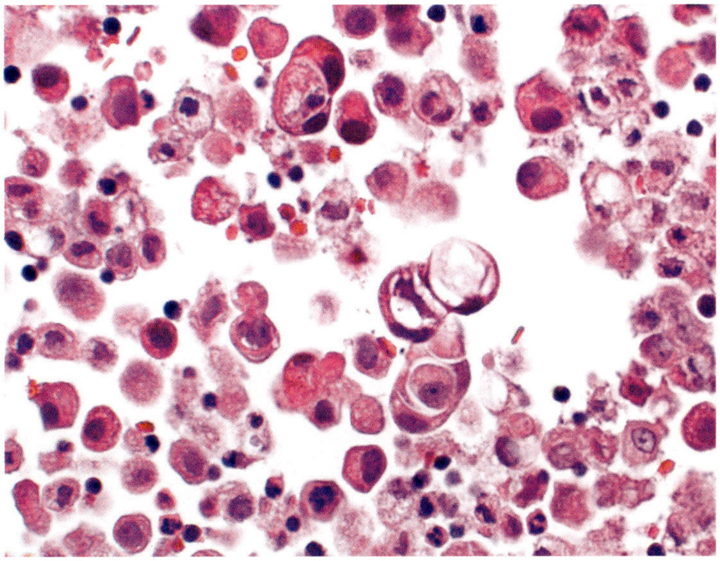

Figure 2.52a — Metastatic Urothelial Carcinoma, Pleural Fluid Cell Block Section [H&E stain; high power]. This field shows an admixture of medium to large cells, some with signet-ring-cell morphology and cell wrapping. Many of the cells contain abundant granular eosinophilic cytoplasm and eccentrically placed round to oval nuclei. The morphologic features of urothelial carcinoma cells in serous fluids tend to overlap with other poorly differentiated carcinomas, including poorly differentiated, nonkeratinizing squamous cell carcinoma. They may be vacuolated, probably because of degeneration, or nonvacuolated, in which they may resemble mesothelial cells.

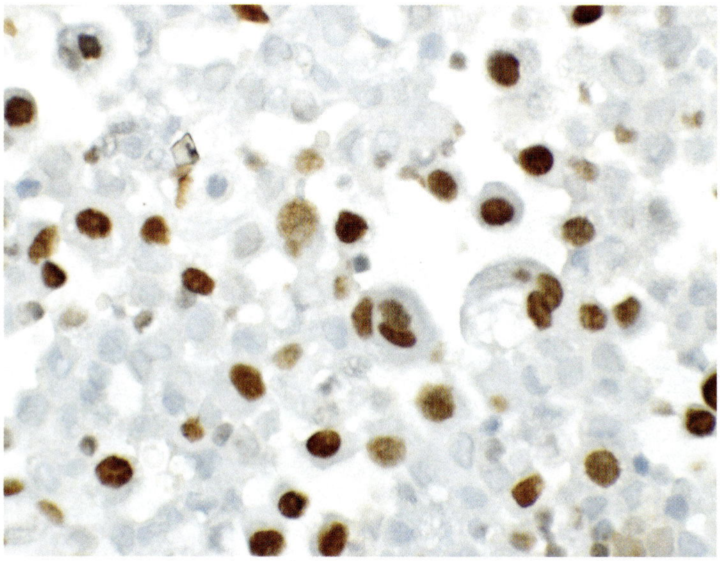

Figure 2.52b — Metastatic Urothelial Carcinoma, Pleural Fluid Cell Block Section [GATA-binding protein 3 (GATA3) immunohistochemical stain; high power]. Although GATA3 is expressed in the majority of urothelial carcinomas (70%–85%), immunoexpression is also seen in ductal and lobular carcinomas of the breast, squamous cell carcinoma, paragangliomas, salivary gland tumors, pancreatic adenocarcinoma, and Renal Cell Carcinoma (RCC). Additional stains typically include uroplakin (60% positive but highly specific) and thrombomodulin (70%–90% positive).

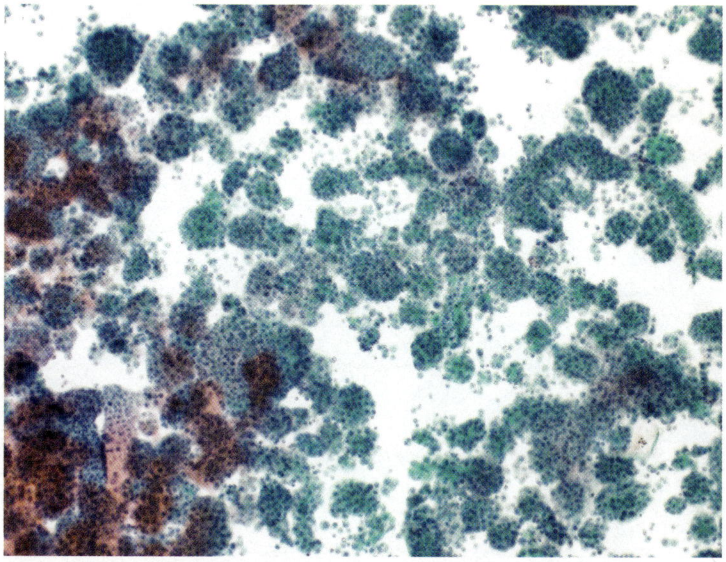

Figure 2.53a — Metastatic Papillary Renal Cell Carcinoma (RCC), Pleural Fluid [Pap stain; low power]. Aspirates of papillary RCC are often very cellular and consist of papillary fragments of cells, as seen here. RCC rarely involves effusions. However, it may be encountered initially in the setting of metastatic disease (e.g., concurrent lung and renal masses) or many years after the initial diagnosis of the primary tumor. Therefore, it is important to recognize the cytomorphologic features of RCC and include it in the differential.

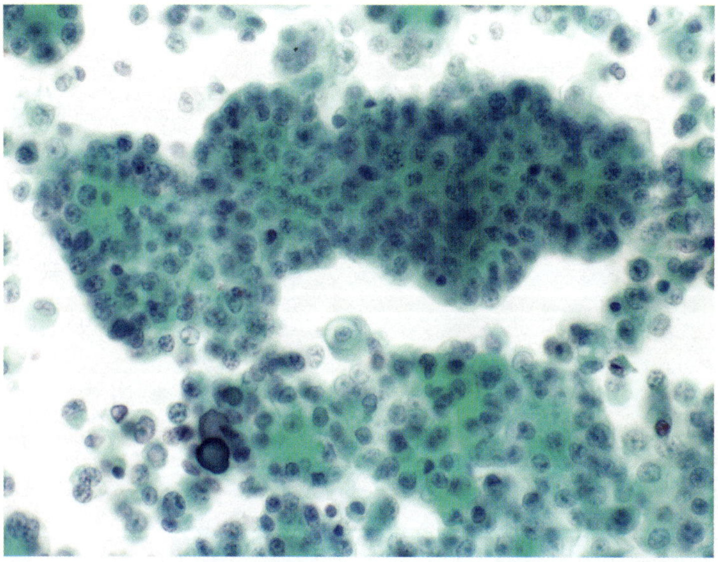

Figure 2.53b — Metastatic Papillary Renal Cell Carcinoma (RCC), Pleural Fluid [Pap stain; high power]. Unlike the short papillae that can occasionally be seen in clear cell RCC, the papillae in papillary RCC are often long with complex branching architecture. In type 1 papillary RCC, the cells are often small to medium and cuboidal, with round, uniform nuclei and scant to moderate amounts of cytoplasm. Depending on the nuclear grade, nucleoli may be inconspicuous (WHO/ International Society of Urological Pathology [ISUP] grade 1), visible but not prominent (grade 2), or conspicuous and eosinophilic (grade 3). Grade 4 tumors typically exhibit overt nuclear pleomorphism, multinucleate giant cells, and rhabdoid or sarcomatoid differentiation. Because cytologic preparations alter cellular size, accurate grading should be performed on H&E sections.

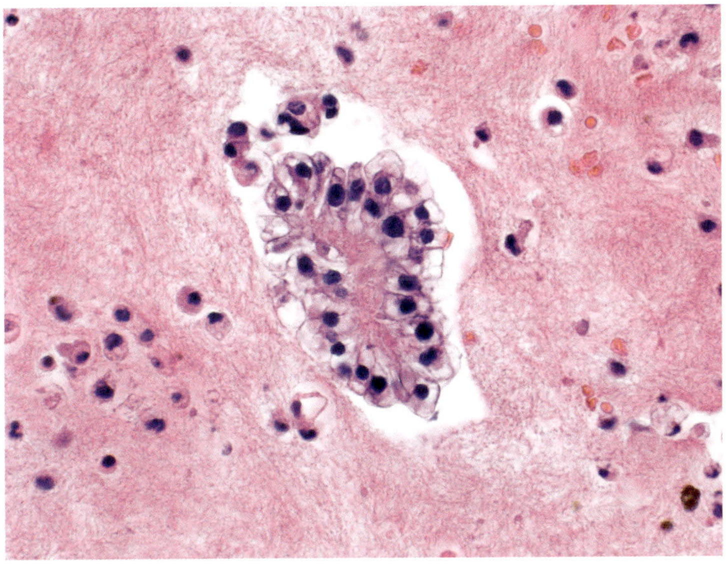

Figure 2.54a — Metastatic Papillary Renal Cell Carcinoma (RCC), Pleural Fluid Cell Block Section [H&E stain; high power]. Papillary architecture with true fibrovascular cores (as seen here), coupled with ancillary immunohistochemical studies, are helpful in confirming a diagnosis of papillary RCC. Another helpful feature is foamy macrophages within papillae, as well as in the background. Hemosiderin deposition in tumor cells and macrophages is also commonly associated with papillary RCC.

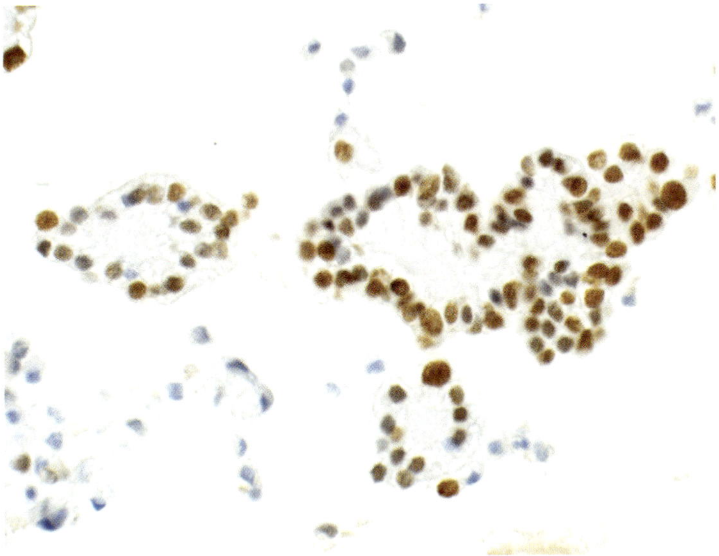

Figure 2.54b — Metastatic Papillary Renal Cell Carcinoma (RCC), Pleural Fluid Cell Block Section [PAX8 immunohistochemical stain; high power]. The paraffin-embedded cell block reveals strong, diffuse nuclear immunoreactivity for PAX8 in the tumor. PAX8 expression, however, is not specific and can be seen in carcinomas of the thyroid gland and female genital tract.

Figure 2.54c — Metastatic Papillary Renal Cell Carcinoma (RCC), Pleural Fluid Cell Block Section [Fumarate hydratase (FH) immunohistochemical stain; high power]. FH expression is normally granular in appearance and localized to the cytoplasm, as shown in this tumor. This patient had two first-degree relatives with RCC, which prompted the pathologist to stain the tumor for FH to screen for the possibility of hereditary leiomyomatosis and renal cell carcinoma (HLRCC) syndrome–associated renal carcinoma. These tumors typically demonstrate high-grade nuclear features with prominent cherry-red nucleoli and show immunohistochemical loss of FH with overexpression of modified cysteine (2SC). Oncogenesis in these tumors is due to autosomal dominant germline *FH* mutations that lead to a defective FH enzyme in the citric acid cycle. Sporadic FH-deficient tumors, which are morphologically and immunophenotypically similar to HLRCC-associated RCC, can also be seen but harbor somatic rather than germline mutations in *FH*.

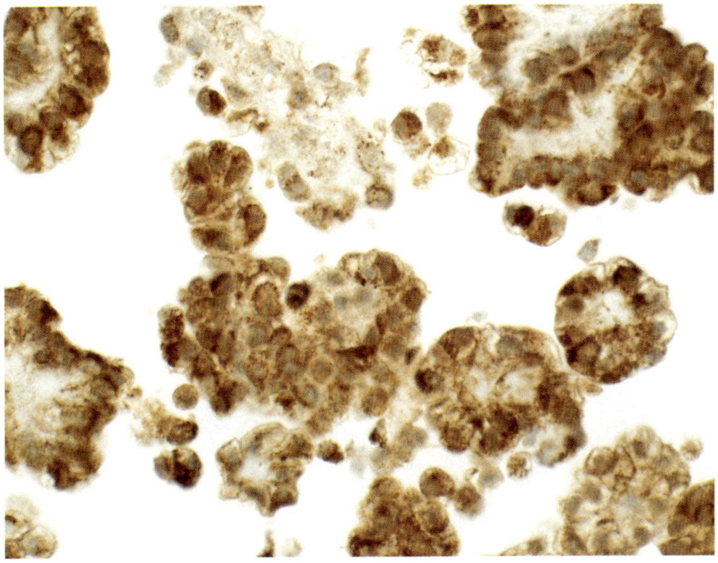

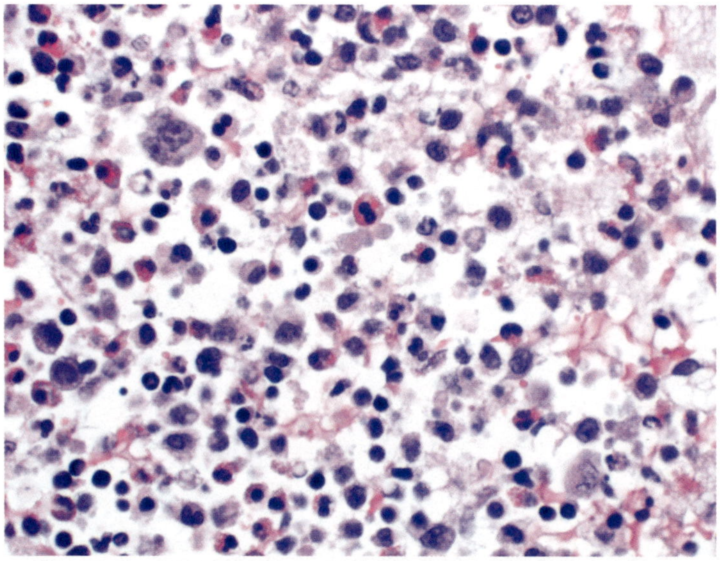

Figure 2.55a — Metastatic Melanoma, Pleural Fluid Cell Block Section [H&E stain; medium power]. A large malignant cell can be seen in the top left corner of the field. Melanoma may blend in with mesothelial cells; some specimens may contain almost entirely melanoma cells, in which case one may assume they are all mesothelial cells and not notice an unusual population of cells. In this case, the cell is binucleate, large, and has prominent nucleoli, features which should suggest melanoma.

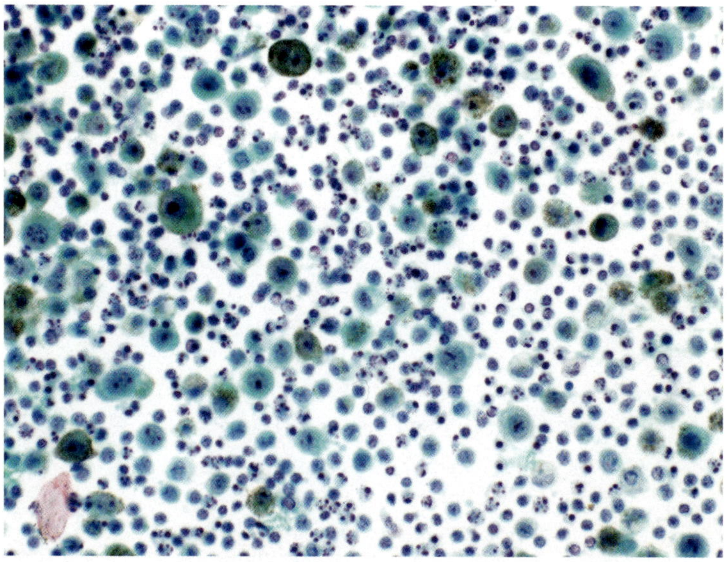

Figure 2.55b — Metastatic Melanoma, Pleural Fluid [Pap stain; medium power]. In this fluid specimen, the melanoma cells stand out as a foreign population; background lymphocytes, mesothelial cells, and macrophages appear very small compared with the larger tumor cells. The coarse green-brown pigment in the cytoplasm should be differentiated from the finer green pigment sometimes seen in macrophages as a result of long-standing degenerated blood in the serous cavity. Even if the pigment were absent, the malignant features of these cells are obvious: large nuclei, prominent nucleoli, and some binucleate forms.

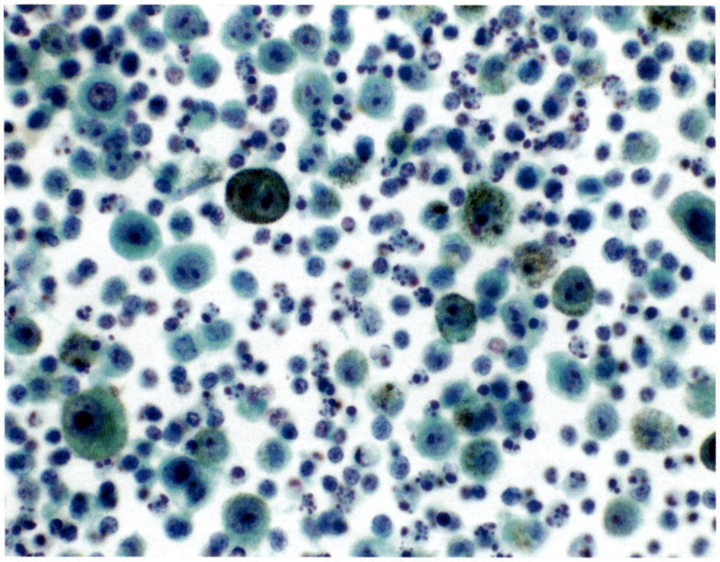

Figure 2.55c — Metastatic Melanoma, Pleural Fluid [Pap stain; high power]. Close examination of the cells reveals at least one cell with all the classical cytomorphologic features of malignant melanoma: binucleation with prominent nucleoli, and coarse cytoplasmic brown-green pigment. In cells with single nuclei, only the nuclei are typically eccentrically located and the cytoplasm is often abundant.

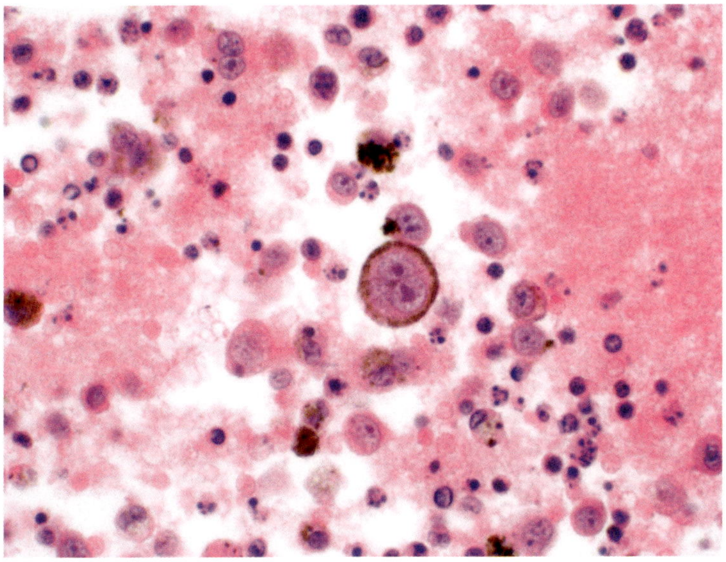

Figure 2.56a — Metastatic Melanoma, Pleural Fluid Cell Block Section [H&E stain; high power]. The pigment in melanoma is not colored secondarily to any one stain preparation and can be seen on even unstained preparations. The malignant cells seen on the cell block preparation recapitulate the morphology seen on the Pap stain preparations. If pigment is absent, immunohistochemical studies can be performed to confirm melanoma, although not all melanoma markers may be expressed. Knowledge of previously expressed markers by the tumor can be used to select markers on the current specimen. Commonly tested markers include S-100 protein, melan-A, HMB-45, and Sox-10 (nuclear).

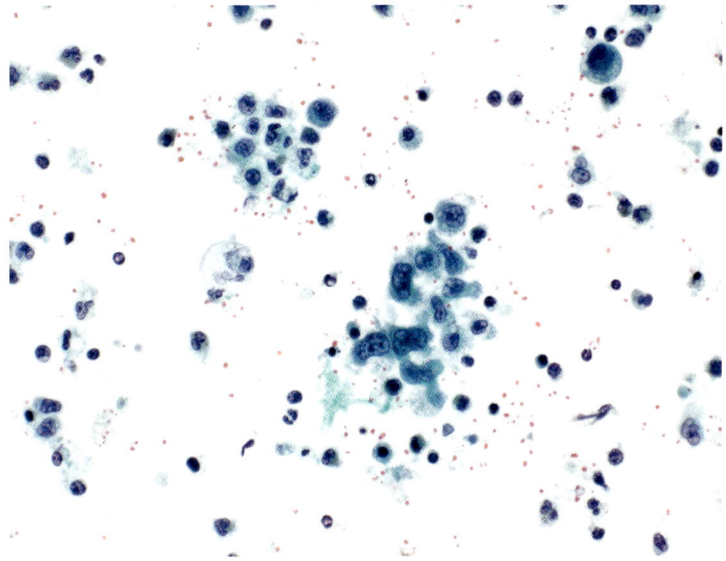

Figure 2.56b — Metastatic Melanoma, Pleural Fluid [Pap stain; intermediate power]. Melanoma can take on many appearances and can look deceivingly bland. Most metastatic melanomas are not pigmented, such as the cells seen here. The cells have irregularly shaped nuclear borders and coarse chromatin; the amount of cytoplasm is not reduced and the cells have a slightly spindled shape. The cells cluster loosely together but do not form fragments; in cytology specimens, melanoma cells should not form fragments, but in some very cellular specimens the cells may be pushed together and appear to be present as a fragment.

Figure 2.57a — Metastatic Melanoma, Biopsy [H&E stain; high power]. Metastatic melanoma is also known as the "great mimicker," because the histology of melanoma can vary widely and mimic many different tumor types. In this biopsy, there are sheets of variably epithelioid to spindled cells with large nuclei and prominent cherry-red macronucleoli, pink cytoplasm, and intracytoplasmic melanin pigment. The differential diagnosis of large pink cells includes melanoma, hepatocellular carcinoma, Renal Cell Carcinoma (RCC), and adrenocortical carcinoma. Both melanoma and hepatocellular carcinoma can contain intracellular pigment, melanin, and bile pigment. Immunostains and a thorough clinical history can help resolve the diagnosis.

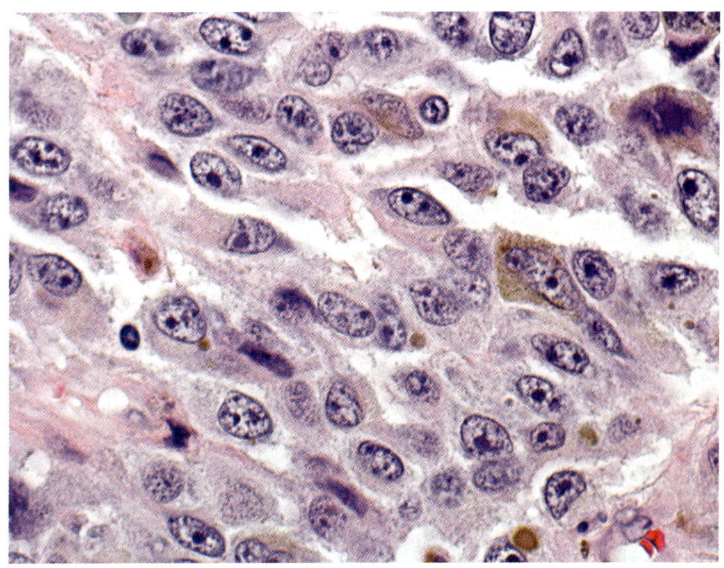

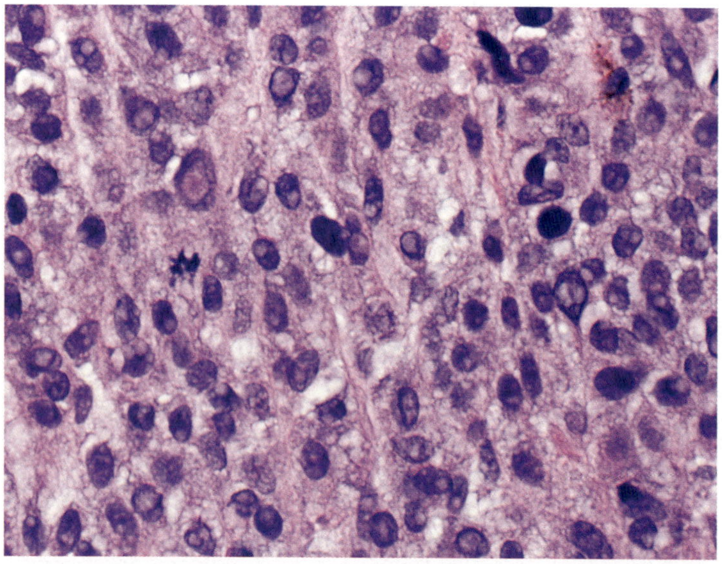

Figure 2.57b — Metastatic Melanoma, Biopsy [H&E stain; high power]. High-power examination of this melanoma specimen reveals sheets of epithelioid cells with abundant eosinophilic cytoplasm. This field also demonstrates intranuclear inclusions, which are a characteristic feature of melanoma, as well as mitotic figures, and intracytoplasmic melanin pigment. It should be noted that other neoplasms, such as melanotic peripheral nerve sheath tumors, can contain melanin pigment. Melanomas are negative for cytokeratin labeling, which is a key feature in distinguishing them from metastatic carcinoma.

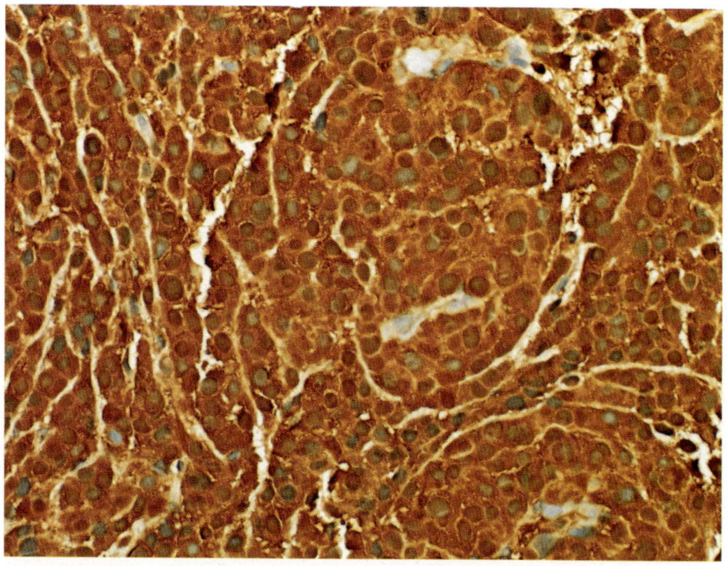

Figure 2.58a — Metastatic Melanoma, Biopsy [S-100 protein immunohistochemical stain; intermediate power]. Most melanomas are strongly and diffusely immunoreactive for S-100 protein. The S-100 protein immunostain labels neural crest–derived cells (such as Schwann cells, melanocytes, and glial cells), as well as myoepithelial cells, dendritic cells, chondrocytes and adipocytes, as well as the neoplasms arising from these cell types. S-100 protein may be the only immunostain that labels desmoplastic (spindled) melanomas. Malignant peripheral nerve sheath tumors (MPNSTs) can be in the histologic differential diagnosis of malignant melanoma, and S-100 protein labeling is typically patchy in MPNSTs in contrast to the typically diffuse labeling in melanomas.

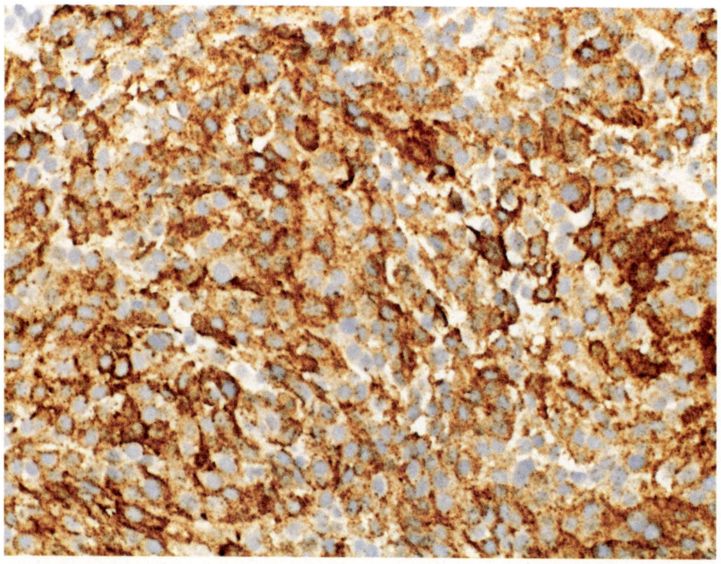

Figure 2.58b — Metastatic Melanoma, Biopsy [HMB-45 immunohistochemical stain; intermediate power]. This metastatic melanoma displays strong and diffuse labeling for HMB-45, which recognizes the melanosomal glycoprotein 100 (gp100). In benign nevi, HMB-45 typically labels the subepithelial component and is lost as the nevocytes mature at the base of the nevus in the deeper dermis. Similarly, HMB-45 is typically negative in the benign nevocytes in nodal nevi (bland, spindled nevocytes present in the lymph node capsule). Thus an immunopanel for Melan-A and HMB-45 can help distinguish nodal nevi (typically Melan-A positive/HMB-45 negative) from metastatic melanoma (typically Melan-A positive/HMB-45 positive).

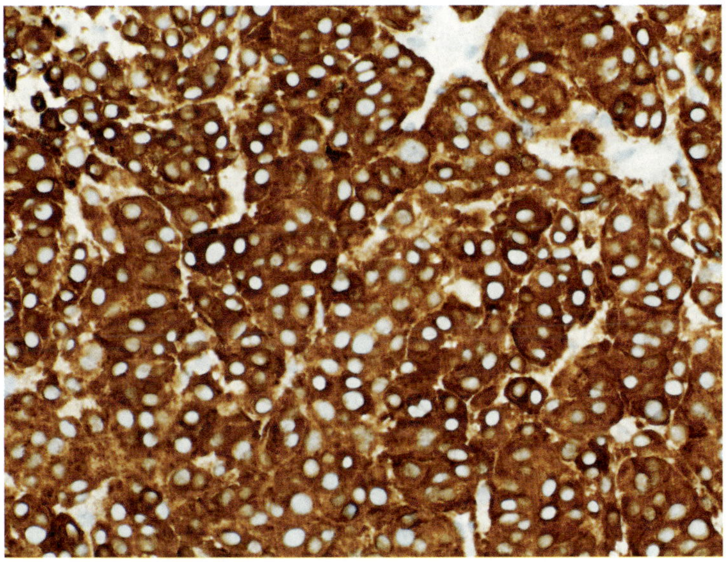

Figure 2.58c — Metastatic Melanoma, Biopsy [Melan-A immunohistochemical stain; intermediate power]. This metastatic melanoma displays strong and diffuse labeling for Melan-A. Melan-A also labels adrenocortical tumors and perivascular epithelioid cell tumors (PEComas), including angiomyolipoma. Melan-A labeling alone should not be used to support a diagnosis of melanoma when adrenocortical carcinoma is also in the differential. In this setting, including other specific markers such as S-100 protein for melanoma and inhibin for adrenocortical carcinoma can aid in the diagnosis.

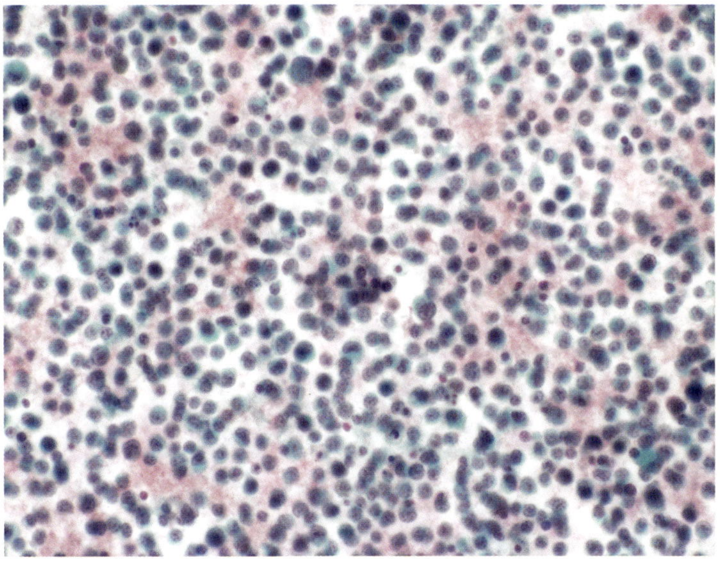

Figure 2.59a — Myelomatous Effusion, Pleural Effusion [Pap stain; medium power]. The presence of immature plasma cells with prominent nucleoli, as seen here, is typical of a plasma cell neoplasm. The differential diagnosis includes florid reactive plasmacytosis associated with HIV or autoimmune disorders, which can be distinguished by normal plasma cell immunophenotype and polytypic cytoplasmic light chain expression. Non-Hodgkin lymphomas with prominent plasmacytic differentiation, most commonly marginal zone lymphomas and lymphoplasmacytic lymphoma, can also be confused for myeloma.

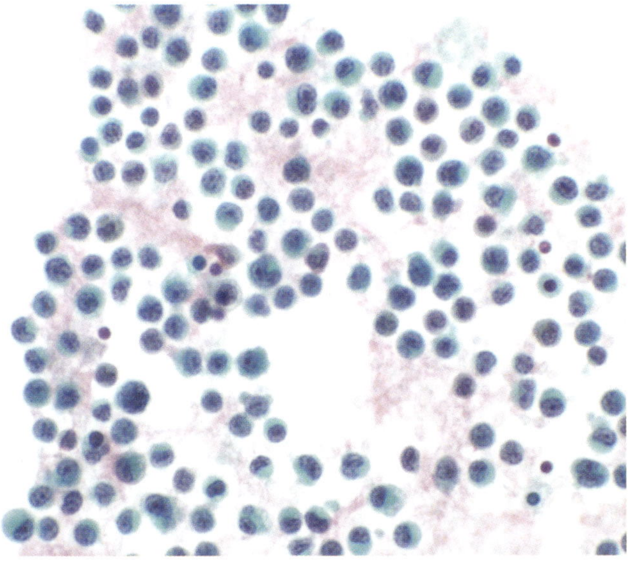

Figure 2.59b — Myelomatous Effusion, Pleural Effusion [Pap stain; high power]. Mature plasma cells are easily identified by their eccentrically placed nuclei with clumped spoke-wheel or clock-face chromatin and perinuclear halo. Neoplastic plasma cells, on the other hand, tend to exhibit greater pleomorphism with immature forms, loose reticular chromatin, and bilobed nuclei. Malignant myelomatous effusions occur in fewer than 1% of patients with multiple myeloma and results from the direct extension of plasma cells from adjacent chest wall lesions or pulmonary lesions, hematogenous spread, or lymphatic obstruction.

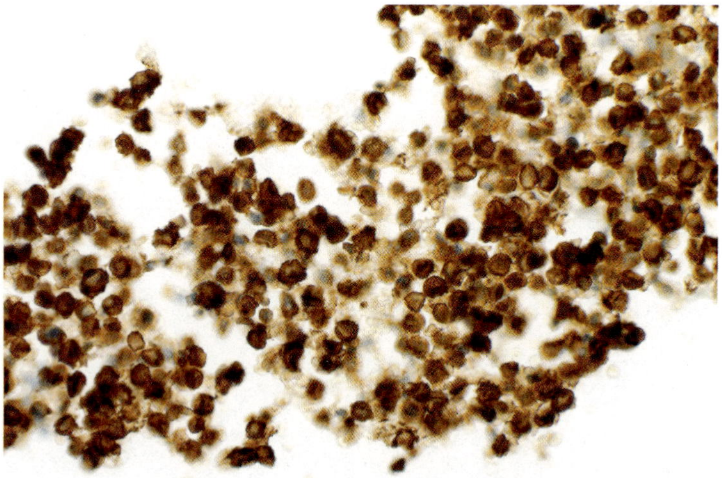

Figure 2.59c — Myelomatous Effusion, Pleural Effusion Cell Block Section [CD138 immunohistochemical stain; high power]. Plasma cells are classically defined immunophenotypically by bright CD138 expression. The CD138 antigen is more specific than CD38 for plasma cells, particularly in flow cytometric analyses. Myeloma cells may also aberrantly express CD56 (~70% of cases), CD117, CD10, and, rarely, cyclin D1. Immunoreactivity for epithelial membrane antigen (EMA; ~70% of cases) and CD33 is also common.

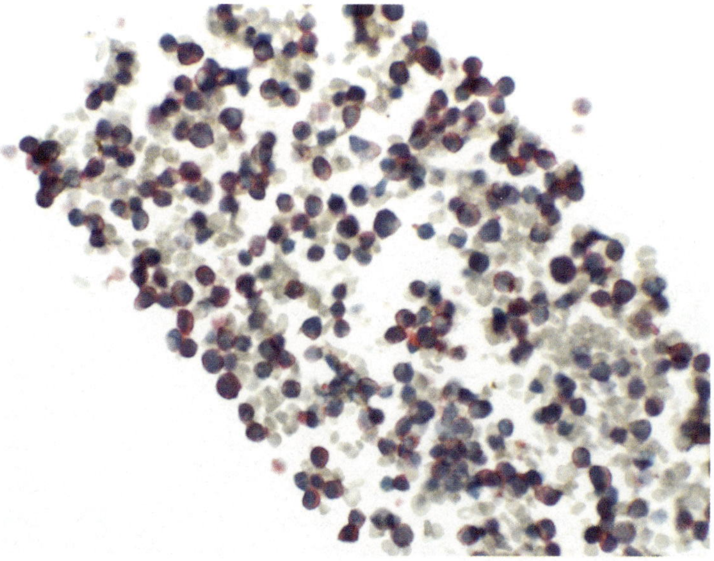

Figure 2.59d — Myelomatous Effusion, Pleural Effusion Cell Block Section [Kappa immunohistochemical stain; high power]. Prominent monotypic expression of kappa or lambda light chains by immunohistochemistry or flow cytometry is characteristic of a clonal plasma cell proliferation. Myeloma can also be classified by the type of monotypic heavy chain produced, with immunoglobulin G (IgG; 50%) being the most common, followed by IgA (20%) and then IgD. Fluorescence in situ hybridization (FISH) and/or cytogenetic testing should be performed in all patients to identify the following high-risk abnormalities: hypodiploidy, complex cytogenetics (≥3 abnormalities), del(1p), del(17p) or monosomy 17, and IgH translocations t(14;16) or t(14;20).

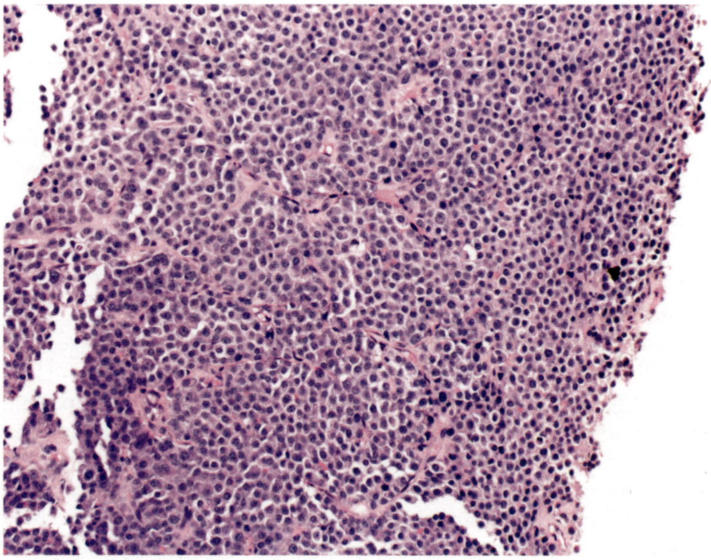

Figure 2.60a — Myeloma, Biopsy [H&E stain; low-intermediate power]. This peritoneal biopsy contains sheets of small, monotonous ovoid cells. From low power, the differential diagnosis includes both reactive inflammatory processes and a malignant process, including a plasmacytoma, lymphoma, melanoma, and carcinoma such as metastatic lobular carcinoma of the breast. An isolated mass of neoplastic, clonal plasma cells is a plasmacytoma. The presence of multiple masses of neoplastic, clonal plasma cells is necessary to diagnose multiple myeloma.

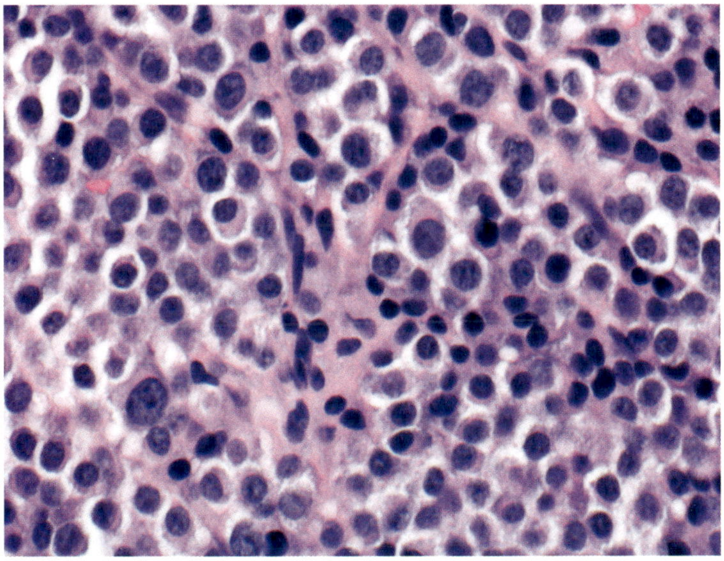

Figure 2.60b — Myeloma, Biopsy [H&E stain; high power]. Plasmacytoma/myeloma cells are characterized by sheets of cells with eccentrically located nuclei, clock-face chromatin with chromatin clumping at the periphery of the nuclei, variably sized nuclei, and the presence of perinuclear clearing (the perinuclear hof, which represents the cell's Golgi apparatus). Neoplastic plasma cells may display binucleate forms. Other common histologic features include the presence of Dutcher bodies that consist of intranuclear rods, the presence of Mott cells that contain blue cytoplasmic inclusions, and the presence of Russell bodies that consist of cherry-red cytoplasmic inclusions.

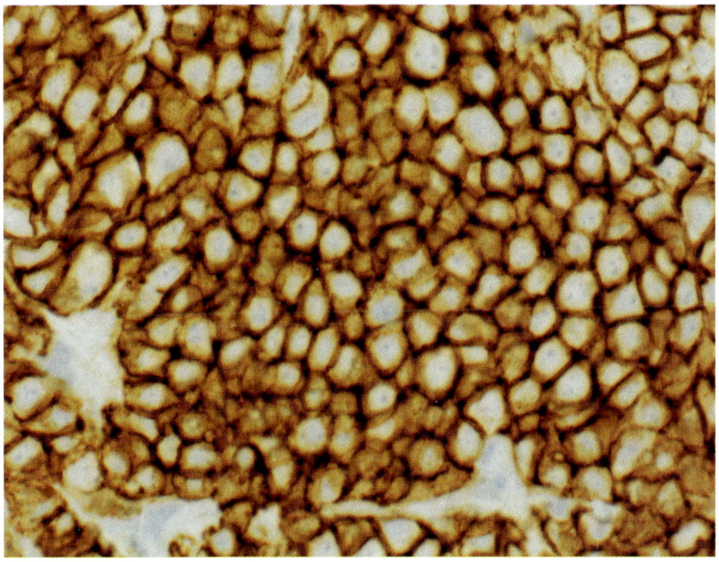

Figure 2.60c — Myeloma, Biopsy [CD138 immunohistochemical stain; high power]. Plasmacytomas/myelomas and non-neoplastic plasma cells label for CD138, as seen here, as well as CD38, and are negative for cytokeratins. Monoclonal, immunoglobulin-producing plasmacytomas/myelomas show labeling for either kappa or lamba, resulting in a designation as kappa or lamba light chain restricted. Plasmacytomas/myelomas can also label for CD56, which should not be overinterpreted as a marker of neuroendocrine differentiation. Of note, CD138 labeling can also be seen in carcinomas, so it should not be used alone to exclude the diagnosis of a carcinoma in a given setting.

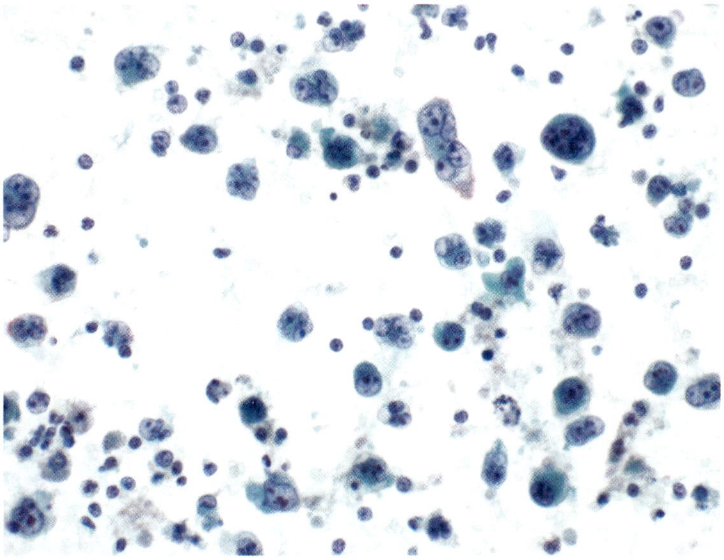

Figure 2.61a — Diffuse Large B-Cell Lymphoma, Pleural Effusion [Pap stain; medium power]. Numerous large, malignant cells can be easily distinguished from background lymphocytes and acute inflammatory cells. The malignant cells are mostly multinucleated with coarse chromatin and exist as individual cells. These pleomorphic cells could represent almost any malignant process—carcinoma, sarcoma, lymphoma—and patient history or immunohistochemical studies would be required to determine the diagnosis.

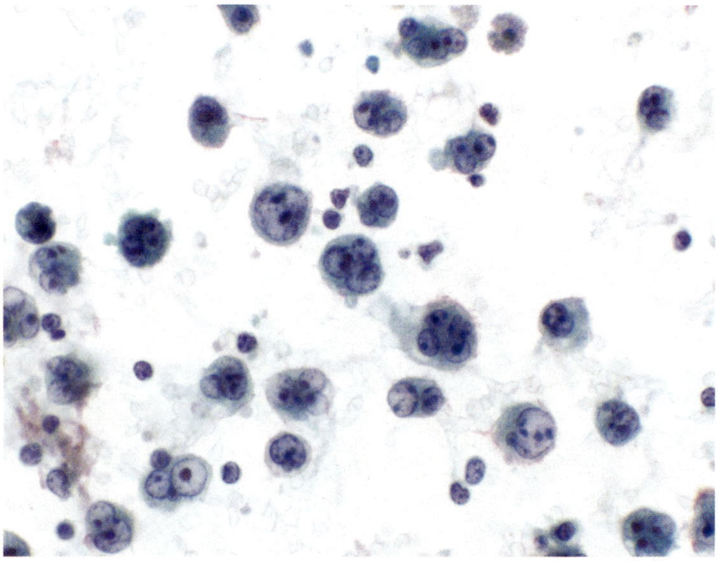

Figure 2.61b — Diffuse Large B-Cell Lymphoma, Pleural Effusion [Pap stain; high power]. The cells are extremely large and have nonspecific cytomorphology. Lymphoglandular bodies, often seen in association with lymphoid populations on conventional smears, are not usually seen on fluid specimens. Although involvement by small cell lymphomas can be difficult to distinguish from reactive lymphocytosis without phenotyping studies, large cell lymphomas appear malignant.

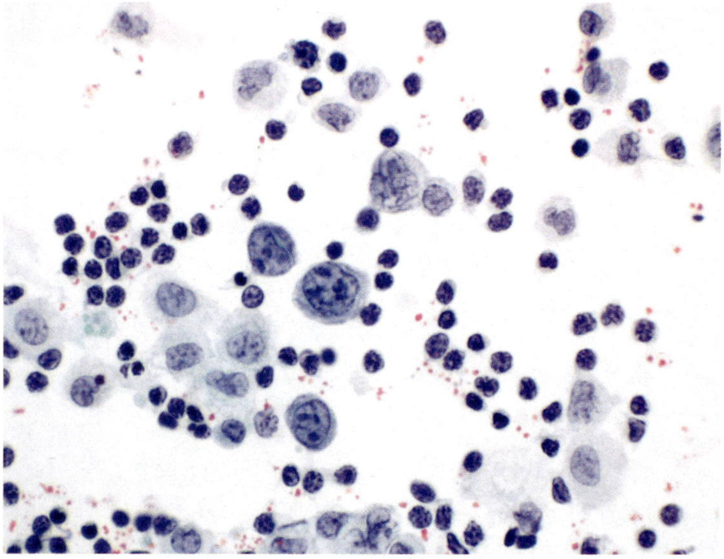

Figure 2.61c — Diffuse Large B-Cell Lymphoma, Pleural Effusion [Pap stain; high power]. Scattered discohesive malignant epithelioid cells are seen interspersed between small polymorphous lymphocytes, mesothelial cells, and histiocytes. The cells are large and have high nuclear-to-cytoplasmic ratios, irregular nuclear borders, and coarse chromatin. In most circumstances, this would likely represent a poorly differentiated adenocarcinoma; however, in this case, the patient had a history of diffuse large B-cell lymphoma. When encountering malignant, discohesive epithelioid cells on a cytology specimen, the initial differential diagnosis should remain broad: poorly differentiated carcinoma, melanoma, sarcoma, and lymphoma. Without a known history, initial immunohistochemical workup might include keratins (carcinoma or mesothelioma), S-100 (melanoma), and CD45 (lymphoma). The absence of any of these markers should cause concern for a sarcoma, an unusual malignancy, or a common malignancy so poorly differentiated that it no longer expresses common markers.

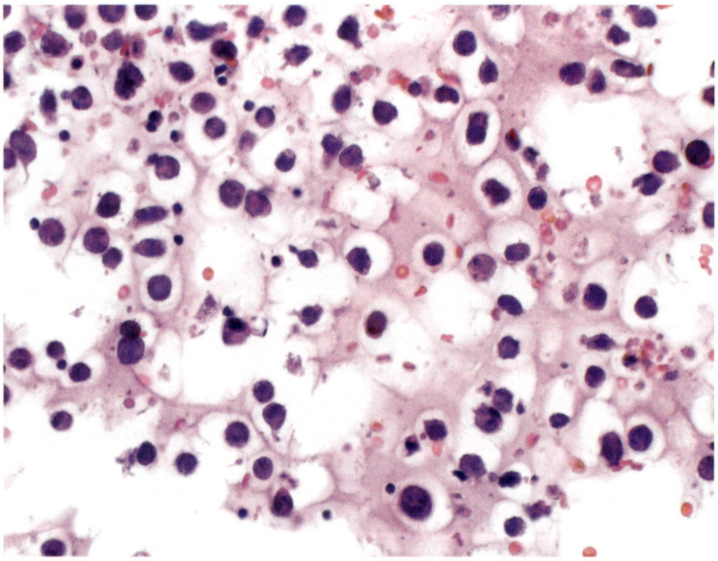

Figure 2.62a — Diffuse Large B-Cell Lymphoma, Pleural Effusion Cell Block Section [H&E stain; medium power]. The lymphoma cells occupy "holes" within the cell block material, caused by preparation artifact. Cell block preparations may allow better visualization of cellular architecture when specimens contain fragments ("mini-biopsies") but are often suboptimal for the examination of cytomorphologic features. In this case, cell block material allows for immunohistochemical studies to be performed to better characterize the malignant cells.

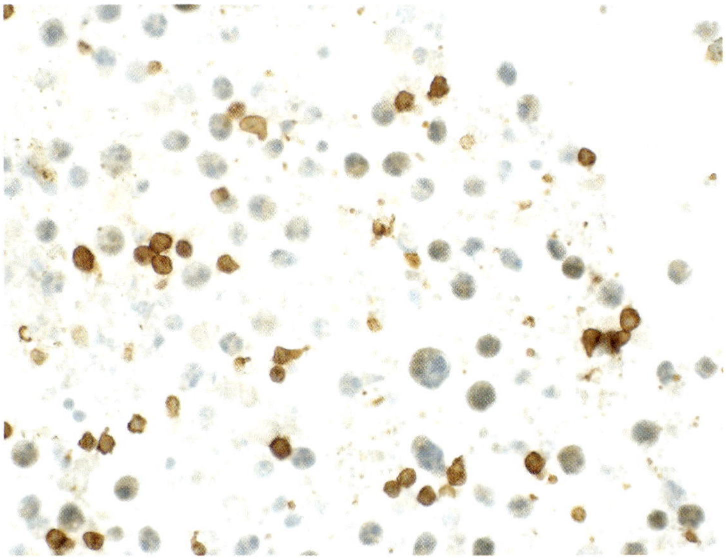

Figure 2.62b — Diffuse Large B-Cell Lymphoma, Pleural Effusion Cell Block Section [CD3 immunohistochemical stain; medium power]. The CD3 immunostain highlights rare background T cells. The larger malignant cells are seen in the background on the counterstain (blue) and thus lack expression of CD3.

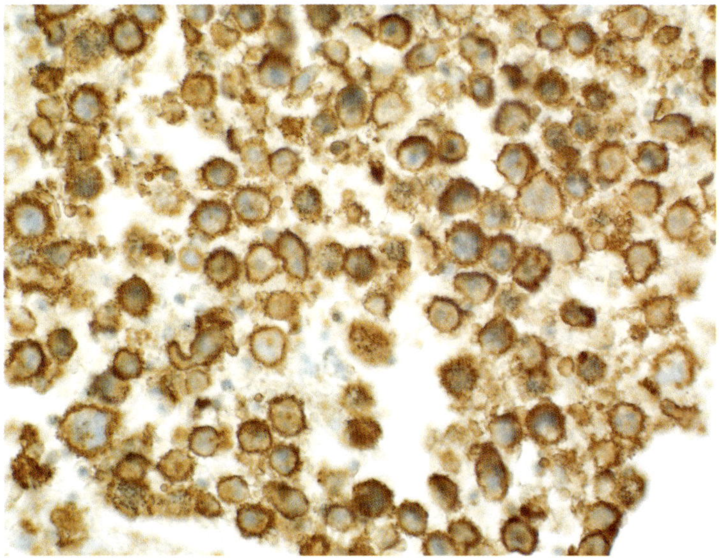

Figure 2.62c — Diffuse Large B-Cell Lymphoma, Pleural Effusion Cell Block Section [CD20 immunohistochemical stain; high power]. Most of the cells in the field demonstrate membranous staining with CD20, a marker for B lymphocytes. This identifies the malignant cells as lymphoma cells and, more specifically, a large B-cell lymphoma. Additional stains for human herpesvirus-8 (HHV-8) and Epstein–Barr virus (EBV) were negative; positivity for either would be suggestive of a primary effusion lymphoma. In this case, the patient did not have known involvement outside of the pleural cavity.

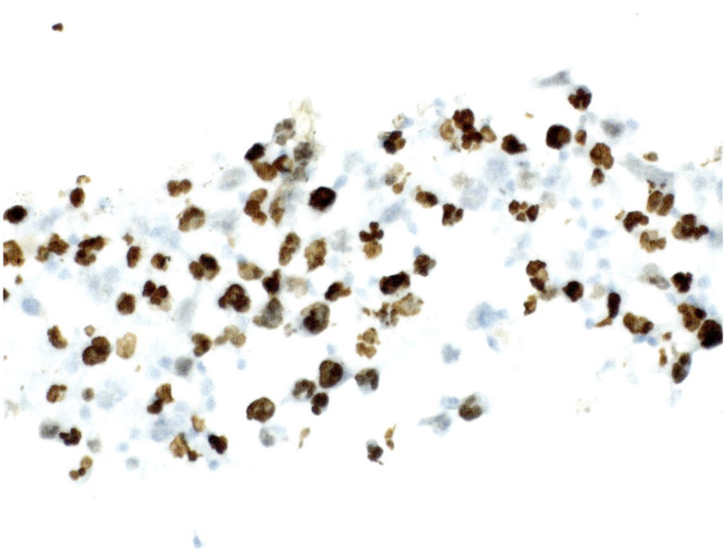

Figure 2.62d — Diffuse Large B-Cell Lymphoma, Pleural Effusion Cell Block Section [Ki-67 immunohistochemical stain; medium power]. The interpretation of Ki-67 is difficult in cell block material, especially in effusion specimens that may contain numerous nonneoplastic cells. Reactive lymphocytes in effusion specimens may also have slightly elevated Ki-67 rates, confounding an accurate estimation. In this case, the lymphoma cell nuclei are bizarre enough to be identified on both the counterstain and immunostain, indicating a high Ki-67 proliferation index.

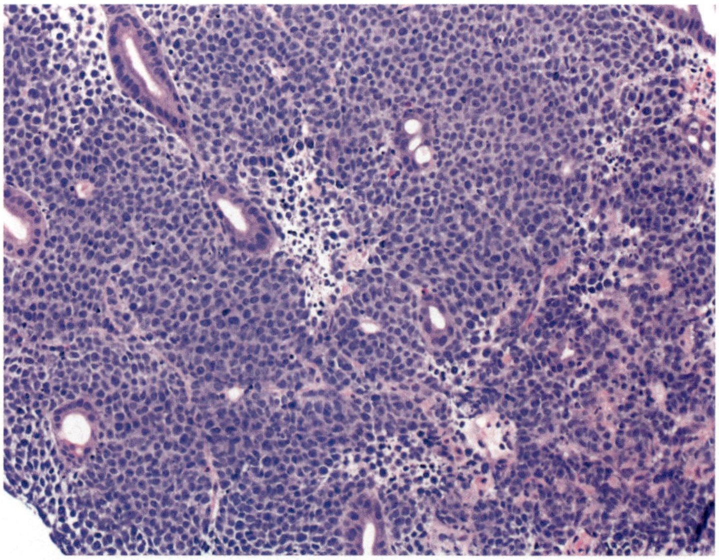

Figure 2.63a — Diffuse Large B-Cell Lymphoma Involving the Stomach, Biopsy [H&E stain; low-intermediate power]. This gastric biopsy shows the lamina propria entirely replaced by uniform cells with hyperchromatic nuclei. From this power, it is not possible to definitively classify the infiltrative cells, but the differential diagnosis includes carcinoma, lymphoma, and melanoma. Primary or secondary lymphomas can involve the stomach. The most common lymphoma to involve the stomach is diffuse large B-cell lymphoma, followed by extranodal marginal zone lymphoma of mucosal-associated lymphoid tissue (MALT) lymphoma. Gastric MALT lymphomas arise in the setting of *Helicobacter pylori* infection and resolve with antibiotic treatment for *H. pylori.*

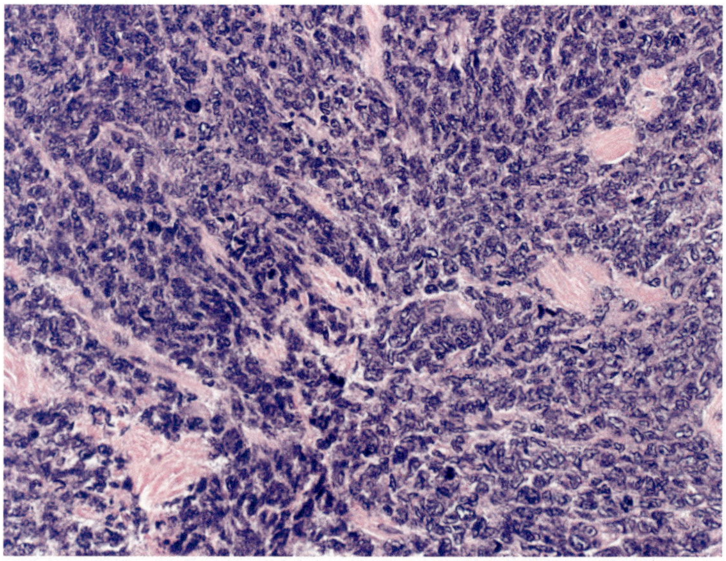

Figure 2.63b — Diffuse Large B-Cell Lymphoma Involving the Stomach, Biopsy [H&E stain; intermediate power]. Areas of this gastric biopsy consist entirely of monotonous cells with enlarged, hyperchromatic nuclei, high nuclear-to-cytoplasmic ratios, nuclear molding, a brisk mitotic rate, and crush artifact. Although this is an example of a diffuse large B-cell lymphoma, the presence of crush artifact and nuclear molding raises the differential diagnosis of small cell carcinoma. Examination of the cells in better-preserved areas can aid in the diagnosis, but it is not uncommon for the entirety of small biopsies to display prominent crush artifact. A targeted immunopanel that includes cytokeratins and hematologic markers can help in this setting, as the crushed cells will still be immunoreactive despite the lack of nuclear or cellular detail on the H&E.

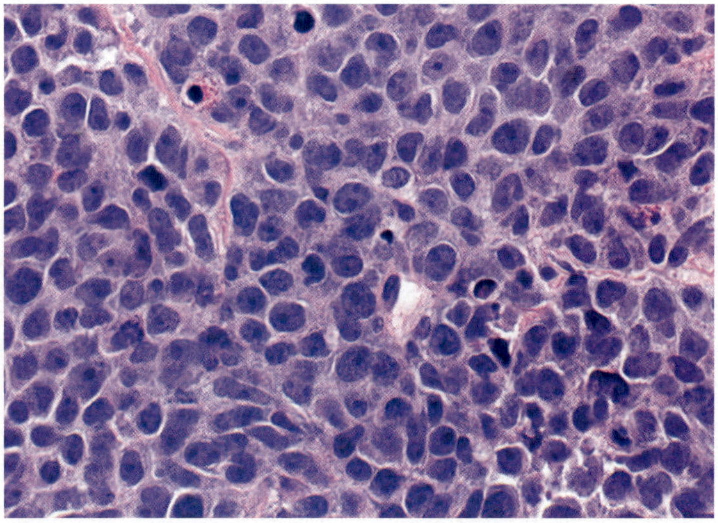

Figure 2.63c — Diffuse Large B-Cell Lymphoma Involving the Stomach, Biopsy [H&E stain; high power]. Diffuse large B-cell lymphomas are characterized by large cells with high nuclear-to-cytoplasmic ratios, irregular nuclear contours, prominent nucleoli, and a brisk mitotic rate. This example also displays scattered eosinophils and apoptotic bodies. Diffuse large B-cell lymphomas can arise de novo secondary to transformation of lower-grade lymphomas, such as follicular lymphoma, or as a post-transplant lymphoproliferative disorder. They are classified as either germinal center B-cell–like or activated (nongerminal center) B-cell–like on the basis of immunolabeling patterns for BLC6, CD10, BLC2, and MUM1.

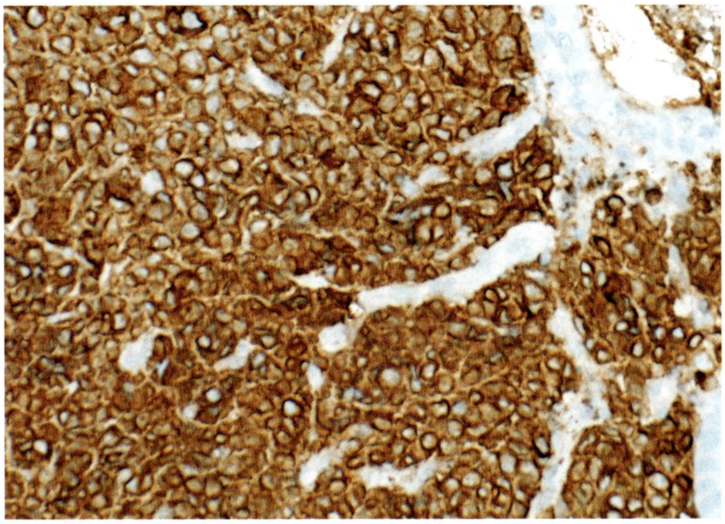

Figure 2.64a — Diffuse Large B-Cell Lymphoma Involving the Stomach, Biopsy [CD20 immunohistochemical stain; intermediate power]. CD20 is a common cell-surface B-cell marker that labels the cell membranes of both benign and malignant B cells. Diffuse large B-cell lymphomas are uniformly immunoreactive for CD20. In contrast, they are negative for cytokeratins (which label carcinomas) and S-100 protein (which labels melanomas). Rituximab is an anti-CD20 chimeric antibody that is used to treat B-cell lymphomas, and CD20 expression can be lost in B-cell lymphomas during and after treatment; however, the expression of other B-cell markers such as CD19 and CD79a is retained.

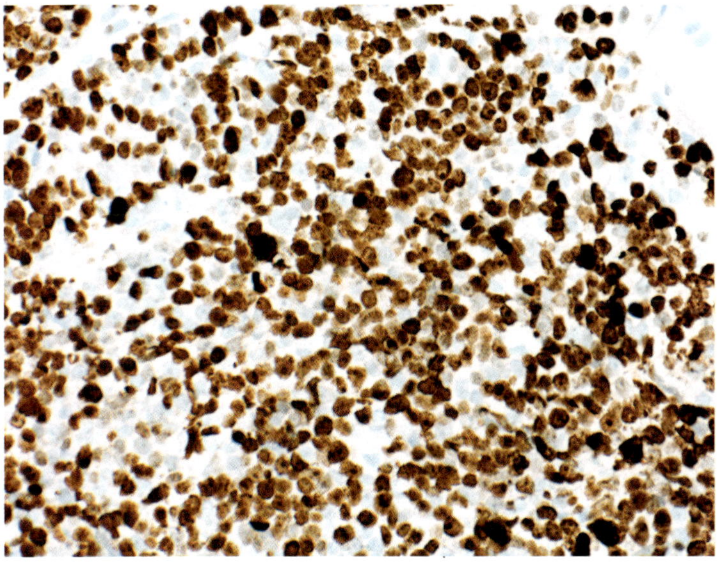

Figure 2.64b — Diffuse Large B-Cell Lymphoma Involving the Stomach, Biopsy [Ki-67 immunohistochemical stain; intermediate power]. Diffuse large B-cell lymphomas display a Ki-67 proliferation index of greater than 90%. Nuclear labeling for Ki-67 (MIB1) detects cells that are actively undergoing cell cycle progression and thus are markers of cell proliferation. The percentage of cells with Ki-67 labeling are higher than the number of mitoses present in a tumor, but the Ki-67 proliferation index parallels the mitotic rate. Small cell carcinomas, which are in the differential diagnosis for diffuse large B-cell lymphoma, also have a high Ki-67 proliferation rate of nearly 100%; however, in contrast to B-cell lymphoma, small cell carcinomas are typically immunoreactive for cytokeratin and negative for CD20. The Ki-67 proliferation index of lower grade lymphomas such as follicular lymphoma is lower.

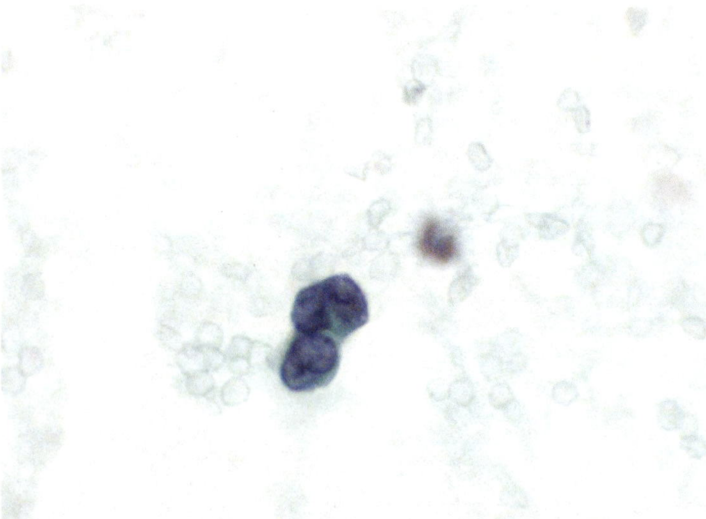

Figure 2.65a — Metastatic Alveolar Rhabdomyosarcoma, Pericardial Fluid [Pap stain; high power]. Tumor cells in alveolar rhabdomyosarcoma tend to be small with scant cytoplasm, round to ovoid nuclei with occasional irregular protrusions or convoluted shapes, and inconspicuous nucleoli. Pleomorphic multilobulated/multinucleated cells with abundant eosinophilic cytoplasm and wreath-like nuclei may also be seen. The differential often includes lymphoma, reactive mesothelial cells, other small round blue cell tumors, and, rarely, small cell carcinoma. A panel of immunohistochemical stains, including myogenin, MyoD1, cytokeratin, CD99, WT-1, synaptophysin, chromogranin, and leukocyte common antigen, helps distinguish alveolar rhabdomyosarcoma from these other entitities. Of note, tumors have been shown to variably express neuroendocrine markers, particularly CD56, as well as smooth muscle actin (SMA), CAM5.2, and activin receptor-like kinase 1 (ALK1).

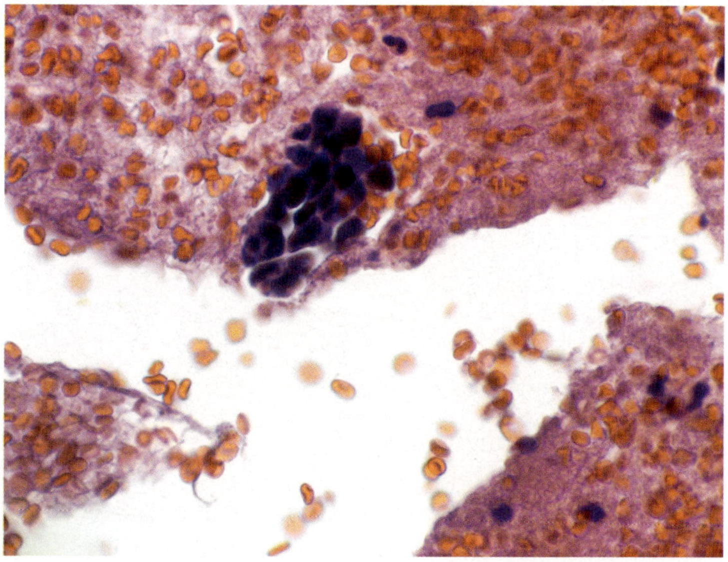

Figure 2.65b — Metastatic Alveolar Rhabdomyosarcoma, Pericardial Fluid Cell Block Section [H&E stain; high power]. Tumor cells may be present in loosely cohesive clusters or as single immature discohesive cells with irregular hyperchromatic nuclei. Although nuclear molding is not typically a feature of alveolar rhabdomyosarcoma, it can be seen in the solid variant of alveolar rhabdomyosarcoma. Most cases of alveolar rhabdomyosarcoma occur sporadically without an apparent genetic predisposition and contain one of two recurrent chromosomal translocations: t(2;13)(q35;q14) or t(1;13)(p36;q14).

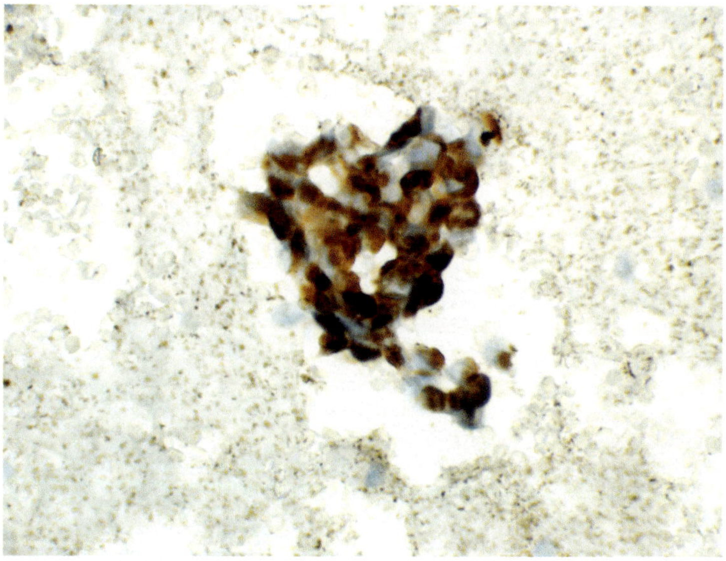

Figure 2.65c — Metastatic Alveolar Rhabdomyosarcoma, Pericardial Fluid Cell Block Section [Desmin immunohistochemical stain; high power]. Desmin is an intermediate filament present near the Z line in sarcomeres of cardiac, smooth, and striated muscle. Cytoplasmic staining can be seen in normal reactive mesothelial cells, myofibroblasts, and smooth muscle cells, as well as a variety of neoplasms. Although generally considered a nonspecific marker of myogenic differentiation, it can be useful as part of a panel in certain differentials such as rhabdomyosarcoma (desmin positive, keratin focal/weakly positive, INI1 positive) versus extrarenal rhabdoid tumor (desmin negative, keratin positive, INI1 loss).

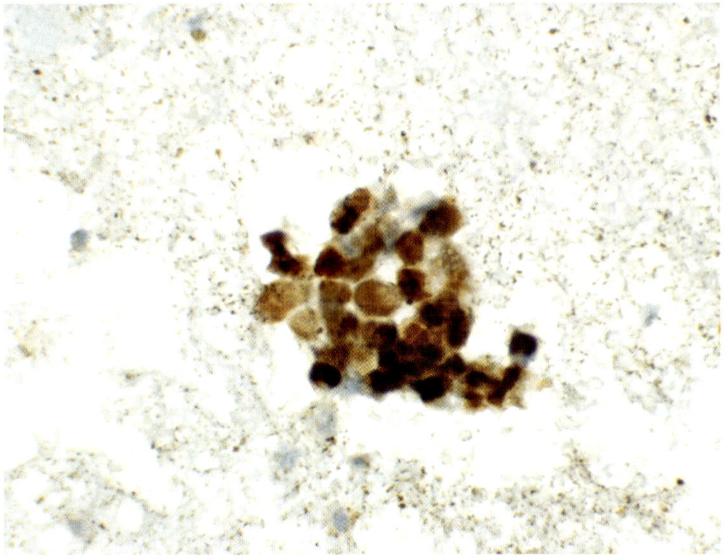

Figure 2.65d — Metastatic Alveolar Rhabdomyosarcoma, Pericardial Fluid Cell Block Section [Myogenin immunohistochemical stain; high power]. Myogenin is a sensitive and specific marker of skeletal muscle differentiation. It is one of several basic myogenic transcription factors, including *MyoD*, *Myf5*, and *MRF4*, that play critical roles in skeletal muscle differentiation and development during embryogenesis. Nuclear positivity for myogenin and MyoD1 is usually strong and diffuse.

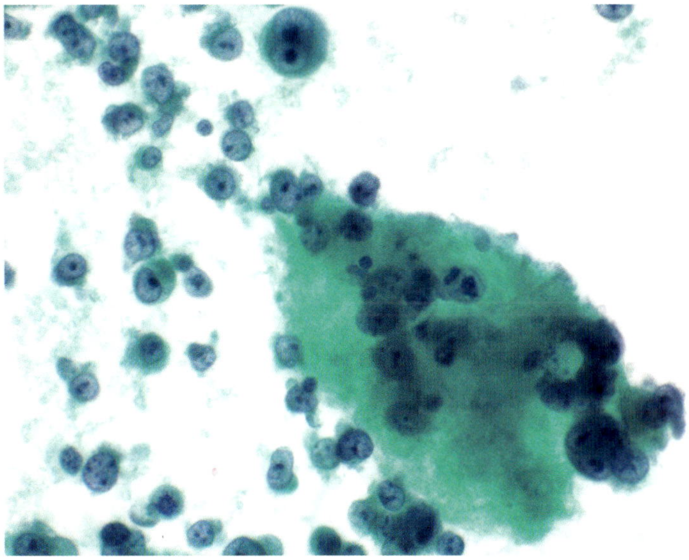

Figure 2.66a — Metastatic Osteosarcoma, Pleural Fluid [Pap stain; high power]. The neoplastic cells are pleomorphic; the only common cytomorphologic feature is the presence of a prominent nucleolus. Many cells have an eccentrically placed, epithelioid nuclei with mildly irregular nuclear borders. The amount of cytoplasm varies between cells and has a green-cyan appearance that matches the color of extracellular matrix material in which several tumor cells are embedded and likely represents osteoid material. Some cells are multinucleated. Aside from translocation sarcomas, which can appear exceedingly bland, high-grade sarcomas often have pleomorphic, large nuclei with prominent nucleoli or coarse chromatin. Extremely large cells may be present with multinucleated or multilobed nuclei, and cells are typically singly dispersed. The presence of matrix material may be a clue. When a fluid specimen is involved, the patient inevitably has a known history.

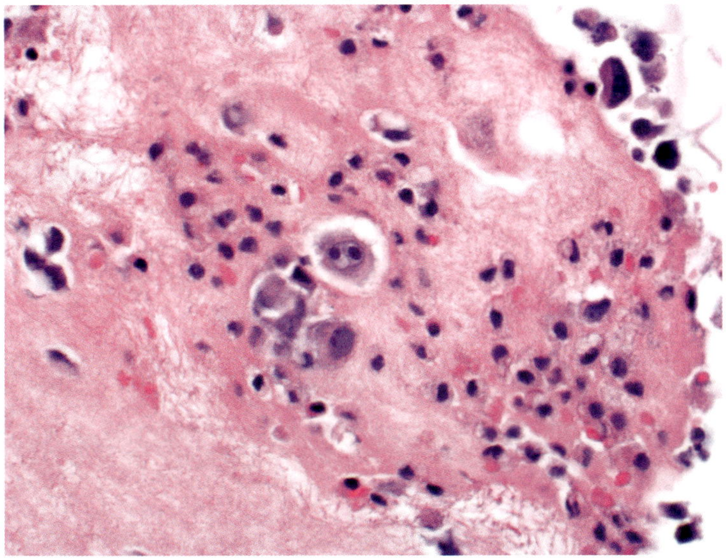

Figure 2.66b — Metastatic Osteosarcoma, Pleural Fluid Cell Block Section [H&E stain; high power]. Several malignant cells are scattered throughout, with a single binucleated tumor cell prominently displayed in the center of the field. The nucleoli possessed by this cell are far too large to be secondary to reactive changes. Note the retraction halo formed around the cell, a visual clue for potentially malignant cells. Other atypical cells can be seen with hyperchromasia and irregularly shaped nuclei.

3

Cerebrospinal Fluid

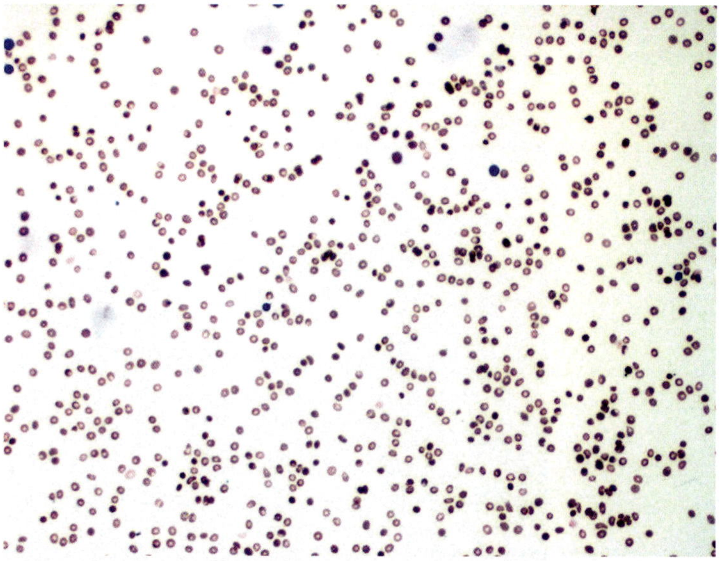

Figure 3.1 — Peripheral Blood Contamination, Lumbar Puncture [Diff-Quik stain; low power]. Traumatic lumbar puncture is the primary cause of peripheral blood contamination in cerebrospinal fluid (CSF) specimens. CSF contaminated by peripheral blood has numerous red blood cells (RBCs) and only scattered white blood cells (WBCs). The ratio of WBC to RBC is similar to that of peripheral blood. In comparison, CSF pleocytosis has a greater WBC:RBC ratio (see Figure 3.3).

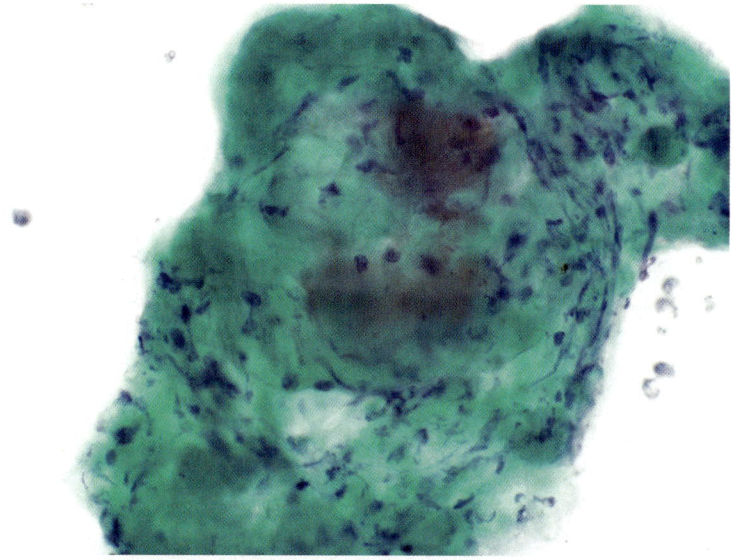

Figure 3.2 — Intervertebral Disk Material, Lumbar Puncture [Pap stain; high power]. This field shows a large fragment of dense, paucicellular, amorphous substance. This finding represents the intervertebral disk material that was accidentally aspirated during the procedure. Intervertebral disks consist of fibrocartilage and hyaline cartilage. The site through which the needle traverses can give some indication of what may contaminate the specimen. For instance, bone marrow elements may occasionally be seen. The presence of these elements often indicates that the procedure did not sample CSF.

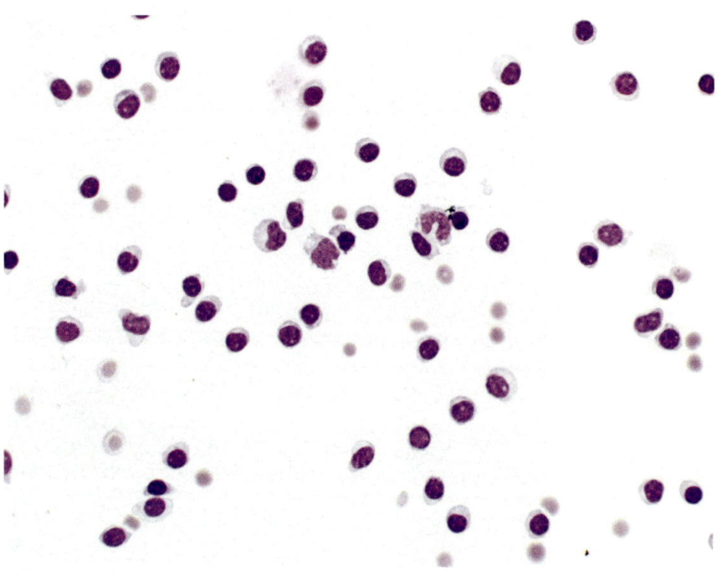

Figure 3.3 — Pleocytosis, Lumbar Puncture [Diff-Quik stain; low power]. *Pleocytosis* is defined as an increased cell count, which usually means an increased WBC count in the CSF. This typically contains a mixture of mononuclear cells and lymphocytes. There are two populations of mononuclear cells in this low-power field: the smaller cells with scant cytoplasm and round to ovoid nuclei are lymphocytes; the larger cells with kidney bean- or horseshoe-shaped nuclei are monocytes.

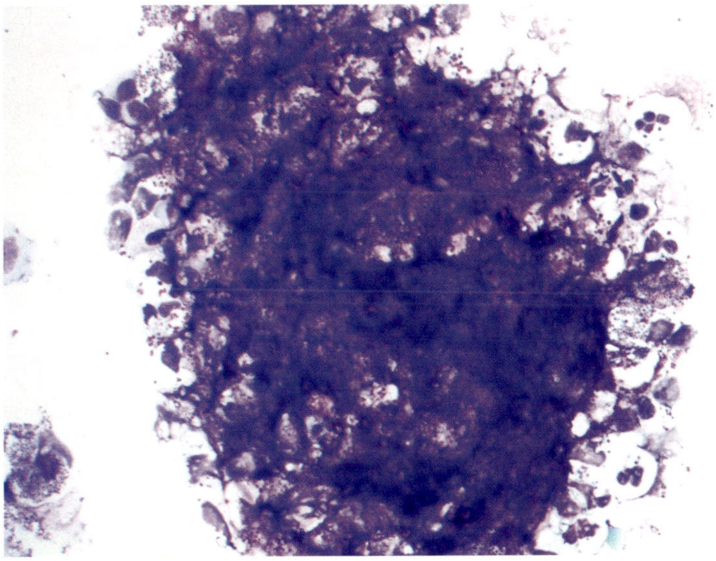

Figure 3.4a — Bacterial Meningitis, Lumbar Puncture [Diff-Quik stain; high power]. This field shows a big clump of fibrin and inflammatory cells. Numerous cocci of 1 to 2 μm can be seen at the edge of the aggregate. The most common pathogens of bacterial meningitis are meningococci (*Neisseria meningitidis*) and pneumococci (*Streptococcus pneumoniae*).

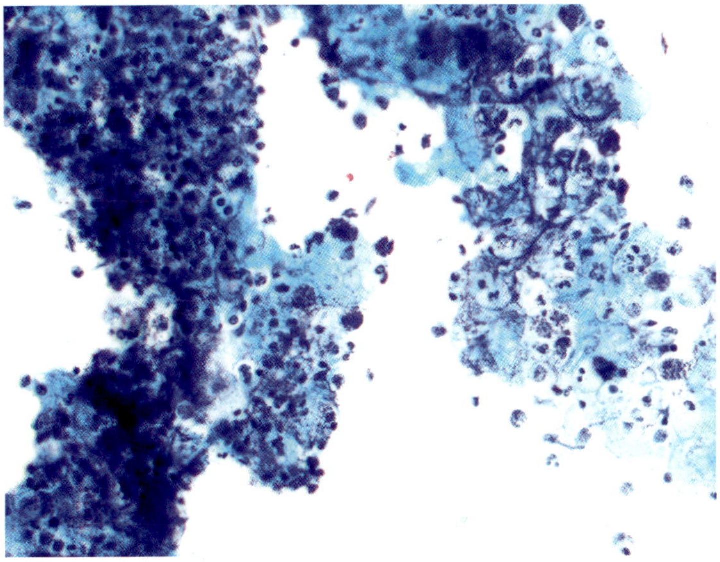

Figure 3.4b — Bacterial Meningitis, Lumbar Puncture [Pap stain; high power]. The inflammatory cells consist predominantly of neutrophils. Abundant cocci can be seen intracytoplasmically. In addition to neutrophil-predominant pleocytosis, bacterial meningitis is often associated with elevated protein and decreased glucose levels in CSF.

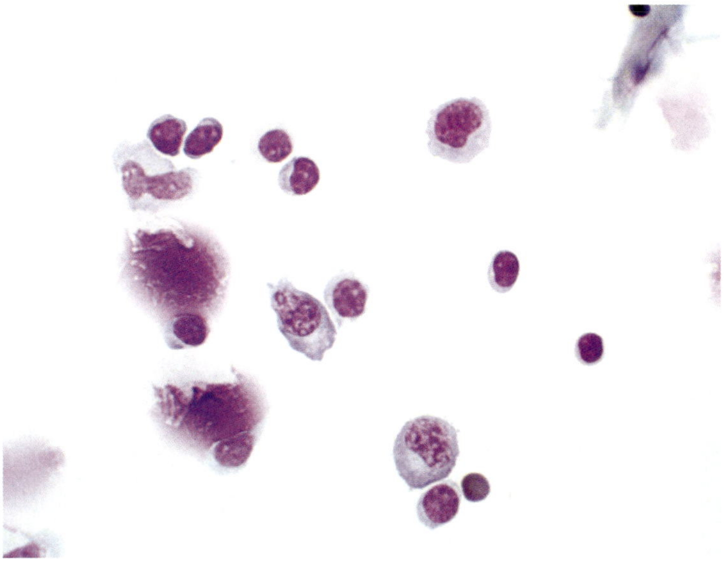

Figure 3.5 — Benign Recurrent Lymphocytic Meningitis, Lumbar Puncture [Diff-Quik stain; high power]. Benign recurrent lymphocytic meningitis, also called *Mollaret meningitis,* is a type of chronic recurrent aseptic meningitis. Herpes simplex virus (HSV) types 1 and 2 and other virus infections have been suspected to be the culprit of the disease. This CSF specimen is characterized by a number of small lymphocytes, along with a few larger Mollaret cells. Mollaret cells are reactive monocytes that contain abundant light basophilic cytoplasm and nuclei in variable shapes, including oval, kidney bean, clefted, and lobulated.

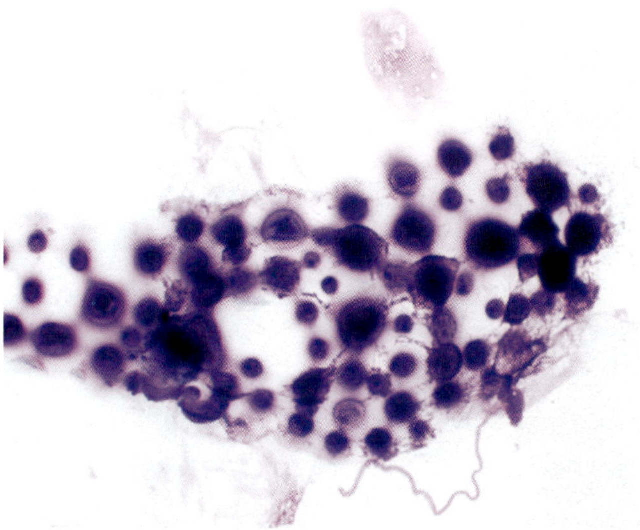

Figure 3.6a—Cryptococcosis, Lumbar Puncture [Diff-Quik stain; high power]. This CSF specimen is remarkable for numerous small spherical structures of slight variability in size. These structures represent encapsulated *Cryptococcus* yeast. The capsule of cryptococci does not stain on Diff-Quik and therefore appears as a clear halo "negative stain." Note the absence of inflammatory cells in the background. Cryptococcosis in the central nervous system (CNS) usually does not cause florid inflammatory reactions.

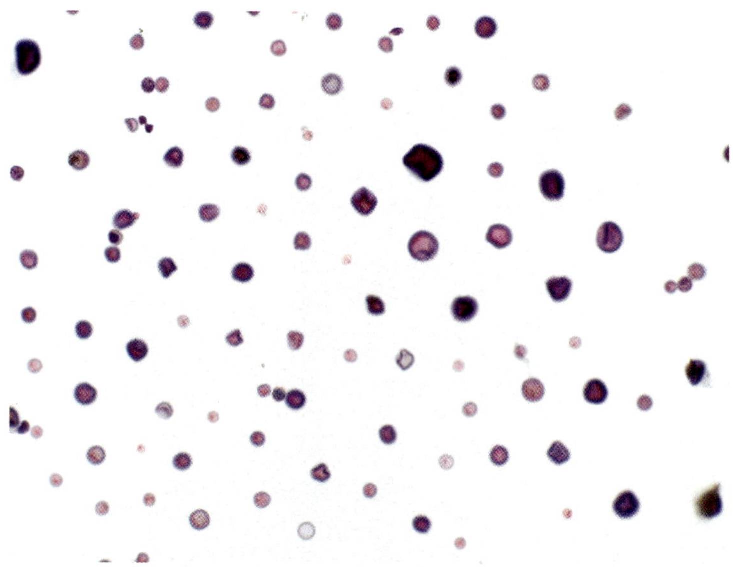

Figure 3.6b — Cryptococcosis, Lumbar Puncture [Pap stain; low power]. Cryptococcosis is an opportunistic infection associated with severe immunosuppression, such as AIDS with extremely low CD4 counts. Although culture is the gold standard for diagnosing cryptococcal infection in the central nervous system (CNS), cerebrospinal fluid (CSF) cytology is also an excellent diagnostic test, given its high sensitivity and specificity. Modern microbiologic laboratory testing has allowed for rapid, specific, and sensitive detection as an adjunct to cytologic examination. Notable in this field are many encapsulated cryptococcal yeasts, some of which demonstrate narrow-based budding. The capsule of *Cryptococcus* is less obvious on Pap stain than Diff-Quik stain.

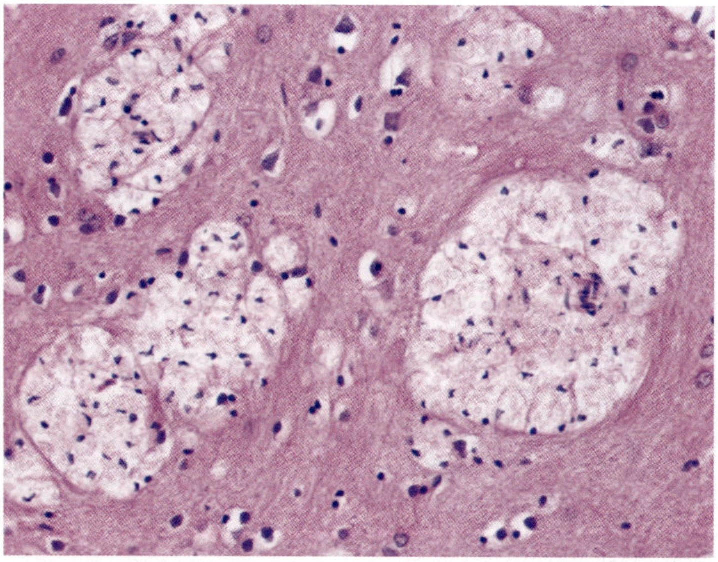

Figure 3.6c — Cryptococcosis, Brain Autopsy [H&E stain; high power]. This tissue section of the brain reveals dilation of perivascular (Virchow–Robin) space. The accumulation of abundant perivascular foamy macrophages gives the brain a soap-bubble appearance. Numerous yeast forms are visible both intracellularly and extracellularly even on the H&E stain. Note that perivascular lymphocytic infiltrates, a common finding of meningitis, are absent in cryptococcal meningitis.

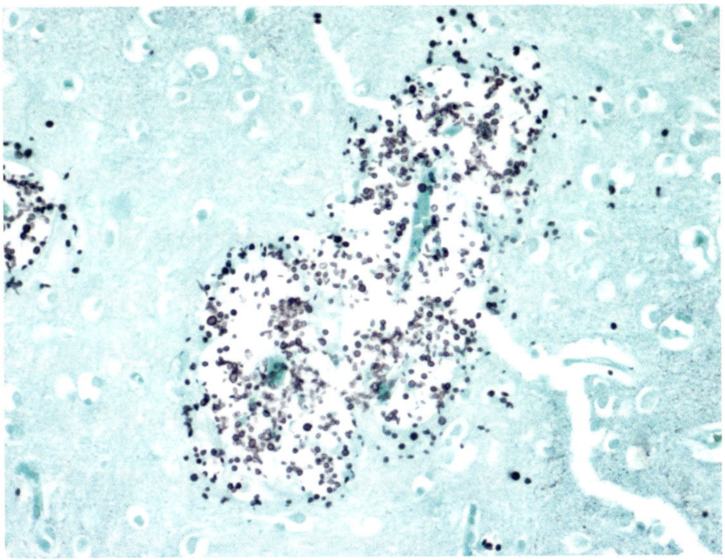

Figure 3.6d — Cryptococcosis, Brain Autopsy [Grocott's methenamine silver (GMS) stain; high power]. Numerous yeasts in the perivascular space are highlighted by GMS fungal stain. The organisms are 4 to 10 μm in diameter. These yeast forms have a thick capsule and demonstrate narrow-neck budding.

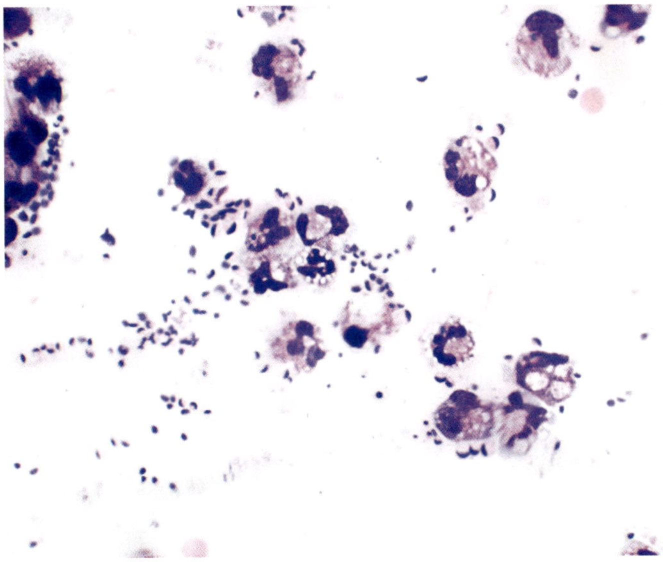

Figure 3.7a — Toxoplasmosis, Lumbar Puncture [Diff-Quik stain; high power]. This image shows a few inflammatory cells and numerous small extracellular organisms called *tachyzoites.* Tachyzoites are a rapidly proliferating form of *Toxoplasma gondii*, which is responsible for expanding the population during the acute phase of infection. The dormant cyst form, bradyzoites, does not exist in CSF.

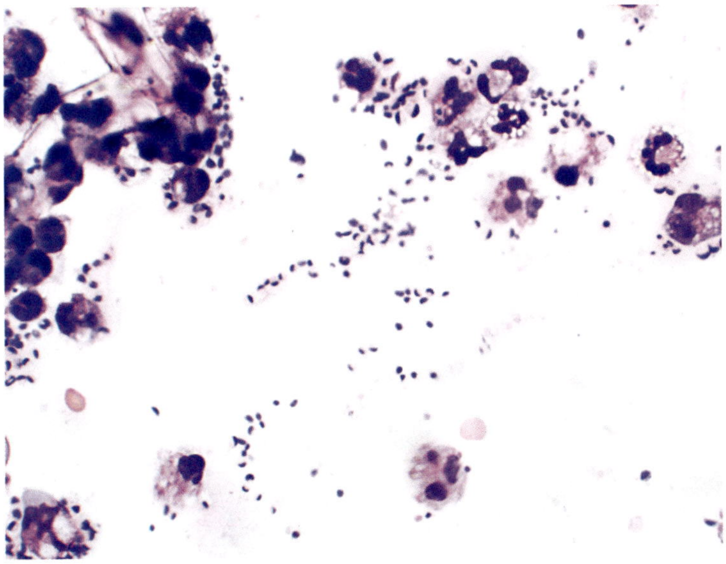

Figure 3.7b — Toxoplasmosis, Lumbar Puncture [Diff-Quik stain; high power]. Toxoplasmosis in the central nervous system (CNS) often elicits marked acute inflammation unless in severely immunocompromised individuals. The inflammatory cells consist of neutrophils, monocytes, and macrophages. Note that tachyzoites are inside some of the inflammatory cells in this field.

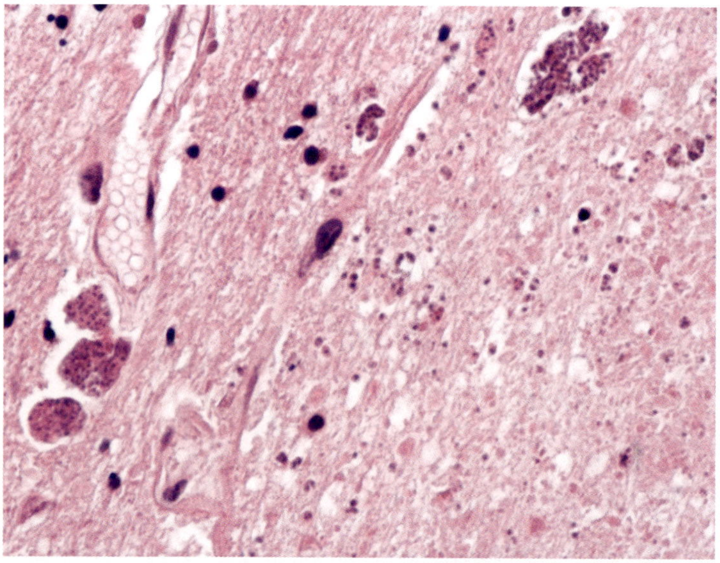

Figure 3.7c — Toxoplasmosis, Brain Autopsy [H&E stain; high power]. The left edge and top right of the image show four bradyzoites, the slowly dividing cyst form of *Toxoplasma gondii*. There are also numerous small microorganisms floating in the adjacent necrotic brain tissue. These are tachyzoites, a rapidly multiplying form of *T. gondii*. Toxoplasmosis often causes "dirty necrosis" in the brain, characterized by large areas of necrosis, abundant cell debris, and many inflammatory cells.

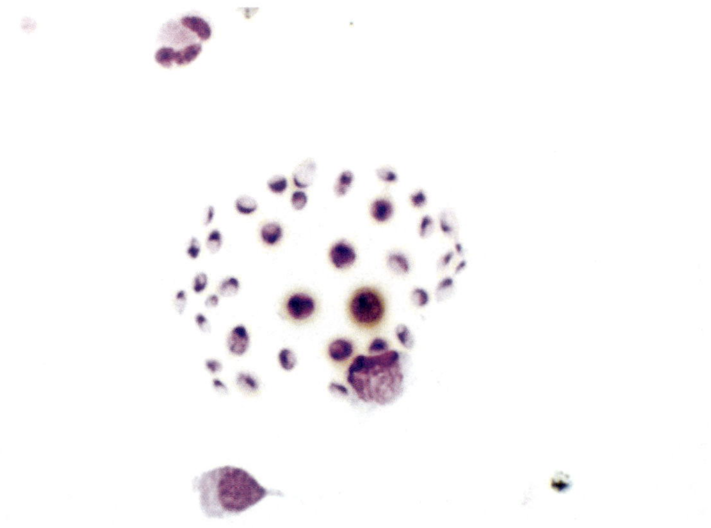

Figure 3.8 — Cryptococcosis and Toxoplasmosis, Lumbar Puncture [Diff-Quik stain; high power]. This field is notable for cells and microorganisms in different shapes and forms. The smallest and most abundant structures are tachyzoites of *Toxoplasma gondii*. In the center, the two larger spherical organisms with vague halos are *Cryptococcus* yeast. In addition, there are three monocytes and neutrophils at the periphery of this microorganism cluster. Concurrent cryptococcal and toxoplasmic infection are unusual but can be seen in severely immunocompromised patients.

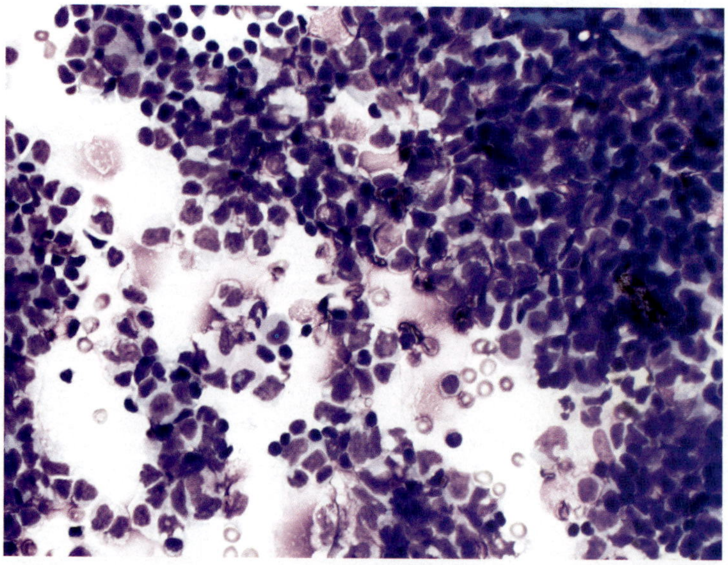

Figure 3.9a — T-Cell Acute Lymphoblastic Leukemia (T-ALL), Lumbar Puncture [Diff-Quik stain; low power]. This is a cellular specimen consisting predominantly of intermediate-sized blasts. The blast cells have scant to moderate amounts of pale blue cytoplasm, irregular nuclei, and occasionally prominent nucleoli. The nuclear-to-cytoplasmic ratio is lower than the background normal lymphocytes.

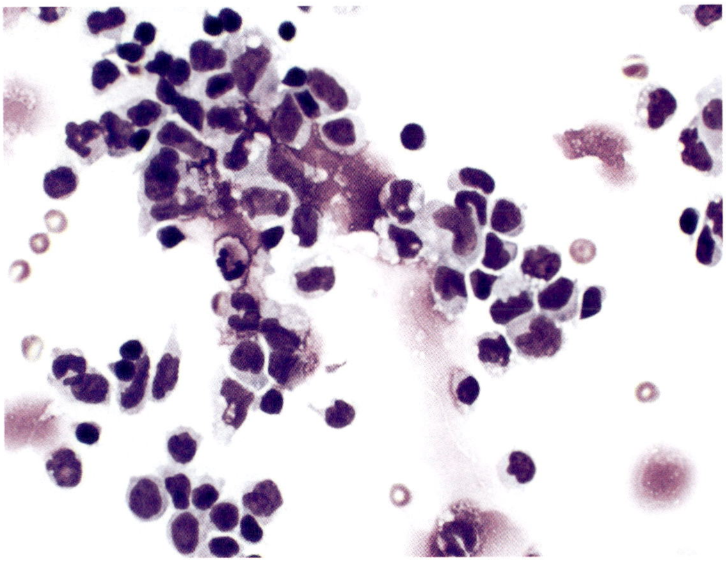

Figure 3.9b — T-Cell Acute Lymphoblastic Leukemia (T-ALL), Lumbar Puncture [Diff-Quik stain; high power]. A lymphoblast is twice the size of a small mature lymphocyte or an RBC, both of which can be found in this field. The blast cells demonstrate slight variability in size and moderate amounts of light basophilic and agranular cytoplasm. Cytoplasmic vacuoles can also be seen.

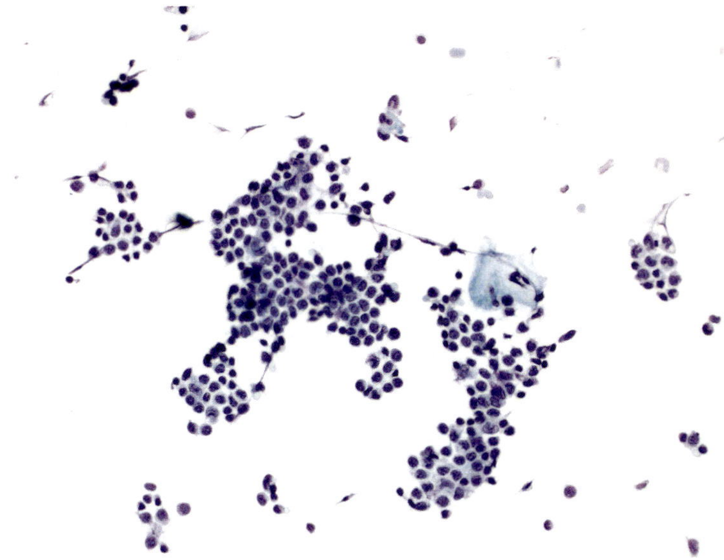

Figure 3.9c — T-Cell Acute Lymphoblastic Leukemia (T-ALL), Lumbar Puncture [Pap stain; low power]. This field shows a relatively monomorphic population of cells. The cells have a small rim of cytoplasm, the amount of which is more than that of normal lymphocytes. Although not apparent, nuclear pleomorphism can be appreciated at low magnification. Nuclear shape is variable: round, oval, indented, or clefted. Nucleoli are better demonstrated on the Pap stain than on the Diff-Quik stain.

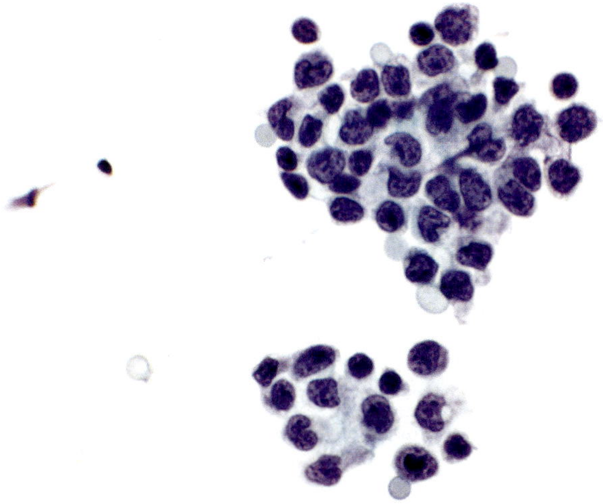

Figure 3.9d — T-Cell Acute Lymphoblastic Leukemia (T-ALL), Lumbar Puncture [Pap stain; high power]. Nuclei of blast cells have irregular or indented contour, slightly coarse chromatin, and multiple small indistinct nucleoli. Conventional cytomorphologic examination of cerebrospinal fluid (CSF) is less sensitive than flow cytometry in detecting leukemic involvement of CSF. To maximize test sensitivity and diagnostic accuracy, cytology is best used in conjunction with flow cytometry.

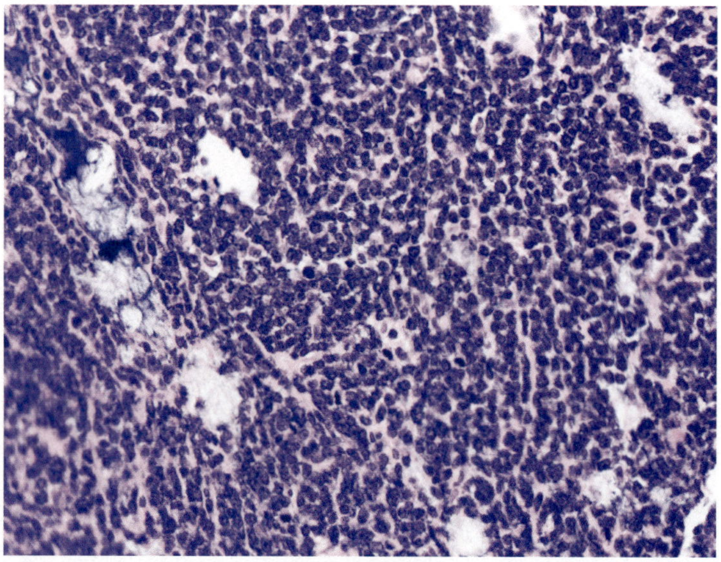

Figure 3.9e — T-Cell Acute Lymphoblastic Leukemia (T-ALL), Mediastinal Biopsy [H&E stain; low power]. This core biopsy shows sheets of small monomorphic cells diffusely infiltrating the soft tissue. Leukemic cells have convoluted or irregular nuclei, finely stippled chromatin and a high nuclear-to-cytoplasmic ratio. Mitotic activity is frequent. Some cells appear to be in a single-file arrangement.

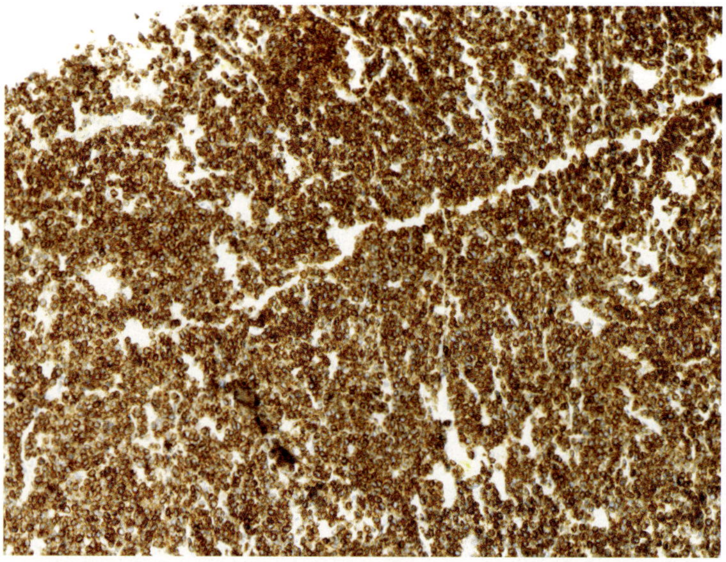

Figure 3.9f — T-Cell Acute Lymphoblastic Leukemia (T-ALL), Mediastinal Biopsy [CD3 immunostain; low power]. As demonstrated on this immunostain, T-ALL is strongly and diffusely positive for CD3, a T-cell marker. This case also shows immunoreactivity for CD4, CD7, CD8, and terminal deoxynucleotidyl transferase (TdT; see Figure 3.9g) and is therefore classified as T-ALL.

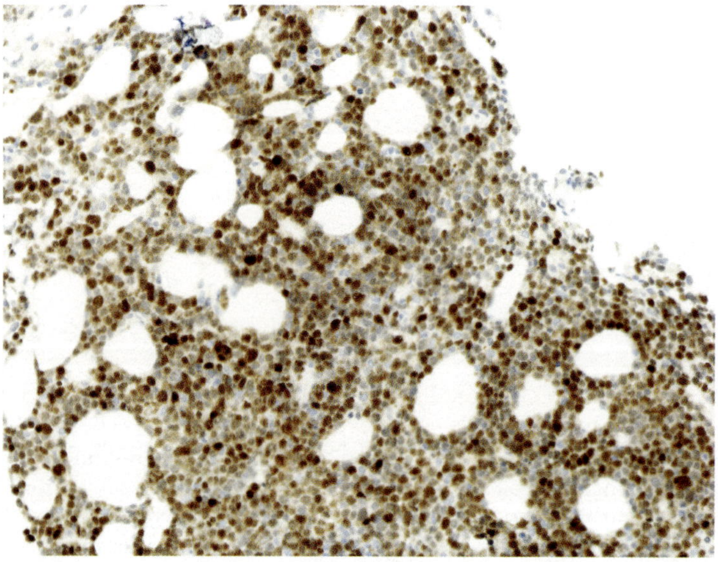

Figure 3.9g — T-Cell Acute Lymphoblastic Leukemia (T-ALL), Mediastinal Biopsy [terminal deoxynucleotidyl transferase (TdT) immunostain; low power]. TdT is a DNA polymerase expressed in pre-B and pre-T cells. It is a marker for ALL (French–American–British [FAB] L1 and L2). TdT-negative ALL is rare but can be seen in T-ALL.

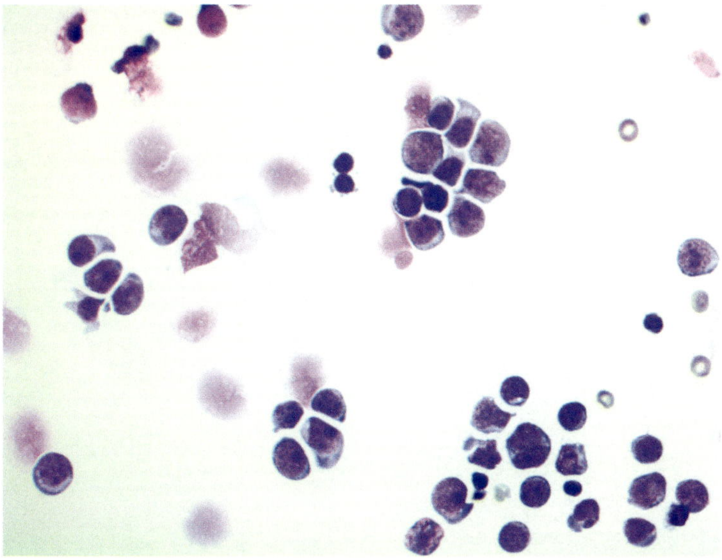

Figure 3.10a — Burkitt Lymphoma, Lumbar Puncture [Diff-Quik stain; high power]. Burkitt lymphoma is the most common type of non-Hodgkin lymphoma in children. It grows rapidly and can be fatal if left untreated. Patients with more extensive disease are more likely to have CSF involvement. This CSF specimen shows numerous cells of variable sizes and shapes. They have moderate amounts of pale basophilic cytoplasm. Nuclei are irregular with coarse chromatin and one to four prominent nucleoli.

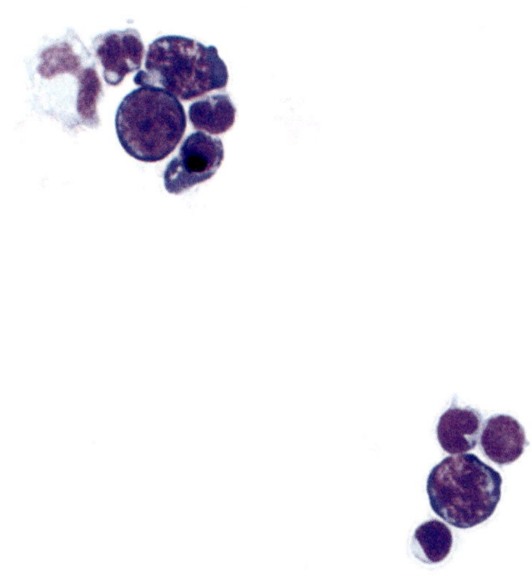

Figure 3.10b — Burkitt Lymphoma, Lumbar Puncture [Diff-Quik stain; high power]. This field is remarkable for a few intermediate-sized cells with large irregular nuclei and intensely basophilic cytoplasm. Also noted in the background are normal lymphocytes and monocytes. Multiple small cytoplasmic vacuoles, a common cytomorphologic feature of Burkitt lymphoma in bone marrow smear, are visible but not obvious in the cerebrospinal fluid (CSF) specimens.

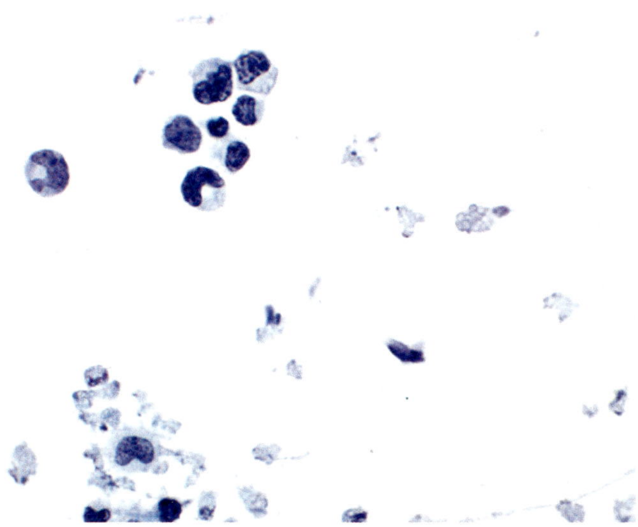

Figure 3.10c — Burkitt Lymphoma, Lumbar Puncture [Pap stain; high power]. Compared with other types of leukemia and lymphomas, Burkitt lymphoma cells in the CSF demonstrate a higher degree of pleomorphism. Their nuclear contour is markedly convoluted. Chromatin is clumped. Nucleoli are conspicuous. The background of this field is also notable for apoptotic cells and debris, likely resulting from treatment.

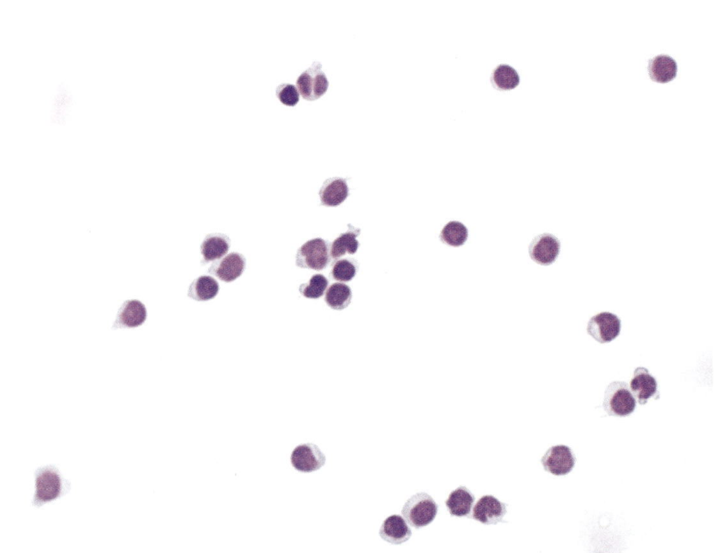

Figure 3.11a — Chronic Lymphocytic Leukemia/Small Lymphocytic Lymphoma (CLL/SLL), Lumbar Puncture [Diff-Quik stain; low power]. CLL/SLL is the most common type of adult leukemia, but CNS involvement is infrequent. This field shows a number of cells that are slightly larger than normal lymphocytes and have scant pale basophilic cytoplasm. Some nuclei are notched, and the chromatin is clumped ("soccer ball like"). One binucleated cell is seen at the top of the image.

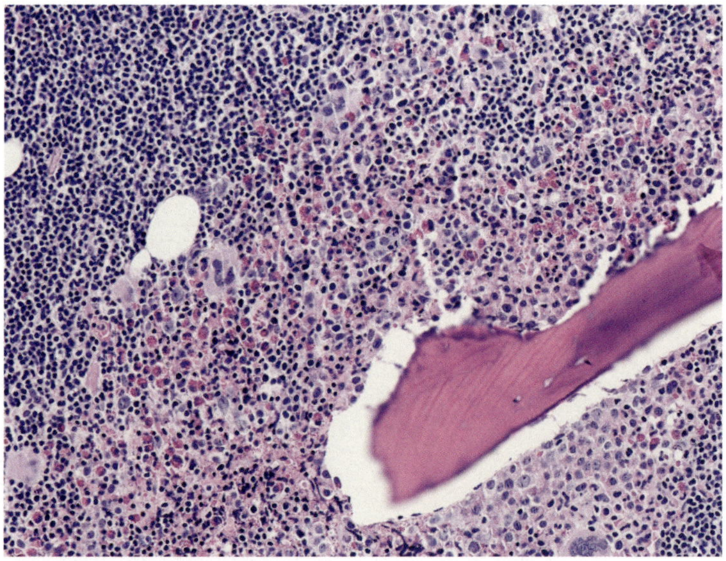

Figure 3.11b — Chronic Lymphocytic Leukemia/Small Lymphocytic Lymphoma (CLL/SLL), Bone Marrow Biopsy [H&E stain; low power]. The top left corner of the image shows a nontrabecular lymphocytic nodule replacing normal hematopoiesis. The nodule consists of small monomorphic lymphocytes with a slightly irregular nuclear contour. This nodular pattern is common in CLL involvement of bone marrow. In addition, individual leukemic cells can be seen infiltrating the marrow, characterized as the "interstitial" pattern of infiltration.

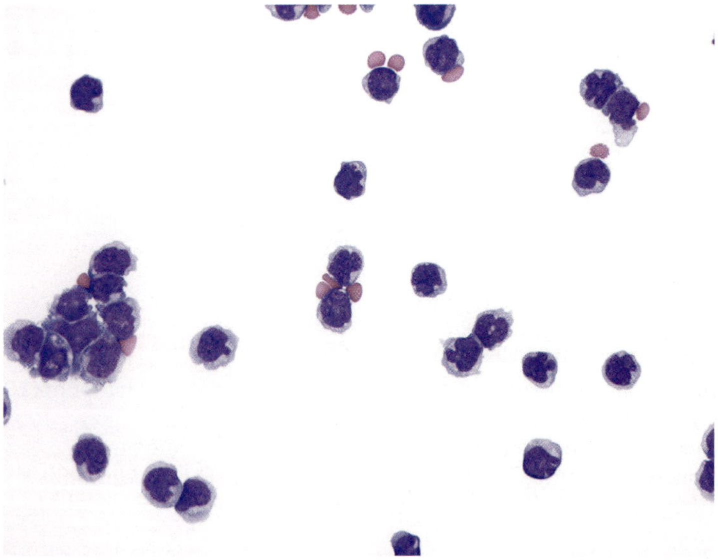

Figure 3.12a — Acute Myeloid Leukemia, Lumbar Puncture [Diff-Quik stain; high power]. This cerebrospinal fluid (CSF) specimen shows numerous intermediate-sized cells with atypical nuclear features. The nuclei are clefted or folded, and the nuclear contour is highly irregular. Nucleoli are prominent. Their cytoplasm is basophilic and slightly vacuolated.

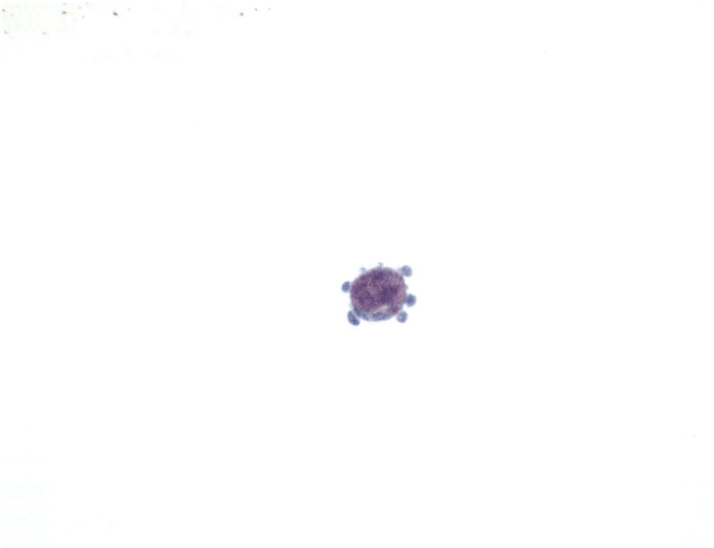

Figure 3.12b — Acute Myeloid Leukemia, Lumbar Puncture [Diff-Quik stain; high power]. The immature myeloblast has a folded nucleus, intensely basophilic cytoplasm, and a high nuclear-to-cytoplasmic ratio. Two conspicuous nucleoli can be seen on the Diff-Quik stain. Cytoplasmic blebs are an important cytomorphologic feature that distinguishes immature myeloblasts from a normal monocyte.

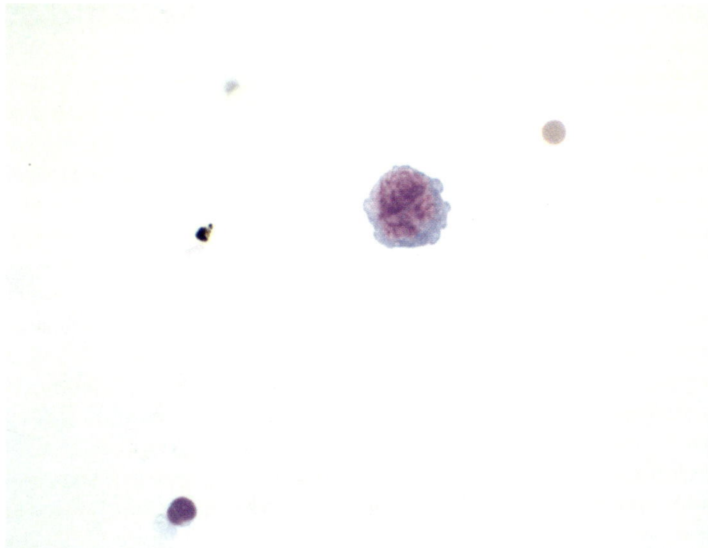

Figure 3.12c — Acute Myeloid Leukemia, Lumbar Puncture [Diff-Quik stain; high power]. Myeloblasts can be very large and harbor a large nucleus. In this field, the myeloblast is almost 10 times the size of the background lymphocyte and red blood cell (RBC). Although not seen in this field, myeloblasts can have azurophilic granules or vacuoles in their cytoplasm.

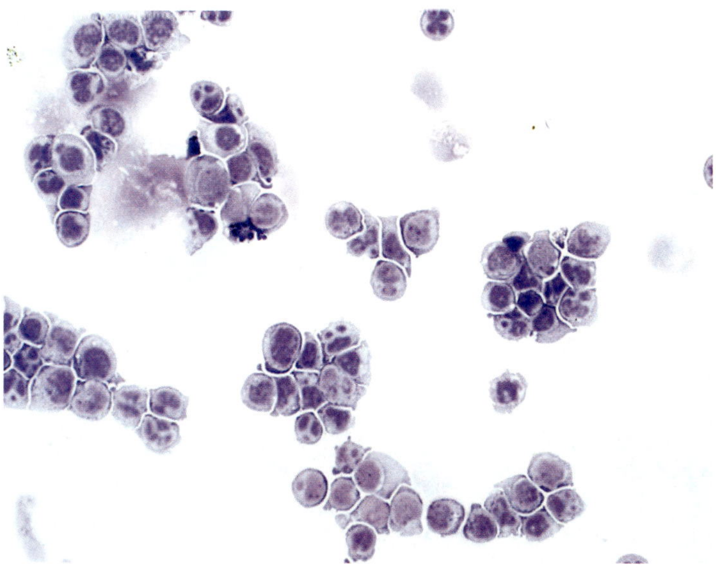

Figure 3.13a — Adult T-Cell Leukemia/Lymphoma (ATLL), Lumbar Puncture [Diff-Quik stain; high power]. ATLL is a mature T-cell malignancy associated with human T-cell lymphotropic virus type 1 (HTLV-1) infection. This hypercellular cerebrospinal fluid (CSF) specimen is composed of pleomorphic lymphoid cells, many of which are undergoing apoptosis (fragmented nuclei), possibly due to treatment effect. The cells bear moderate amounts of basophilic cytoplasm. A signature "flower cell" is seen at the top of the image.

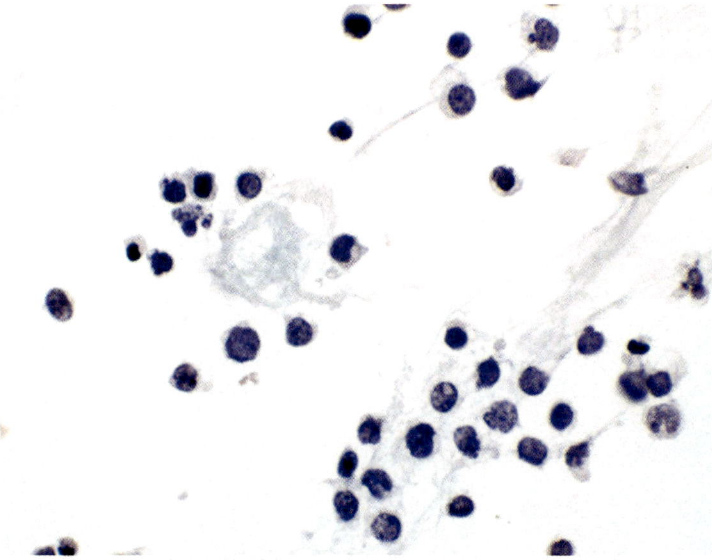

Figure 3.13b — Adult T-Cell Leukemia/Lymphoma (ATLL), Lumbar Puncture [Pap stain; high power]. ATLL cells often have markedly irregular nuclear membranes or multilobed nuclei. Nuclear pleomorphism is more prominent than in other types of leukemia/lymphoma that spread to the cerebrospinal fluid (CSF). Chromatin is coarsely clumped. There are a few apoptotic cells on the left of the image.

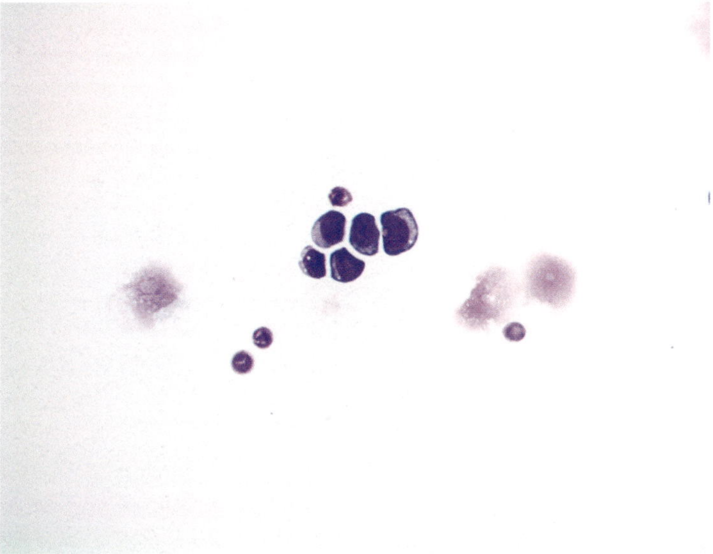

Figure 3.14a — Primary Central Nervous System (CNS) B-Cell Lymphoma, Lumbar Puncture [Diff-Quik stain; low power]. This field shows a handful of atypical lymphoid cells that are three to five times the size of a red blood cell (RBC). The cells have large, irregular nuclei and moderate amounts of delicate cytoplasm. Prominent nucleoli are an important cytomorphologic feature of primary CNS B-cell lymphoma, which is predominantly a diffuse large B-cell lymphoma.

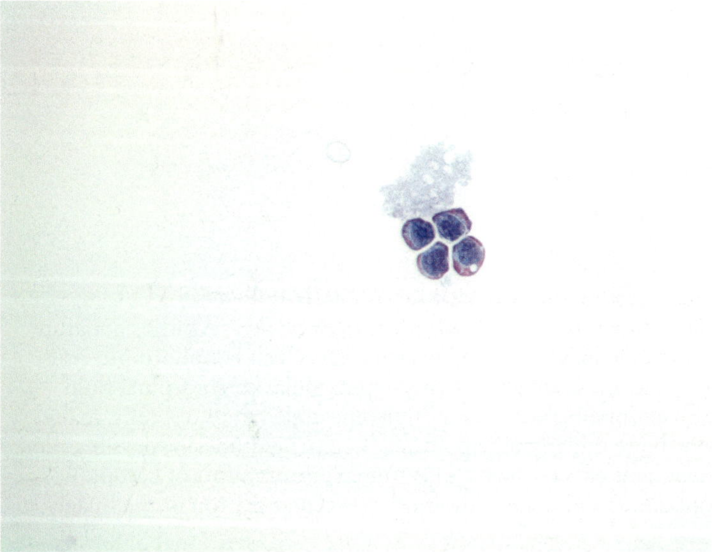

Figure 3.14b — Primary Central Nervous System (CNS) B-Cell Lymphoma, Lumbar Puncture [Pap stain; low power]. Approximately, a quarter of the cases of primary CNS lymphomas have positive cerebrospinal fluid (CSF) cytologic findings. Nucleoli, an important feature of primary CNS B-cell lymphoma, are more easily visualized on the Pap stain. The cytoplasm of the leukemic cells is occasionally vacuolated.

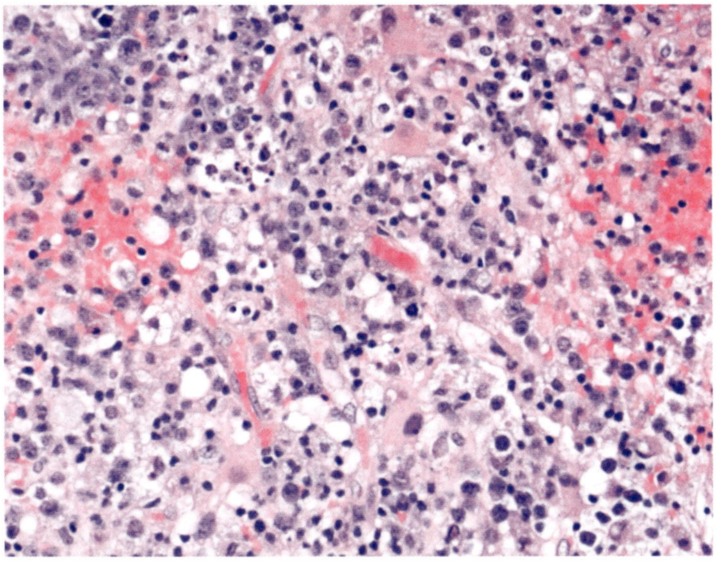

Figure 3.14c — Primary Central Nervous System (CNS) B-Cell Lymphoma, Brain Biopsy [H&E stain; low power]. This brain biopsy specimen shows a mixture of cells, including foamy macrophages, reactive astrocytes, small lymphocytes, and some atypical lymphoid cells with fine chromatin and conspicuous nucleoli. There is apoptotic debris in the background. Macrophage infiltrate is a common finding of primary CNS lymphoma after treatment with steroids, which is often used to decrease intracranial pressure in the setting of a mass lesion. Although not seen in this image, primary CNS lymphoma cells frequently demonstrate angiocentricity.

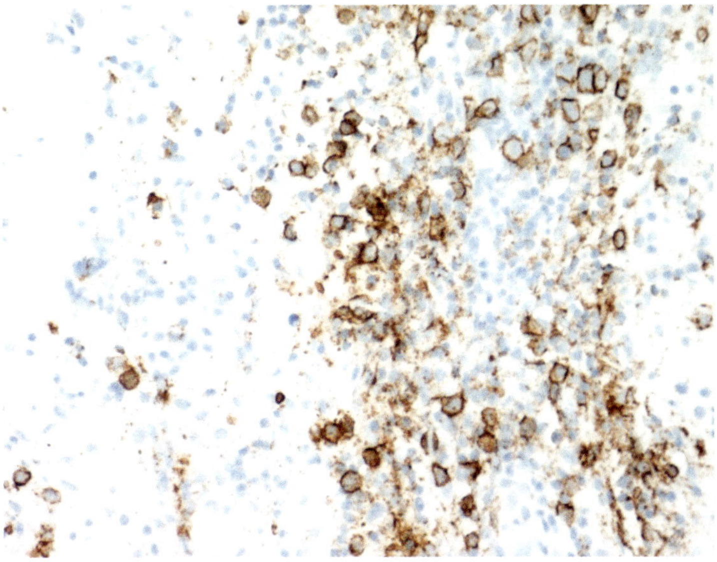

Figure 3.14d — Primary Central Nervous System (CNS) B-Cell Lymphoma, Brain Biopsy [CD20 immunostain; low power]. The majority of the primary CNS lymphomas are B-cell lymphomas. Therefore CD20 is a useful marker for diagnosing primary CNS lymphomas. In this image, CD20 immunostain highlights the atypical lymphoid cells among other benign cells. Note that the CD20-positive cells are larger compared with the CD20-negative cells. Primary CNS lymphomas tend to affect the elderly or immunocompromised individuals. The cases associated with immunodeficiency are usually positive for Epstein–Barr virus (EBV).

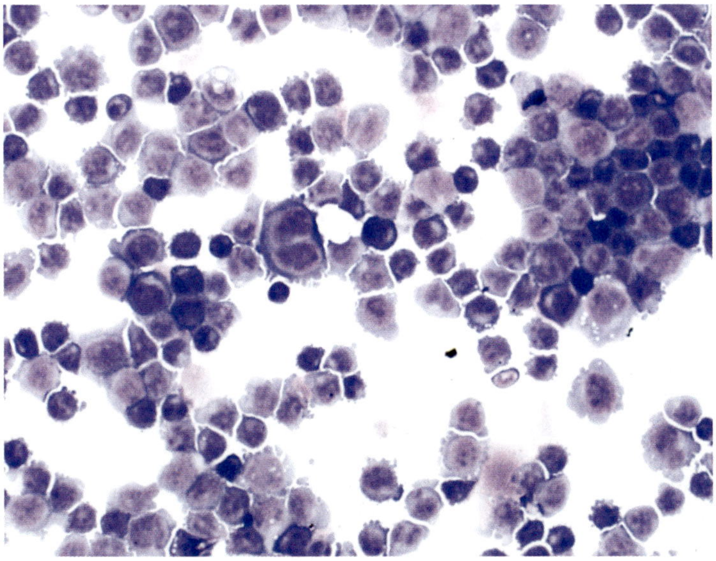

Figure 3.15a — Multiple Myeloma, Lumbar Puncture [Diff-Quik stain; high power]. The specimen is remarkably cellular with many medium to large cells. Some cells have abundant basophilic cytoplasm and a perinuclear hof (clearing) that displaces the nucleus. These are morphologic features of mature plasma cells. The atypical plasma cells demonstrate paler cytoplasm, loss of hof, and occasional binucleation.

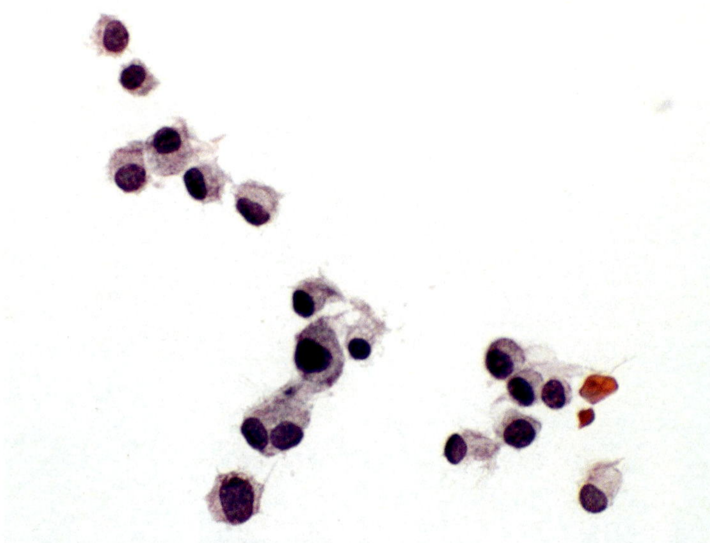

Figure 3.15b — Multiple Myeloma, Lumbar Puncture [Pap stain; high power]. Eccentric nuclei and clumpy chromatin are important cytomorphologic features of plasma cells. The presence of plasma cells in the cerebrospinal fluid (CSF) is more often caused by inflammatory or infectious etiologies, such as multiple sclerosis or meningoencephalitis. It is uncommon in plasma cell neoplasms.

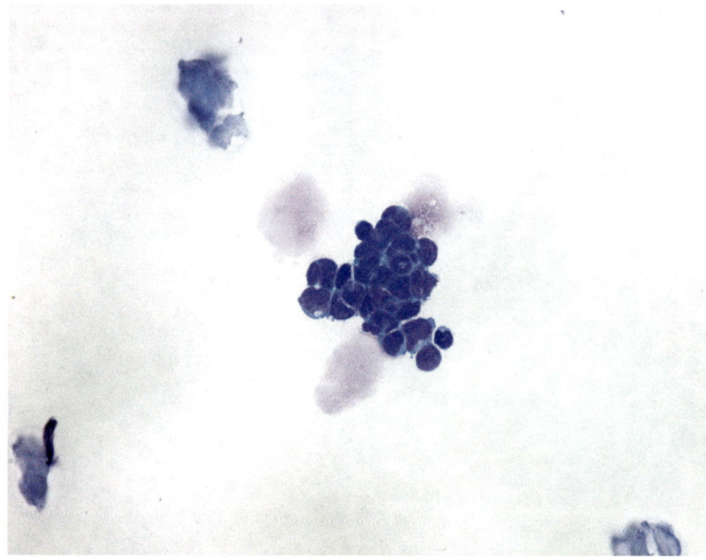

Figure 3.16a — Medulloblastoma, Lumbar Puncture [Giemsa stain; low power]. Cerebrospinal fluid (CSF) dissemination through leptomeningeal extension occurs in approximately 30% of medulloblastomas at diagnosis. CSF cytology is considered a method of choice for detecting metastatic diseases because of its excellent correlation with imaging findings. In the CSF, metastatic medulloblastoma most frequently appears as medium to large clusters of cells. Individual cells are uncommon. In this field, tumor cells harbor scant to moderate amounts of basophilic cytoplasm and mildly pleomorphic polygonal nuclei. There is nuclear molding. Two cells on the left have a single prominent cytoplasmic vacuole.

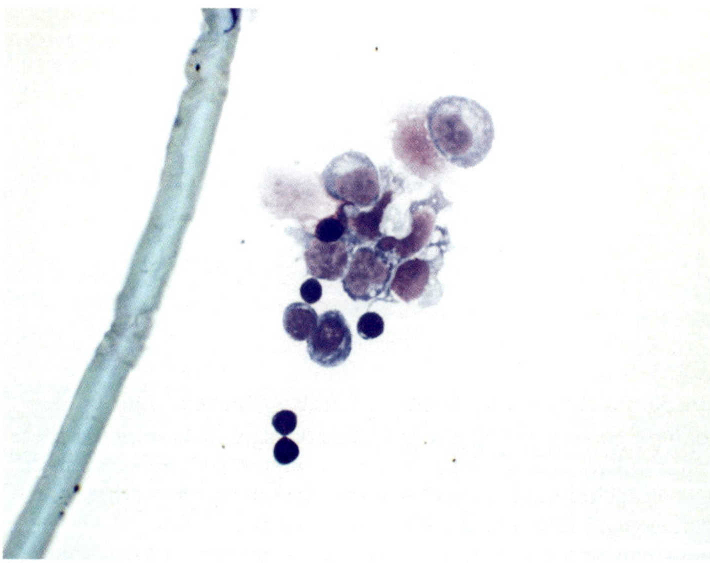

Figure 3.16b — Medulloblastoma, Lumbar Puncture [Diff-Quik stain; high power]. Although medulloblastomas are considered a small blue round cell tumor, the disseminated tumor cells in the cerebrospinal fluid (CSF) are often medium to large, regardless of the histologic subtype. In this field, tumor cells are 5 to 10 times the size of a normal lymphocyte. Their cytoplasm has abundant small vacuoles. Nuclei are convoluted with multiple conspicuous, small nucleoli.

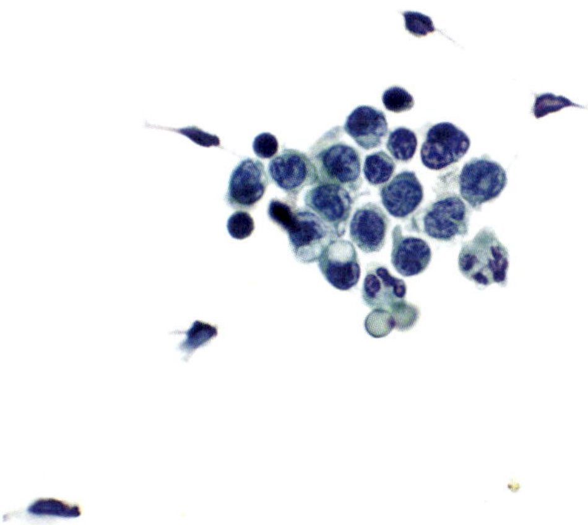

Figure 3.16c — Medulloblastoma, Lumbar Puncture [Pap stain; high power]. Medulloblastoma cells have a tendency to cluster when disseminating in the cerebrospinal fluid (CSF). The sizes of tumor cells in this field range from a normal lymphocyte to a neutrophil (the bottom two cells with multilobed nuclei). The Pap stain highlights the convoluted nuclear contour and coarsely granular chromatin. Two tumor cells in the left corner are notable for large cytoplasmic vacuoles.

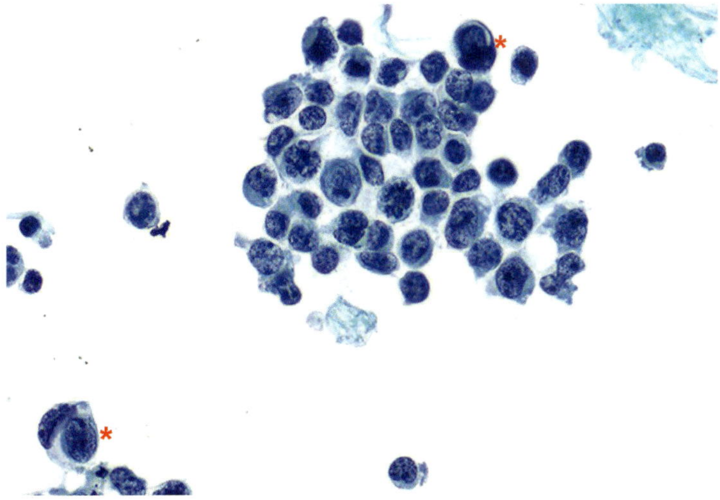

Figure 3.16d — Medulloblastoma, Lumbar Puncture [Pap stain; high power]. Tumor cells in this field demonstrate cytologic features of large cell/anaplastic medulloblastomas: prominent nucleoli (large cell feature) and cell wrapping (anaplastic feature, indicated by red asterisks). Large cell/anaplastic medulloblastomas are associated with early leptomeningeal dissemination and a dismal prognosis. Therefore the rate of positive cerebrospinal fluid (CSF) cytology is the highest in this histologic subtype.

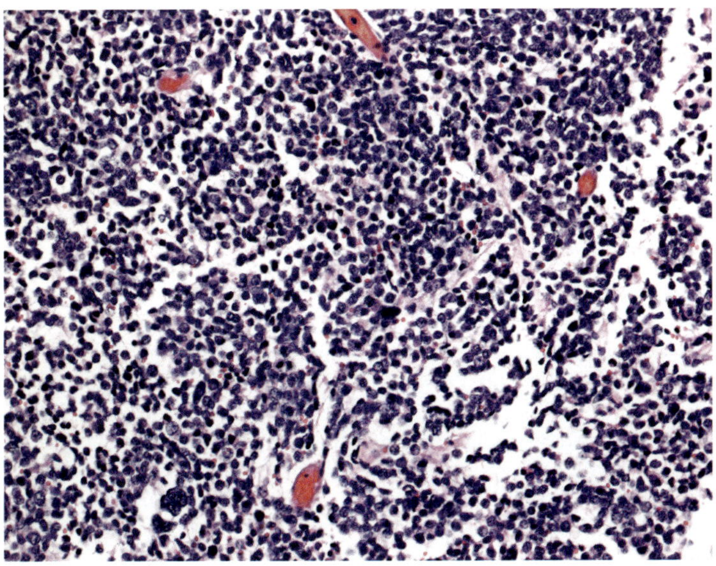

Figure 3.16e — Medulloblastoma, Classic Type, Brain Resection [H&E stain; high power]. Medulloblastoma is the most common malignant brain tumor of childhood. It occurs exclusively in the posterior fossa. The classic histologic type of medulloblastoma has the typical morphology of a small round blue cell tumor. The tumor cells have slightly open chromatin. Nuclear atypia and pleomorphism are mild to moderate.

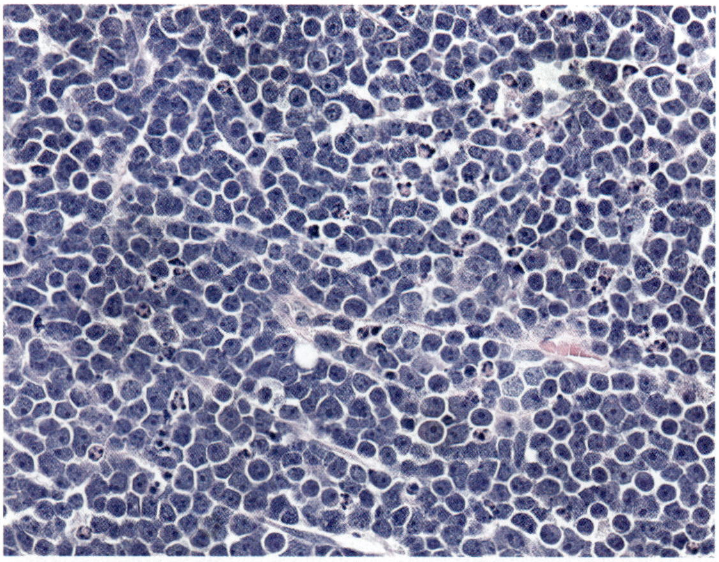

Figure 3.16f — Medulloblastoma, Large Cell Variant, Brain Resection [H&E stain; high power]. Large cell variant of medulloblastoma is characterized by larger tumor cell size (at least twice the size of a classic medulloblastoma cell), prominent nucleoli, cell wrapping, nuclear molding, and abundant apoptosis. It is important to recognize this histologic variant due to its frequent involvement of cerebrospinal fluid (CSF) and association with an aggressive clinical course. Large cell medulloblastomas often demonstrate *MYC* gene amplification.

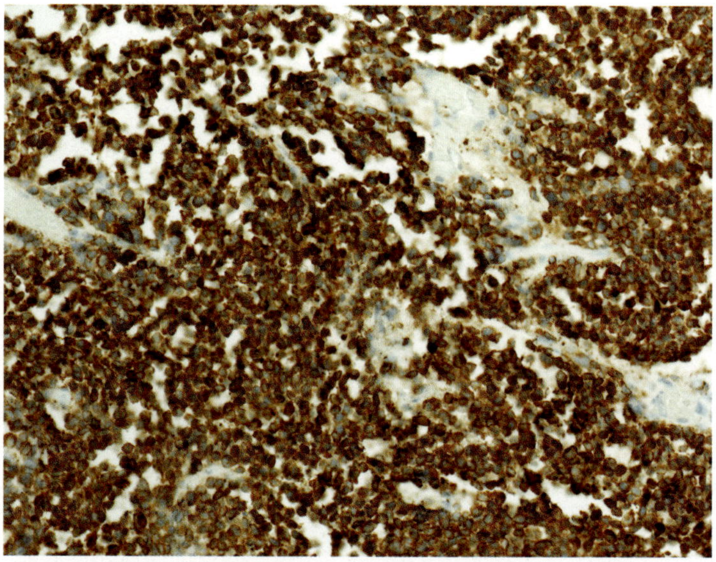

Figure 3.16g — Medulloblastoma, Brain Resection [Synaptophysin immunostain; high power]. Synaptophysin is a marker of neuronal differentiation. It is commonly used for diagnosis of medulloblastoma, which is a primitive neuronal neoplasm. The majority of medulloblastomas show strong, diffuse immunoreactivity for synaptophysin.

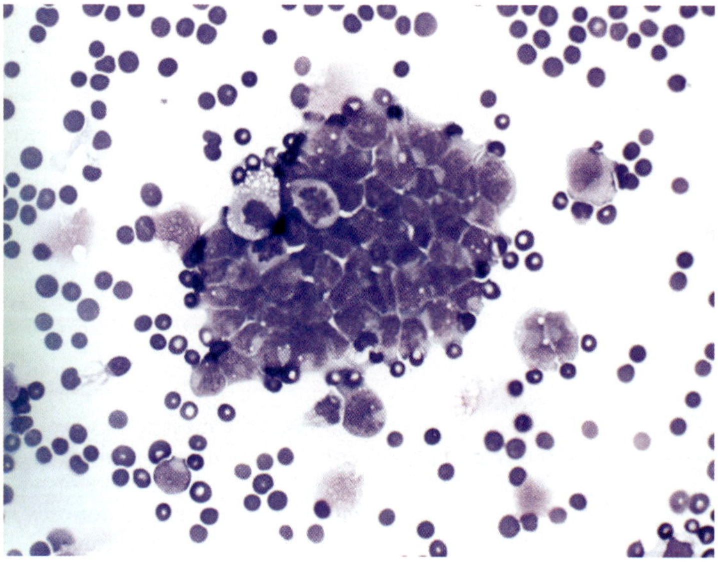

Figure 3.17a — Central Nervous System (CNS) Embryonal Tumor, Lumbar Puncture [Diff-Quik stain; high power]. CNS embryonal tumor is the supratentorial counterpart of medulloblastoma. The terminology has recently been changed from *central primitive neuroectodermal tumor (PNET)* to this name. The cytomorphologic features of CNS embryonal tumor are very similar to those of medulloblastoma. CNS embryonal tumor cells also tend to form large clusters in the cerebrospinal fluid (CSF). In this field, tumor cells are medium to large, with moderate amounts of lavender cytoplasm and large, angulated nuclei. Cytoplasmic vacuoles are abundant. Two cells demonstrate mitotic figures.

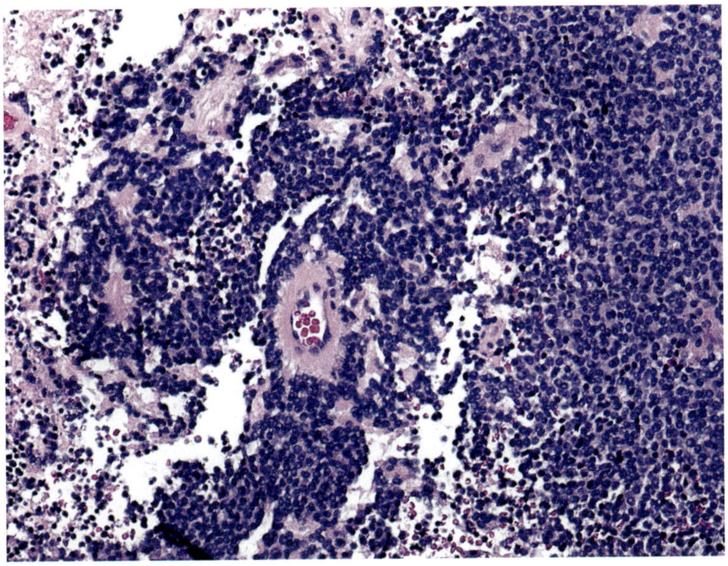

Figure 3.17b — Central Nervous System (CNS) Embryonal Tumor, Brain Resection [H&E stain; low power]. CNS embryonal tumors represent a morphologically and genetically heterogeneous group of primitive neoplasms. Some tumors have the appearance of a small round blue cell tumor, as shown in this image. Other tumors may demonstrate secondary structures such as a rosette formation.

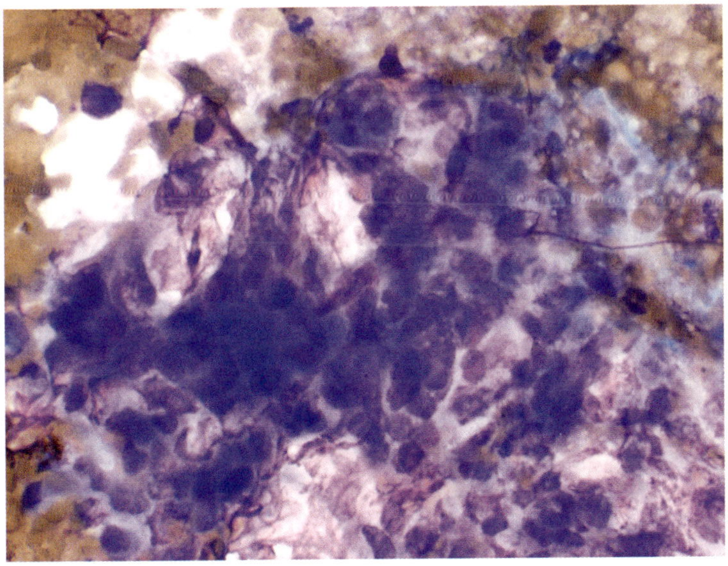

Figure 3.18a — High-Grade Astrocytoma, Lumbar Puncture [Diff-Quik stain; high power]. High-grade gliomas include WHO grade III and IV (glioblastoma) tumors. As the most common type of central nervous system (CNS) neoplasms in adults, high-grade gliomas surprisingly seldom have cerebrospinal fluid (CSF) involvement. In the CSF, high-grade astrocytoma cells can appear singly or in clusters. This field shows a large cluster or fragment of cells with irregular, hyperchromatic nuclei, and scant cytoplasm. Nuclear pleomorphism is not apparent in this case.

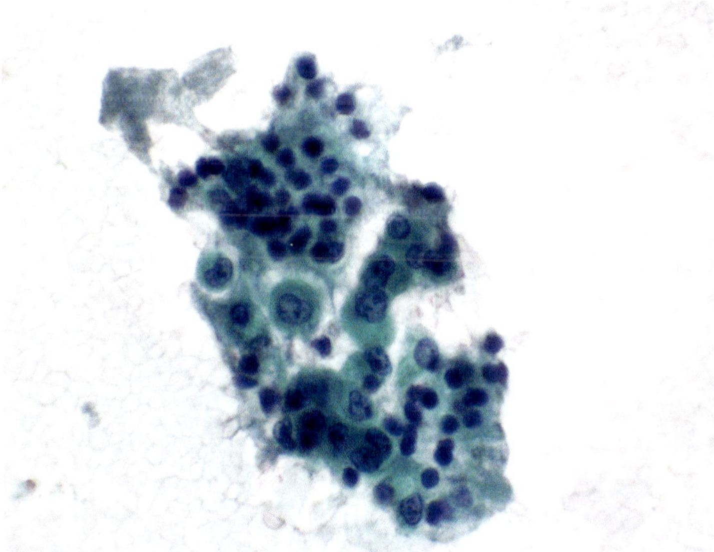

Figure 3.18b — High-Grade Astrocytoma, Lumbar Puncture [Pap stain; high power]. This cluster of high-grade astrocytoma cells demonstrates size variation, nuclear pleomorphism, and nuclear overlapping. There are a few large cells with abundant cytoplasm. The smaller cells have hyperchromatic nuclei, whereas the large cells have vesicular chromatin with indistinct nucleoli. In addition to these nuclear features, nuclear contour irregularity is another diagnostic finding of high-grade astrocytoma.

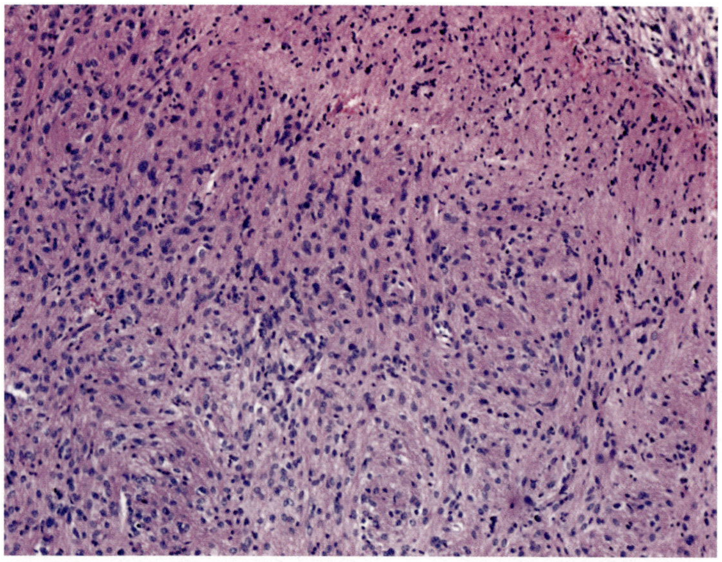

Figure 3.19a — Anaplastic Astrocytoma, Brain Biopsy [H&E stain; low power]. This image shows a widely infiltrative glioma of moderate to high cellularity. Tumor cells exhibit nuclear atypia such as irregular nuclear contour and coarse but open chromatin. Nuclear atypia and pleomorphism are usually more obvious in high than low-grade infiltrating gliomas.

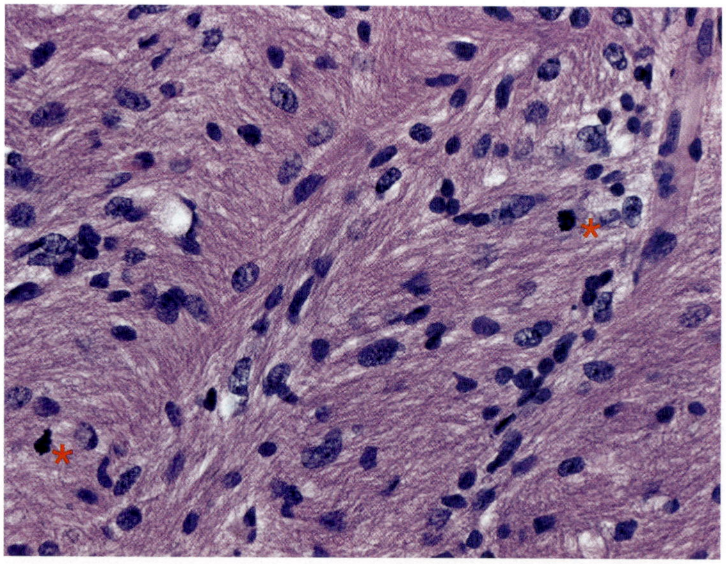

Figure 3.19b — Anaplastic Astrocytoma, Brain Biopsy [H&E stain; high power]. Anaplastic astrocytoma is a WHO grade III glioma that has notable mitotic activity. This high-magnification image shows two mitotic figures, denoted by the red asterisks. There are two populations of cells in this field: Tumor cells have larger oval or polygonal nuclei with open chromatin, and nonneoplastic glia have round nuclei with condensed chromatin.

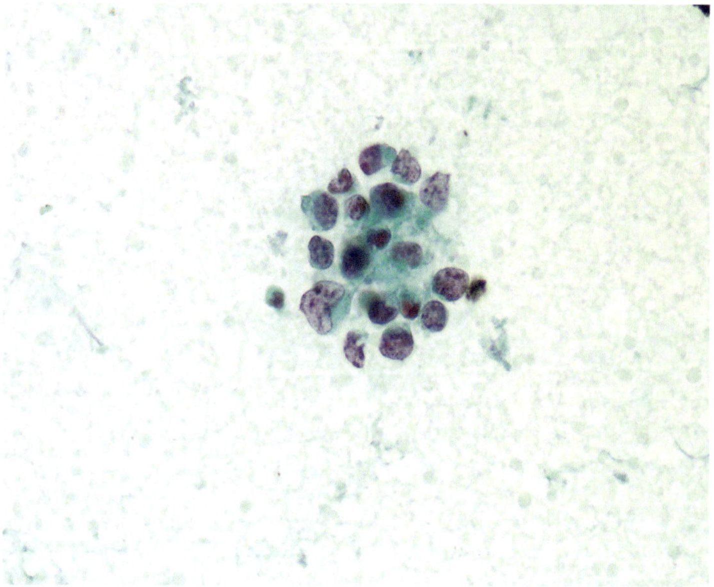

Figure 3.20a — Glioblastoma, Lumbar Puncture [Pap stain; low power]. This field demonstrates a cluster of medium to large atypical cells with a high nuclear-to-cytoplasmic ratio. The chromatin is open, with several small nucleoli. The cell in the lower left corner is binucleated. Binucleation and multinucleation are frequently seen in glioblastomas. Note the irregularity of the nuclear membrane.

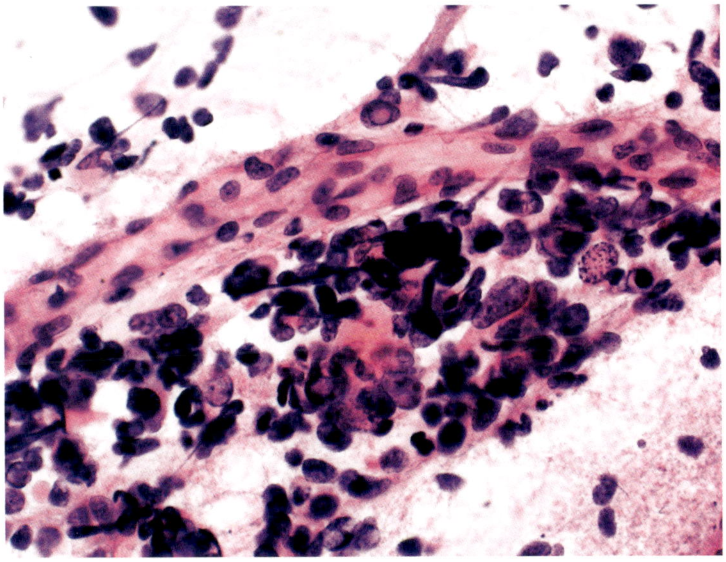

Figure 3.20b — Glioblastoma, Lumbar Puncture [H&E stained cell block; high power]. This cell block is hypercellular and contains many atypical cells. Eosinophilic glial processes, a characteristic finding of glial neoplasms, are well presented in this image. The majority of the tumor cells here have an ovoid nucleus with stippled coarse chromatin. Nuclear pseudoinclusions, although uncommon, can occasionally be observed in high-grade astrocytomas.

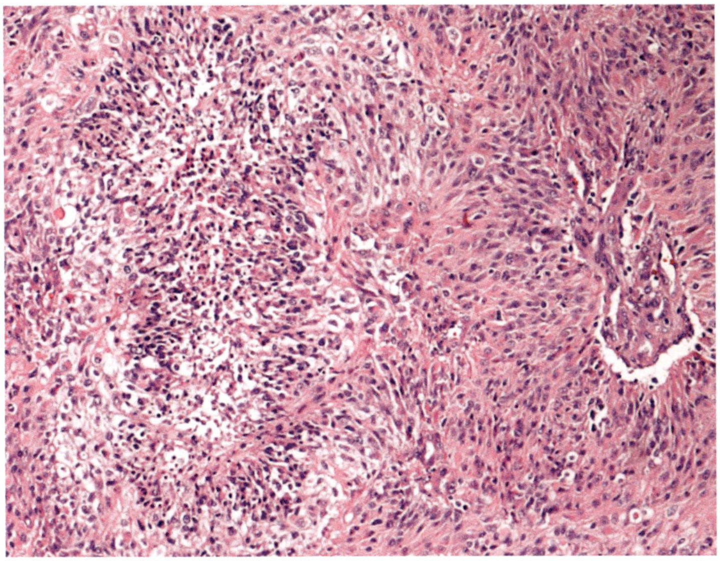

Figure 3.20c — Glioblastoma, Brain Resection [H&E stain; low power]. This image demonstrates a hypercellular infiltrating glioma with cytologic atypia and mitotic activity. This tumor is also notable for pseudopalisading necrosis on the left side of the image and vascular proliferation on the opposite side. Necrosis and vascular proliferation are important grading features of glioblastoma, the highest grade glioma (WHO grade IV).

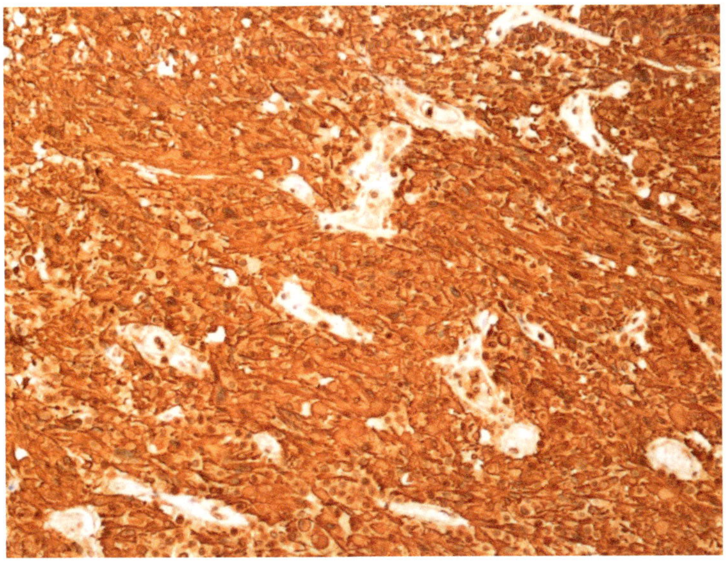

Figure 3.20d — Glioblastoma, Brain Resection [Glial fibrillary acidic protein (GFAP) immunostain; low power]. This glioblastoma is strongly and diffusely positive for GFAP. GFAP is a glial marker that is usually positive in glioblastomas, even though the immunoreactivity can be quite focal. Occasionally, glioblastomas can lose the GFAP expression. Another glial marker, oligodendrocyte transcription factor (OLIG2), can be useful in this setting to demonstrate glial differentiation.

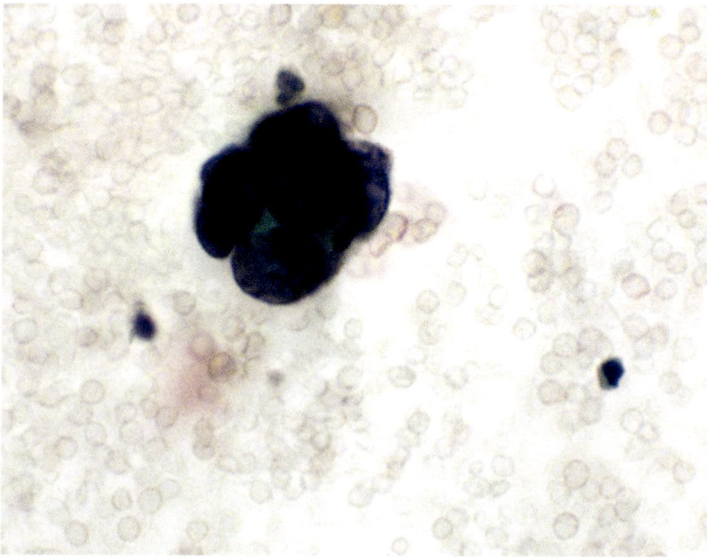

Figure 3.21a — Giant Cell Glioblastoma, Lumbar Puncture [Pap stain; high power]. Giant cell glioblastoma is a histologic variant of glioblastoma. With marked nuclear pleomorphism and severe cytologic atypia, the prognosis of giant cell glioblastoma is surprisingly better than that of classical glioblastomas. This field shows a multinucleated tumor giant cell with huge, hyperchromatic nuclei.

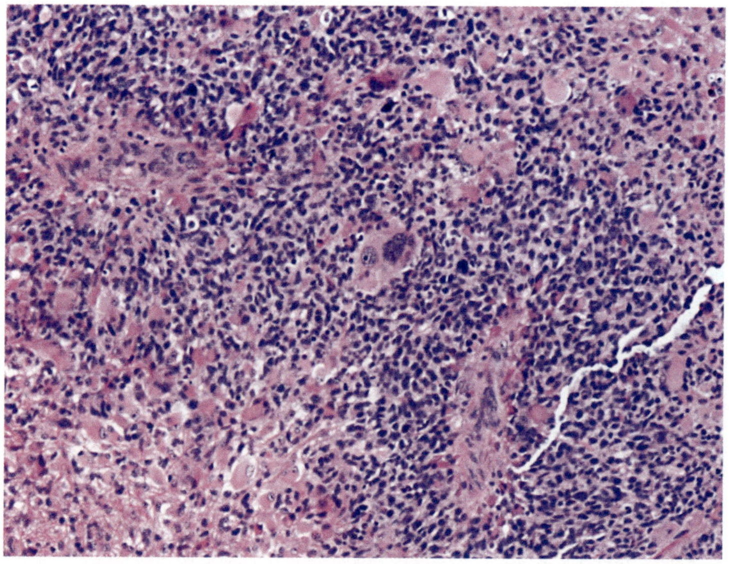

Figure 3.21b — Giant Cell Glioblastoma, Brain Resection [H&E stain; low power]. Giant cell glioblastoma is associated with a better prognosis among all histologic types of glioblastomas. This case represents a glioblastoma with both giant cell and small cell features. There are clusters of multinucleated tumor giant cells with abundant pink cytoplasm and prominent nucleoli. Some of the tumor giant cells (e.g., the one in the center) can have bizarre morphology.

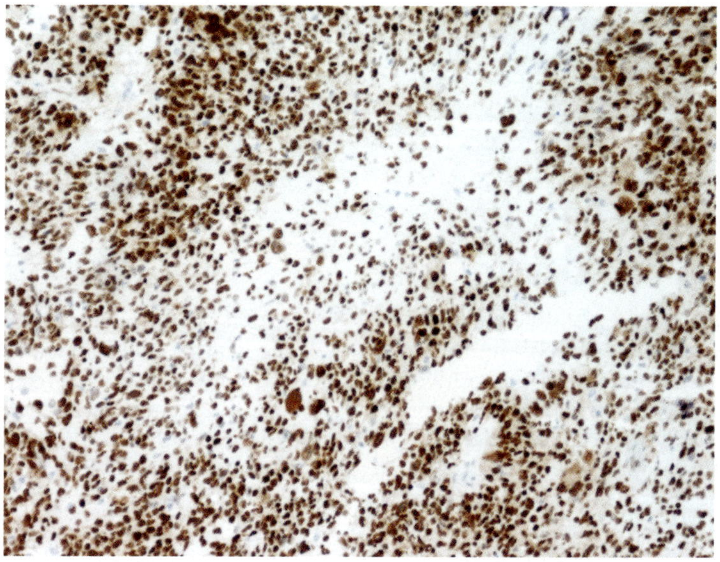

Figure 3.21c — Giant Cell Glioblastoma, Brain Resection [p53 immunostain; low power]. Approximately 80% of the giant cell glioblastomas carry *TP53* mutation, which in gliomas can be detected by immunohistochemistry. This image shows strong, diffuse immunoreactivity for p53. Despite the presence of *TP53* mutation, the prognosis of giant cell glioblastoma is better than that of glioblastoma with classic morphology.

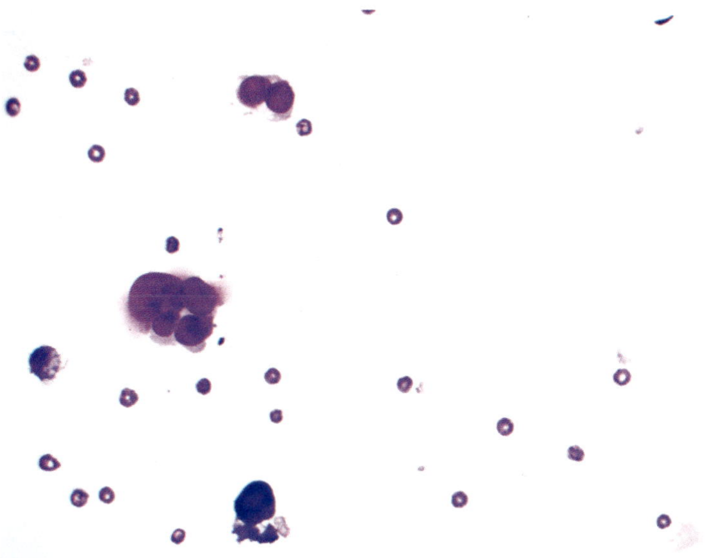

Figure 3.22a — Low-Grade Astrocytoma, Lumbar Puncture [Diff-Quik stain; low power]. *Low-grade astrocytoma* is a generic term that in most instances refers to WHO grade I pilocytic astrocytoma. Although uncommon, low-grade astrocytomas can disseminate along the neuraxis and result in positive cerebrospinal fluid (CSF) cytology. This CSF specimen is notable for two to three gigantic cells, one of which is binucleated and the other multinucleated. Although the size of the cells is concerning, the nuclear membrane is smooth, suggesting a benign nature of the tumor.

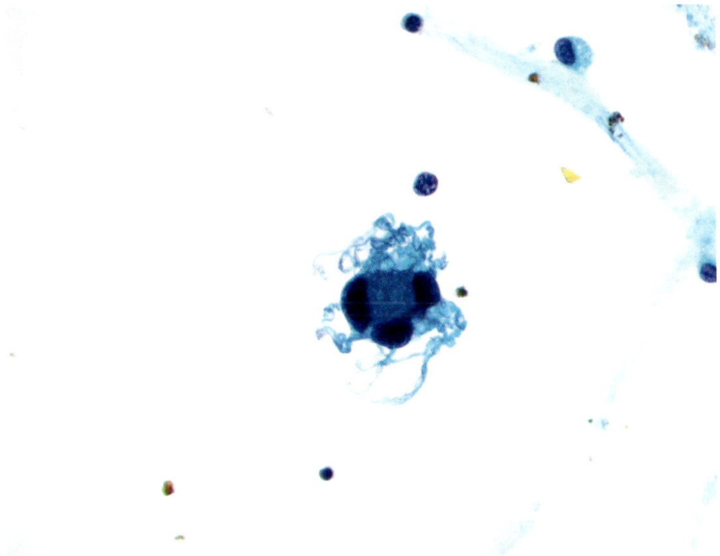

Figure 3.22b — Low-Grade Astrocytoma, Lumbar Puncture [Pap stain; high power]. The giant multinucleated cell in the center of the field shows multiple glial processes on the Pap stain, a diagnostic cytologic feature of gliomas. The cell immediately above the tumor giant cell has a naked nucleus and stippled chromatin. These are important features that distinguish gliomas from reactive gliosis on smears of tissue during intraoperative evaluation.

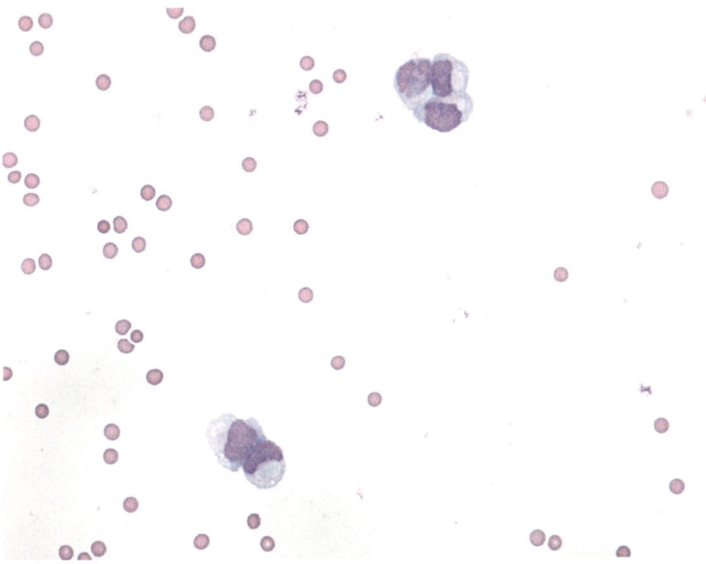

Figure 3.23a — Atypical Teratoid/Rhabdoid Tumor (AT/RT), Lumbar Puncture [Diff-Quik stain; low power]. AT/RT is an aggressive embryonal tumor that has a 15% to 30% rate of leptomeningeal disease. Different from most primary CNS tumors which tend to form large clusters in the cerebrospinal fluid (CSF), AT/RT often appears singly or in small aggregates. This image demonstrates two small clusters of tumor cells. These cells are large and have abundant amphophilic cytoplasm; large, irregular nuclei; and visible nucleoli. The eosinophilic cytoplasmic inclusion is especially prominent, giving the cells a rhabdoid appearance. Small cytoplasmic vacuoles can also be observed.

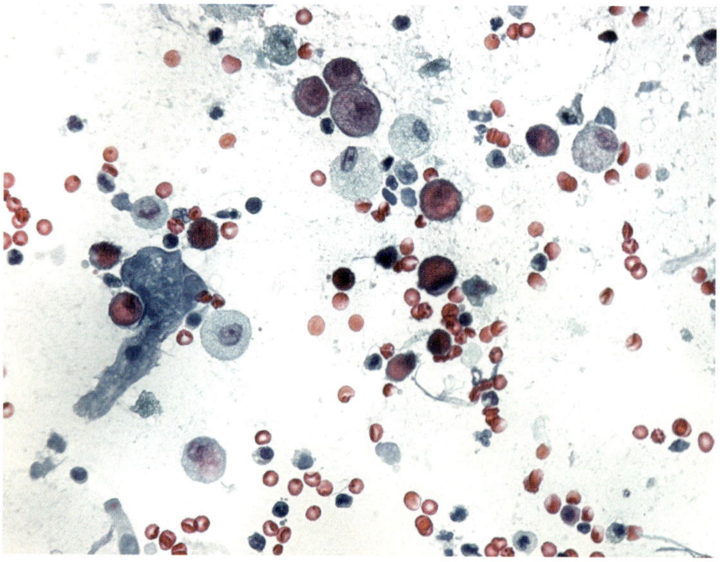

Figure 3.23b — Atypical Teratoid/Rhabdoid Tumor (AT/RT), Lumbar Puncture [Pap stain; high power]. Because of the ample cytoplasm, the nuclear-to-cytoplasmic ratio of AT/RT tumor cells can be quite low. With the low nuclear-to-cytoplasmic ratio and conspicuous nucleoli, these cells can look very similar to macrophages, which are often prominent in the post-treatment cerebrospinal fluid (CSF). Cell size variation, the presence of cytoplasmic inclusion, and rhabdoid morphology are key cytomorphologic features that distinguish AT/RT tumor cells from macrophages.

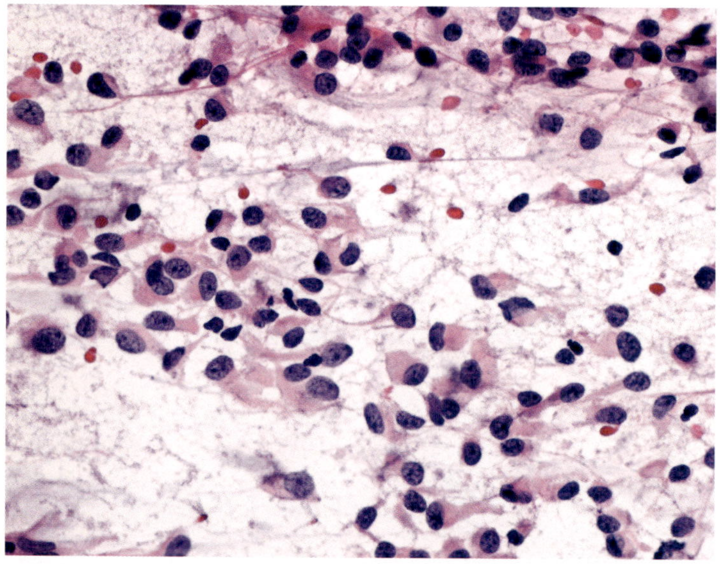

Figure 3.23c — Atypical Teratoid/Rhabdoid Tumor (AT/RT), Brain Resection, Intraoperative Smear [H&E stain; high power]. This image is taken from an H&E-stained intraoperative cytologic smear of a surgical specimen. The smear is hypercellular and consists of relatively monomorphic, intermediate-sized cells. Tumor cells have abundant eosinophilic cytoplasm and ovoid, eccentrically located nuclei. Nucleoli are visible but not prominent. Note the blue-tinged mucin in the background. This finding is rarely seen in other embryonal tumors or high-grade gliomas.

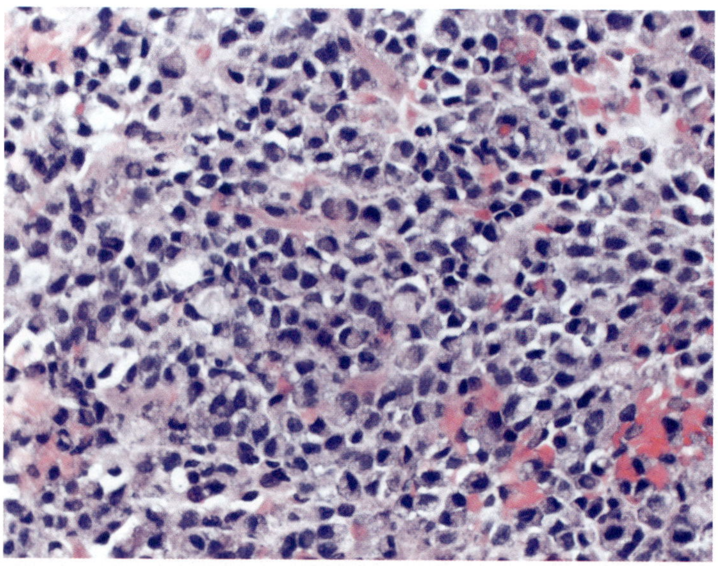

Figure 3.23d — Atypical Teratoid/Rhabdoid Tumor (AT/RT), Brain Resection [H&E stain; low power]. Tumor cells in this case are discohesive and demonstrate classic rhabdoid morphology: eccentric nuclei with ample pale pink cytoplasm. Cytoplasmic inclusions can be vaguely observed. Nuclei are hyperchromatic with indistinct nucleoli. Although not seen in this field, necrosis is common in AT/RT.

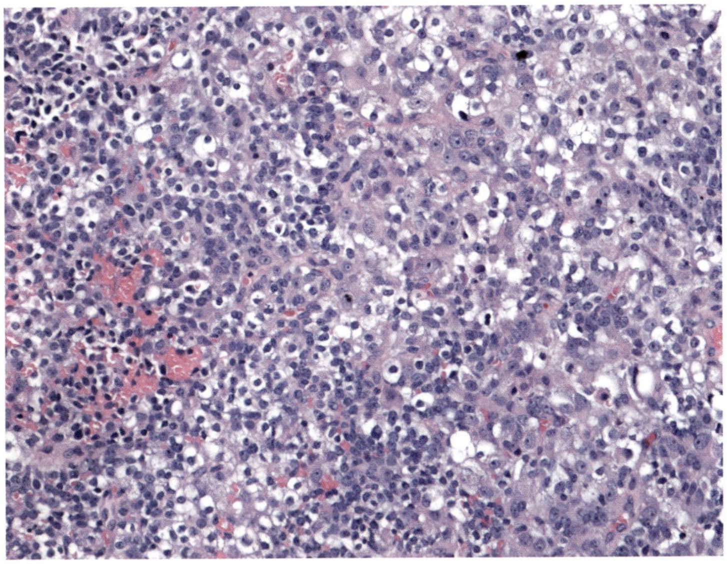

Figure 3.23e — Atypical Teratoid/Rhabdoid Tumor (AT/RT), Brain Resection [H&E stain; low power]. Tumor cells in this case appear more pleomorphic than in the previous case: There are tumor cells with epithelioid morphology, pink cytoplasm, and prominent nucleoli admixed with tumor cells, with primitive small blue round cell morphology. Also observed are scattered clear cells with perinuclear halos. The presence of nucleoli and clear cells are important histologic features that distinguish AT/RT from other embryonal tumors, especially medulloblastoma. Among different histologic subtypes of medulloblastomas, only the large cell variant harbors nucleoli. Clear cells are rare in medulloblastomas.

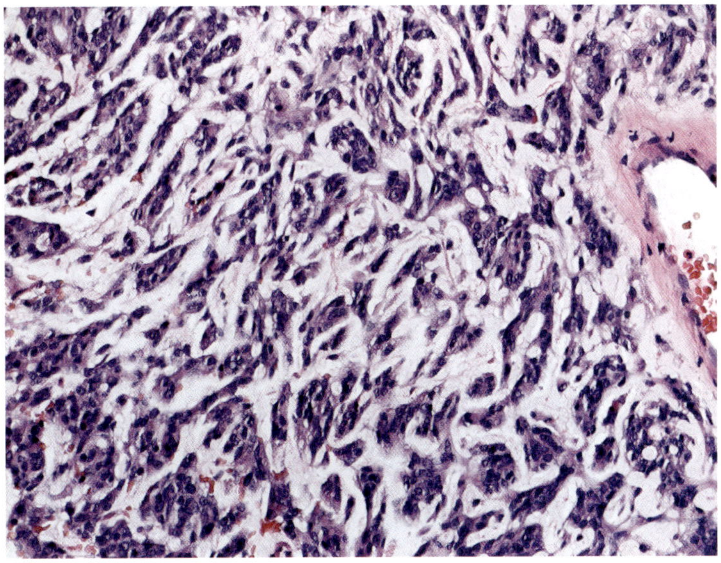

Figure 3.23f — Atypical Teratoid/Rhabdoid Tumor (AT/RT), Brain Resection [H&E stain; low power]. Occasionally, AT/RT tumor cells are arranged in cords or trabeculae in a background of myxoid matrix. This architecture is an important clue that leads to the diagnosis of AT/RT because it is lacking in the differential diagnoses of AT/RT, including medulloblastoma, central nervous system (CNS) embryonal tumor, and glioblastoma. Note the prominent nucleoli in a fraction of the cells.

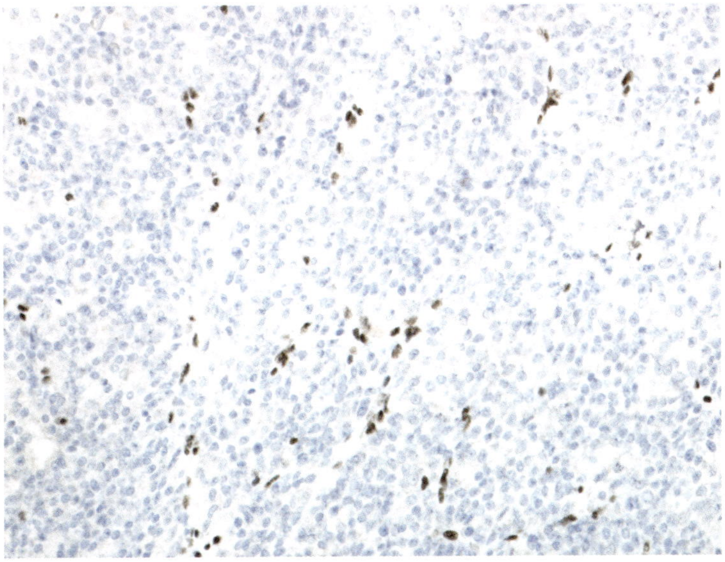

Figure 3.23g — Atypical Teratoid/Rhabdoid Tumor (AT/RT), Brain Resection [INI1 immunostain; low power]. AT/RT is characterized by deletion or mutation of the gene *SMARCB1/INI1*, which results in loss of gene expression. Loss of INI1 expression can be detected by immunohistochemical staining. This image demonstrates absence of staining in AT/RT tumor cells. Note positive INI1 staining in the endothelial cells of capillaries, which serve as the internal positive control for the stain.

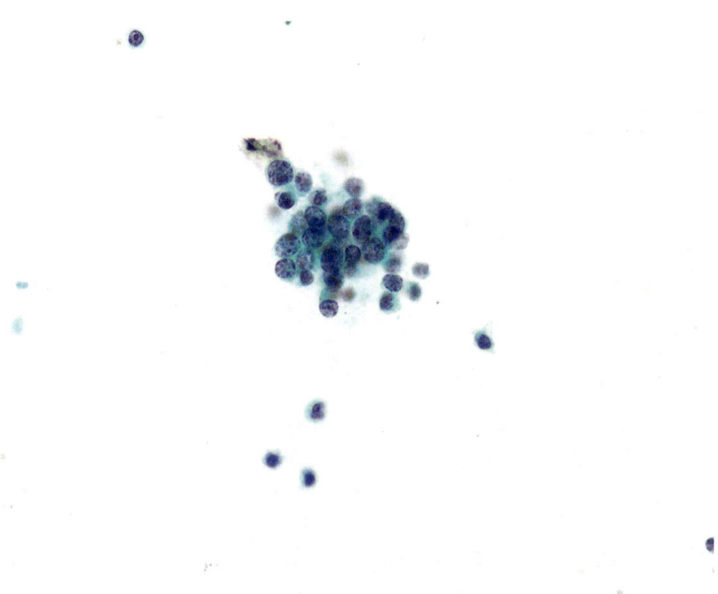

Figure 3.24a — Ependymoma, Lumbar Puncture [Pap stain; low power]. Cerebrospinal fluid (CSF) involvement by ependymoma is rare but can occur through "drop metastasis." This field shows a three-dimensional fragment consisting of monomorphic small to medium cells. The nuclei are round to oval shaped with smooth contour and finely stippled chromatin. Cytoplasm is scant.

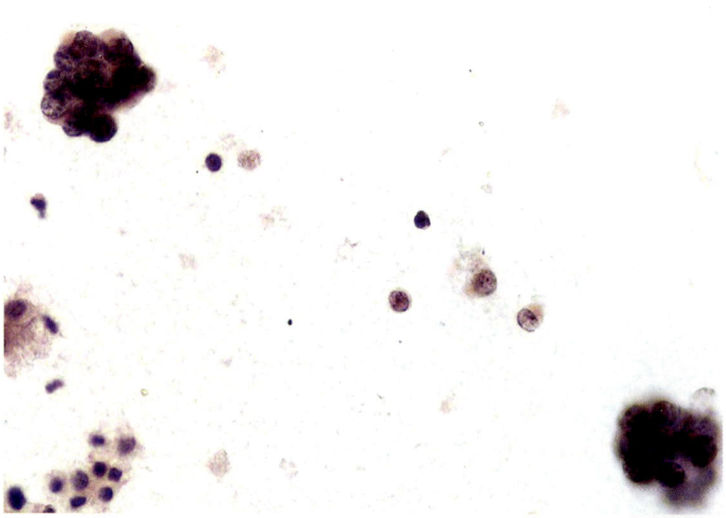

Figure 3.24b — Ependymoma, Lumbar Puncture [Pap stain; high power]. Two small fragments of ependymoma are seen in the field surrounded by scattered tumor cells. The ependymoma cells are in a vague rosette formation. True (ependymal) rosettes are an important diagnostic feature of ependymoma. They are formed by tumor cells surrounding a true lumen.

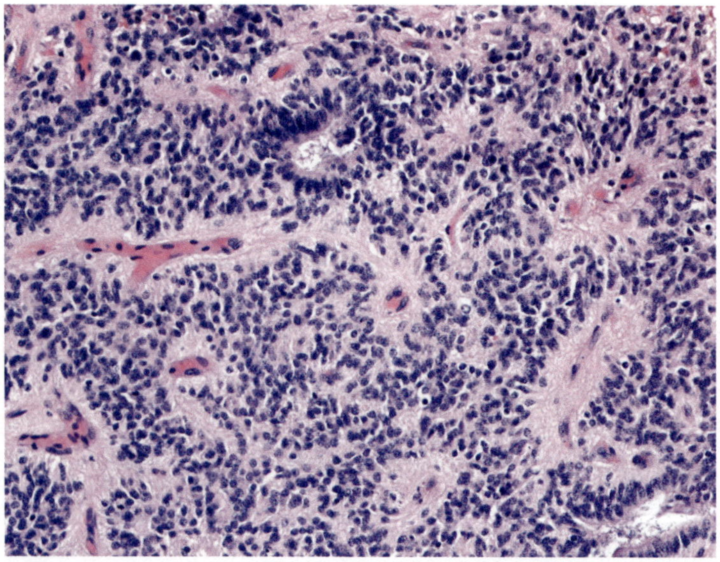

Figure 3.24c — Ependymoma, Brain Resection [H&E stain; low power]. Ependymoma is a well-circumscribed glioma of WHO grade II or III. It can occur in the supratentorial region, posterior fossa, or spinal cord. Histologically, ependymoma is characterized by abundant perivascular pseudorosettes and rare true rosettes (one at the top center of the image). Grade III (anaplastic) ependymomas show frequent mitotic activity and/or microvascular proliferation. Of note, grading of ependymomas does not correlate with the metastasis rate or prognosis.

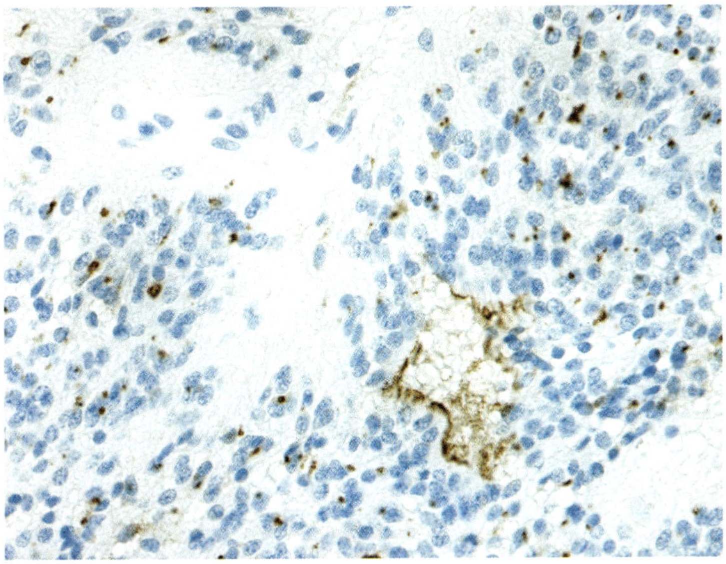

Figure 3.24d — Ependymoma, Brain Resection [Epithelial membranous antigen (EMA) immunostain; high power]. EMA is a useful marker for the diagnosis of ependymoma. It highlights the microlumina, which appear as perinuclear dot-like activity on the stain. The linear pattern of stain in this image represents the luminal surface, another diagnostic finding of ependymomas.

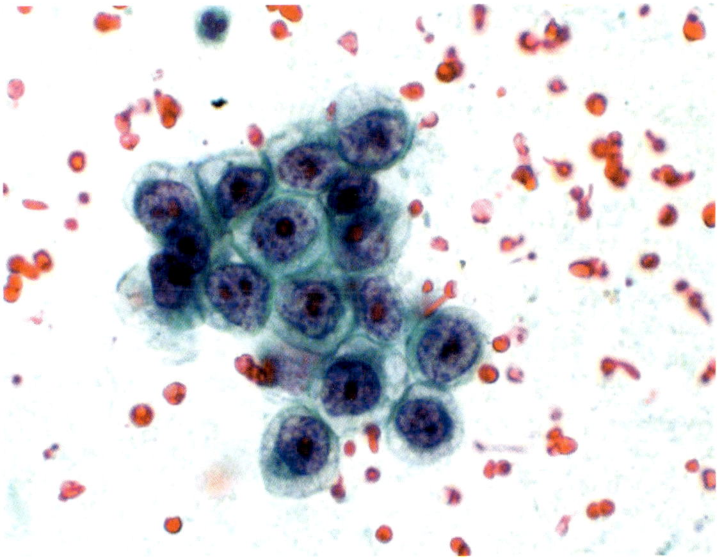

Figure 3.25a — Germinoma, Lumbar Puncture [Pap stain; high power]. Germinoma is the most common type of germ cell tumor in the central nervous system (CNS). Cerebrospinal fluid (CSF) cytology is routinely performed to detect spinal dissemination of the tumor. Germinoma cells have distinctive cytomorphology: Cells have ample clear cytoplasm, a large central nucleus, and one to three prominent eosinophilic nucleoli. In addition to cytology, germ cell markers such as alpha-fetoprotein (AFP) and beta–human chorionic gonadotropin (beta-HCG) in CSF are often used in diagnostic workup before biopsy of the tumor.

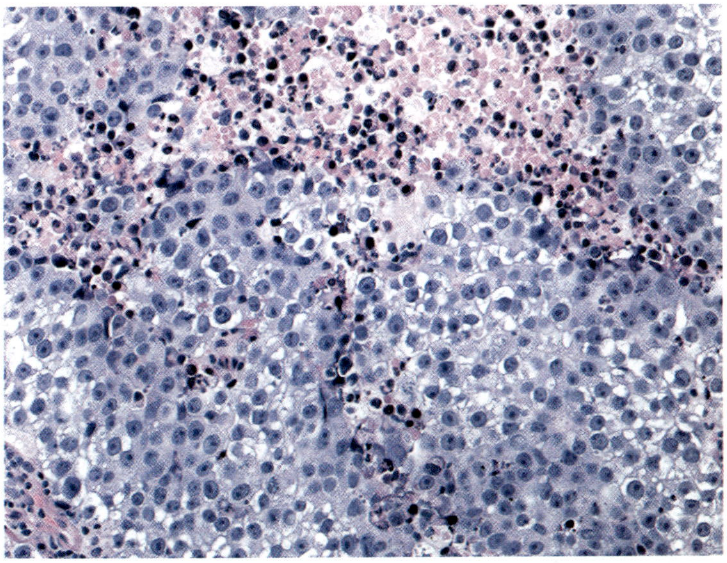

Figure 3.25b — Germinoma, Sellar Biopsy [H&E stain; low power]. Histologically, germinoma appears as sheets of clear cells with a large central nucleus and one to four distinct nucleoli. This case also shows focal necrosis. Although not present in this case, lymphocytic infiltrates are often prominent in the tumor.

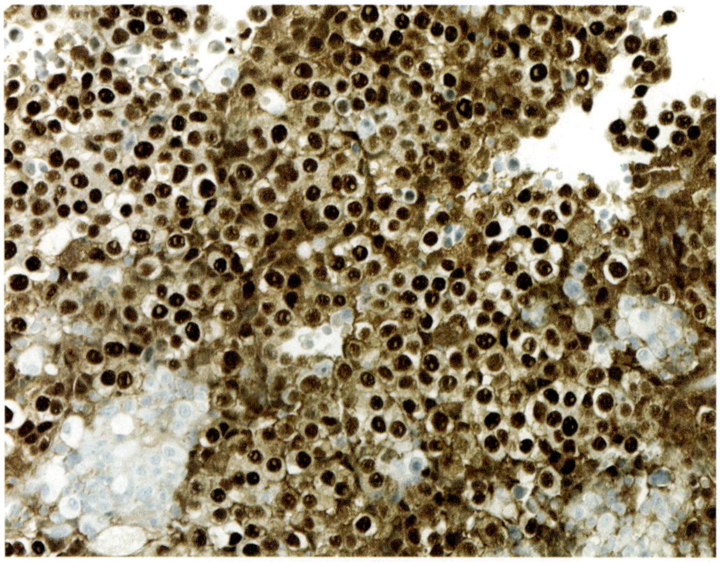

Figure 3.25c — Germinoma, Sellar Biopsy [Oct-3/4 immunostain; low power]. Oct-3/4 is a marker for germ cell tumors. It shows strong nuclear positivity in this case. For diagnosis of germinoma, other useful immunostains include SALL4 and c-kit. Similar to Oct-3/4, SALL4 is a nuclear stain. c-Kit is a cytoplasmic stain but a good marker for germinomas, since brain parenchyma, inflammatory cells, and other common brain tumors do not express c-kit.

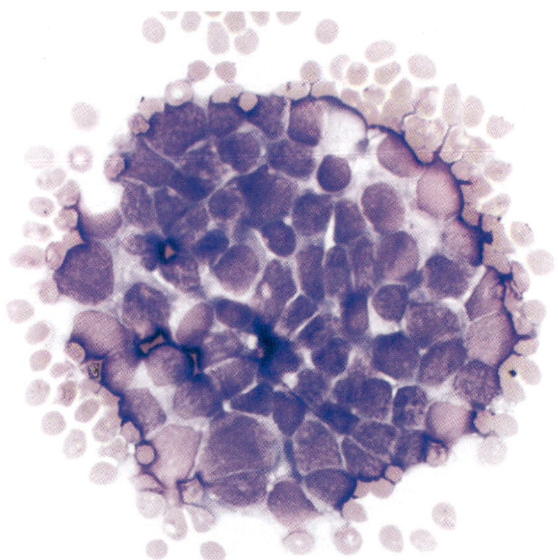

Figure 3.26a — Retinoblastoma, Lumbar Puncture [Diff-Quik stain; high power]. Retinoblastoma is considered an embryonal tumor and is therefore morphologically similar to other embryonal tumors such as medulloblastoma and central nervous system (CNS) embryonal tumor in the cerebrospinal fluid (CSF). The retinoblastoma cell clusters, however, are often larger than medulloblastoma. This field demonstrates a very large aggregate of large tumor cells with scant amphophilic cytoplasm and irregular nuclei. Nuclear molding and slight nuclear pleomorphism are also noted.

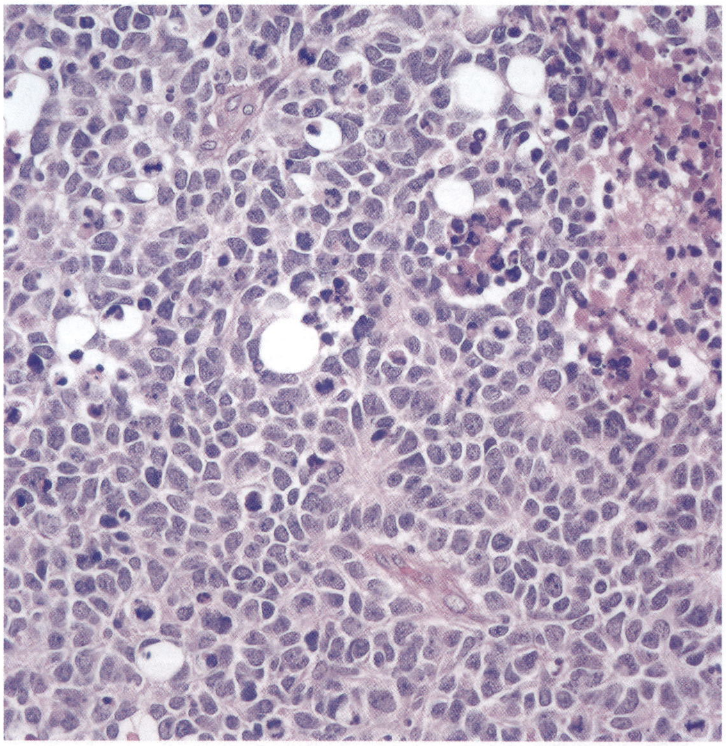

Figure 3.26b — Retinoblastoma, Eye Enucleation [H&E stain; low power]. The histologic section of retinoblastoma demonstrates a malignant embryonal neoplasm with brisk mitotic activity and focal necrosis. Tumor cells have polygonal hyperchromatic nuclei. Two Flexner-Wintersteiner pseudorosettes, the spoke-and-wheel type of structures formed by cytoplasmic extensions, are seen off center.

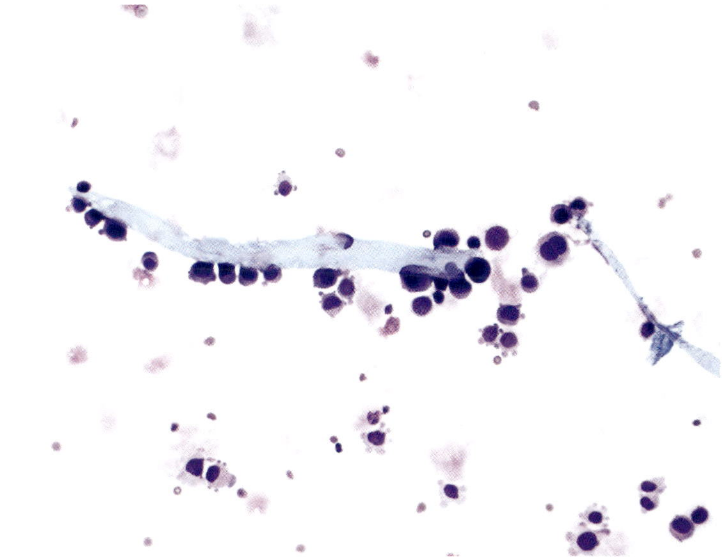

Figure 3.27a — Metastatic Lung Adenocarcinoma, Lumbar Puncture [Diff-Quik stain; low power]. Lung cancer is the most common source of brain metastases, accounting for approximately 50% of all cases. In this moderately cellular specimen, tumor cells contain a round to ovoid nucleus and abundant amphophilic cytoplasm, demonstrating a fried-egg or plasmacytoid morphology. Cytoplasmic blebs are frequent.

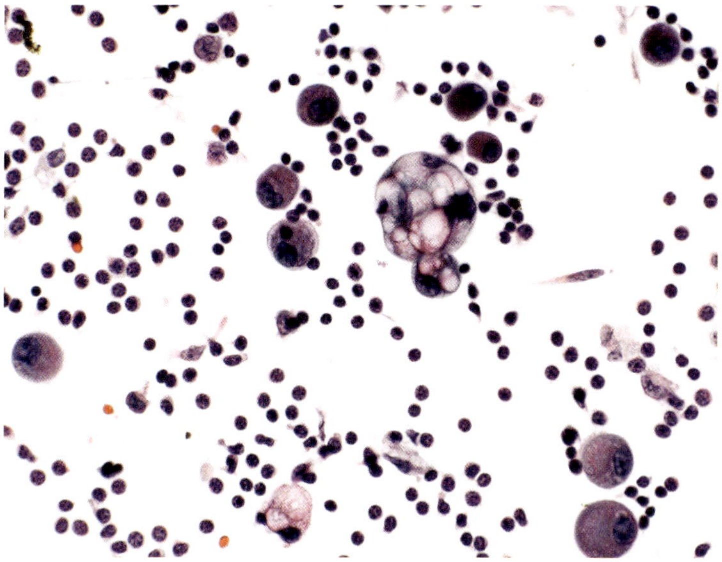

Figure 3.27b — Metastatic Lung Adenocarcinoma, Lumbar Puncture [Pap stain; low power]. This hypercellular cerebrospinal fluid (CSF) specimen shows several large cells scattered among abundant small cells. The large cells have peripherally located nuclei, and some are binucleated. Nucleoli are visible but indistinct. The two cells in the center harbor multiple mucin vacuoles in the cytoplasm. The background small tumor cells have a high nuclear-to-cytoplasmic ratio and are morphologically reminiscent of lymphocytes, but the chromatin is more clumped.

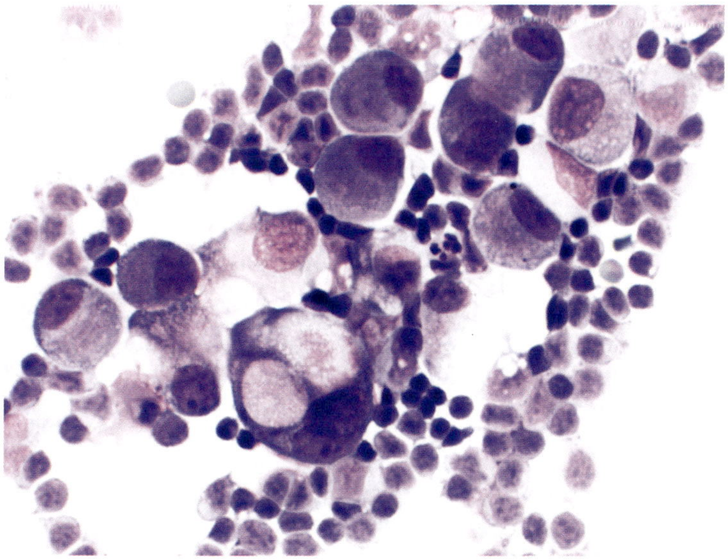

Figure 3.27c — Metastatic Lung Adenocarcinoma, Lumbar Puncture [Diff-Quik stain; high power]. This is a high-magnification image of the previous image but stained with Diff-Quik. Like the previous image, this field is also notable for two populations of cells: multiple small cells with a high nuclear-to-cytoplasmic ratio and clumped chromatin and scattered gigantic cells with ample basophilic cytoplasm and clear cytoplasmic vacuoles. Occasional binucleated cells are also observed.

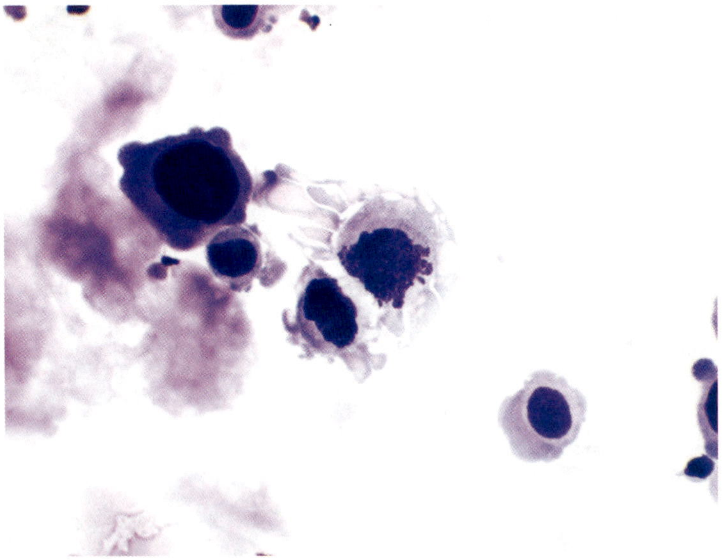

Figure 3.27d — Metastatic Lung Adenocarcinoma, Lumbar Puncture [Diff-Quik stain; high power]. Tumor cells in this field are highly pleomorphic and atypical: Their nuclei demonstrate irregular contour, hyperchromasia, and condensed prophase chromatin. Cytoplasmic blebs are prominent. Mucin vacuoles can be seen inside the blebs in the two cells in the center. In addition, small pools of mucin are present in the background.

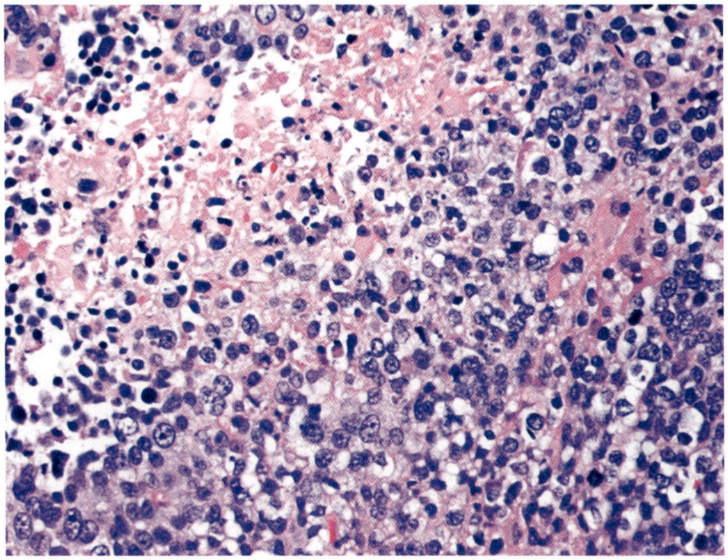

Figure 3.27e — Metastatic Lung Adenocarcinoma, Brain Resection [H&E stain; low power]. On the histologic section, metastatic lung adenocarcinoma demonstrates features of malignancy, including brisk mitotic activity and large areas of necrosis. The geographic necrosis seen in metastatic carcinoma is distinct from the pseudopalisading necrosis, a histologic hallmark of glioblastoma. Note the conspicuous nucleoli in the tumor cells. Nucleoli can be seen in glioblastomas but are relatively uncommon.

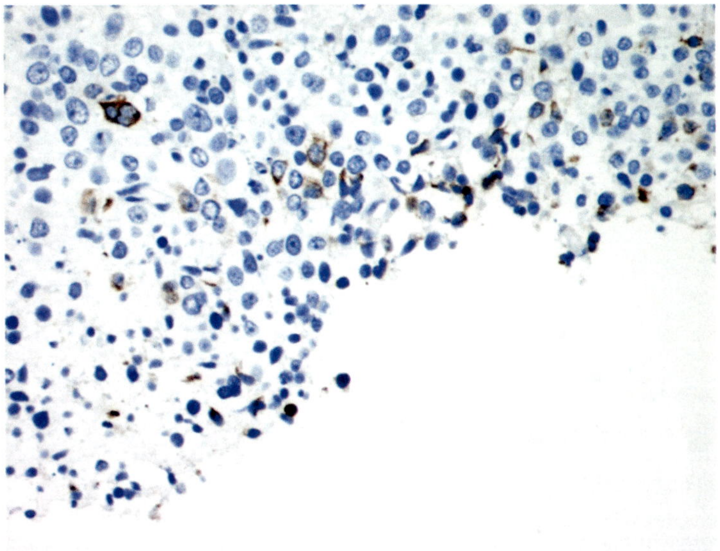

Figure 3.27f — Metastatic Lung Adenocarcinoma, Brain Resection [CAM5.2 immunostain; low power]. CAM5.2 is an antibody against low-molecular-weight cytokeratin. It is an excellent antibody for detecting metastatic carcinomas in the central nervous system (CNS). Pan cytokeratin antibodies often cross-react with reactive astrocytes, and therefore their use in the CNS tissue should be avoided. The metastatic lung adenocarcinoma cells in this image show focal immunoreactivity for CAM5.2. Cytokeratin 8/18 is a good alternative if CAM5.2 antibody is not available.

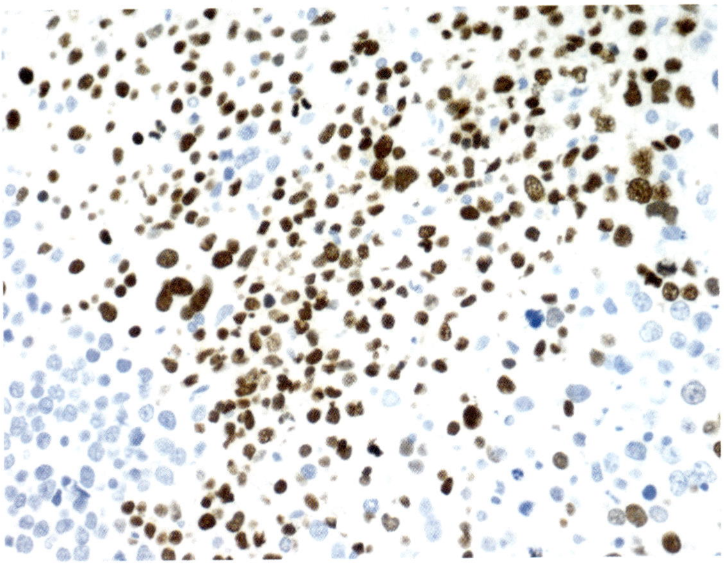

Figure 3.27g — Metastatic Lung Adenocarcinoma, Brain Resection [Thyroid Transcription Factor-1 (TTF-1) immunostain; low power]. TTF-1 is the most commonly used marker for lung adenocarcinomas. Its expression is detected in approximately 80% lung adenocarcinomas. In this image, at least half of the tumor cells demonstrate nuclear immunoreactivity for TTF-1. In addition to TTF-1, Napsin A is another good marker for lung adenocarcinomas. Compared with the nuclear localization of TTF-1, Napsin exhibits cytoplasmic distribution.

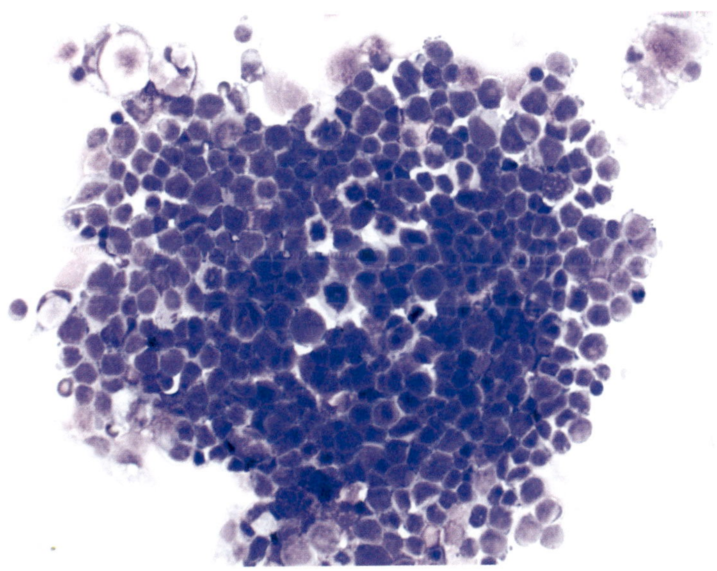

Figure 3.28a — Metastatic Small Cell Carcinoma, Lumbar Puncture [Diff-Quik stain; low power]. This field demonstrates a very large cluster of small to medium cells with oval to polygonal nuclei and scant granular cytoplasm. Nuclear molding is an important cytomorphologic feature of small cell carcinoma that can also be seen in the tumor cells in cerebrospinal fluid (CSF). Mitotic figures are frequent.

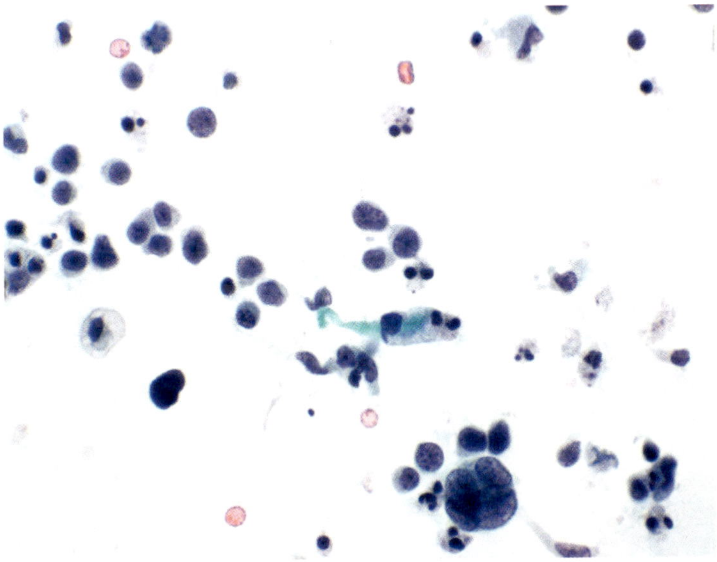

Figure 3.28b — Metastatic Small Cell Carcinoma, Lumbar Puncture [Pap stain; low power]. Small cell carcinoma cells have a high nuclear-to-cytoplasmic ratio. Some resemble "naked nuclei" because of the paucity of the cytoplasm. Their nuclei have the characteristic salt-and-pepper chromatin. Although small cell carcinoma cells are relatively monotonous, occasional large or multinucleated cells can be seen. Note the frequent apoptotic cells in the background.

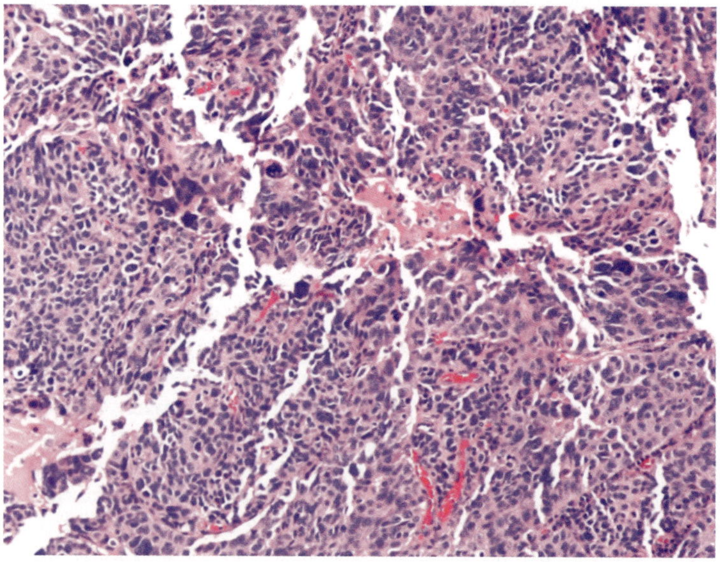

Figure 3.28c — Metastatic Small Cell Carcinoma, Brain Resection [H&E stain; low power]. This image demonstrates sheets of epithelioid tumor cells of high cellularity. There is no obvious glandular formation. Tumor cells of this metastatic small cell carcinoma have a fair amount of cytoplasm, which is uncommon in the primary tumors. Nuclear irregularity and occasional large hyperchromatic cells are histologic clues to the diagnosis of small cell carcinoma.

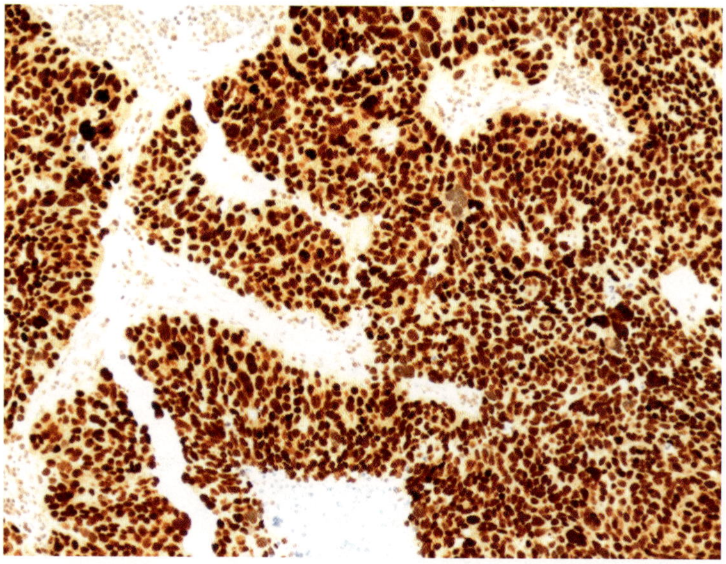

Figure 3.28d — Metastatic Small Cell Carcinoma, Brain Resection [Thyroid transcription factor-1 (TTF-1) immunostain; low power]. Similar to lung adenocarcinomas, metastatic small cell carcinomas express the nuclear marker TTF-1. TTF-1 expression is common in pulmonary small cell carcinomas but can also be seen in small cell carcinomas from extrapulmonary sites. Selected primary brain tumors, such as pituicytomas and subependymal giant cell astrocytomas, also express TTF-1.

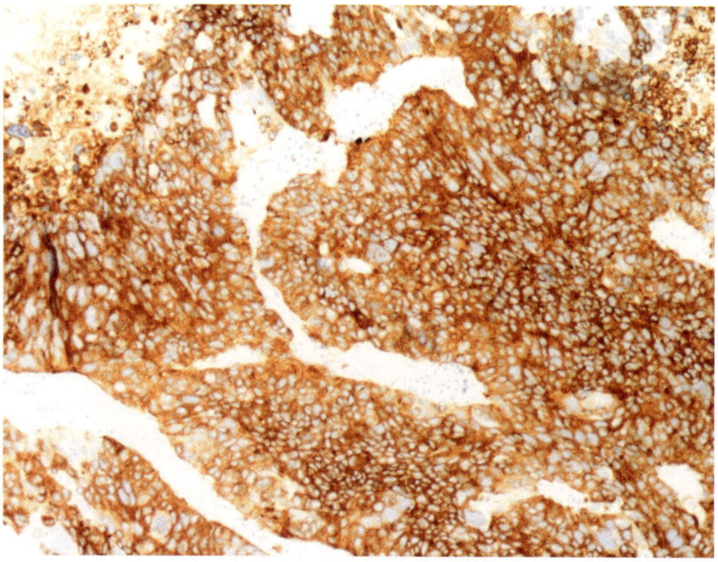

Figure 3.28e — Metastatic Small Cell Carcinoma, Brain Resection [Synaptophysin immunostain; low power]. This case of metastatic small cell carcinoma is strongly and diffusely immunoreactive for synaptophysin, a neuroendocrine and neuroectodermal marker. Synaptophysin stain needs to be interpreted with caution in brain tissue because normal gray matter and many primary brain tumors of neuronal lineage also show strong expression.

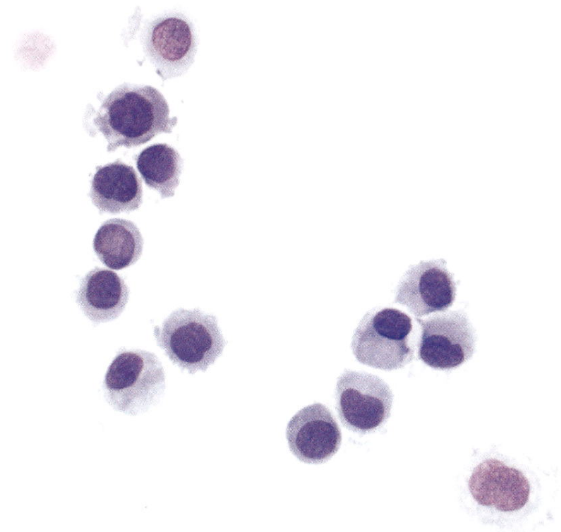

Figure 3.29a — Metastatic Breast Carcinoma, Lumbar Puncture [Diff-Quik stain; low power]. Metastatic breast carcinoma cells in this cerebrospinal fluid (CSF) are fairly monomorphic. These medium-sized cells have moderate amounts of delicate cytoplasm and a round to oval nucleus, giving them a fried-egg or plasmacytoid appearance. The cytoplasm is often ruffled at the edge. Nuclei are folded or notched.

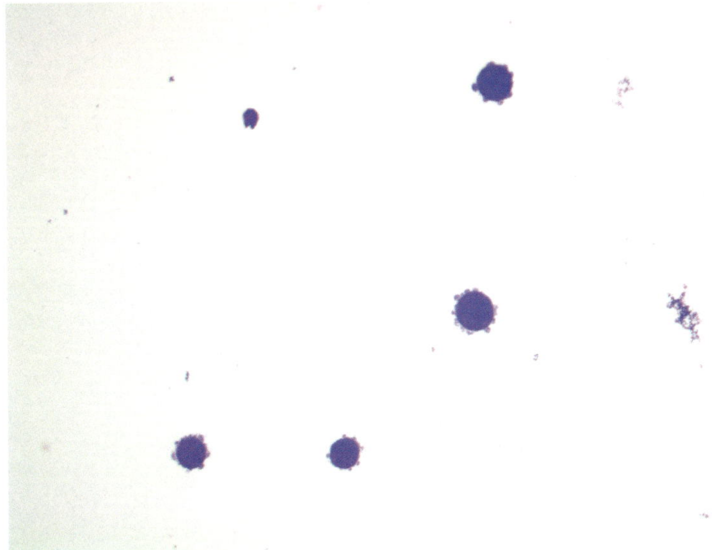

Figure 3.29b — Metastatic Breast Carcinoma, Lumbar Puncture [Diff-Quik stain; low power]. This field demonstrates individual atypical cells with a high nuclear-to-cytoplasmic ratio. The cells have a large, round nucleus and scant cytoplasm. The nuclear contour is quite smooth. The cytoplasm is notable for multiple small blebs.

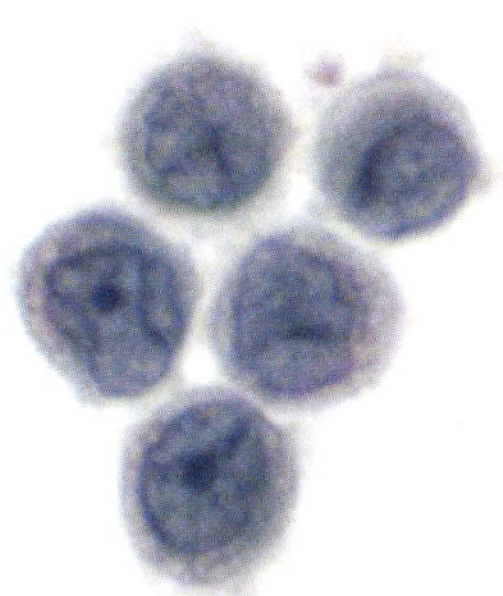

Figure 3.29c — Metastatic Breast Carcinoma, Lumbar Puncture [Pap stain; high power]. Breast carcinoma is common among central nervous system (CNS) metastases, comprising 10% to 20% of cases. Breast carcinoma cells in this field contain a large nuclei with markedly irregular nuclear membranes and a single conspicuous nucleoli. Cytoplasmic blebs/protrusions are vaguely visible on the Pap stain.

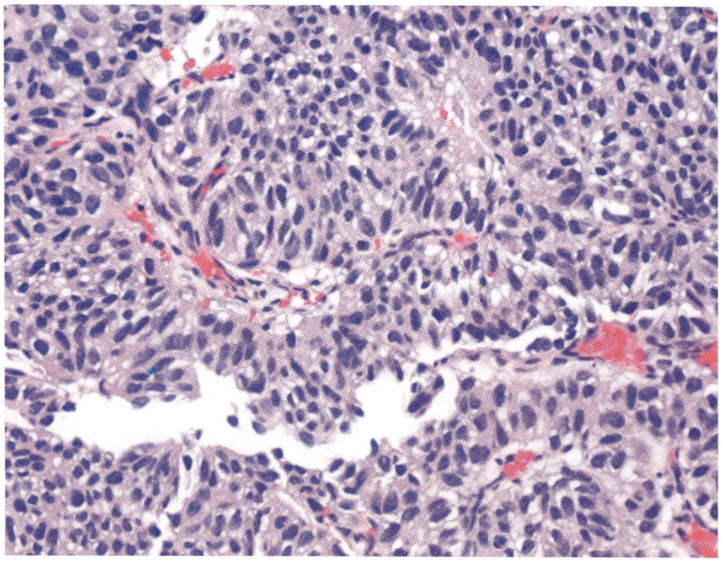

Figure 3.29d — Metastatic Breast Carcinoma, Spine Resection [H&E stain; low power]. This metastatic ductal breast carcinoma has complex papillary and glandular arrangements. Tumor cells show mild nuclear pleomorphism and nuclear hyperchromasia. Loss of nuclear polarity is also noted.

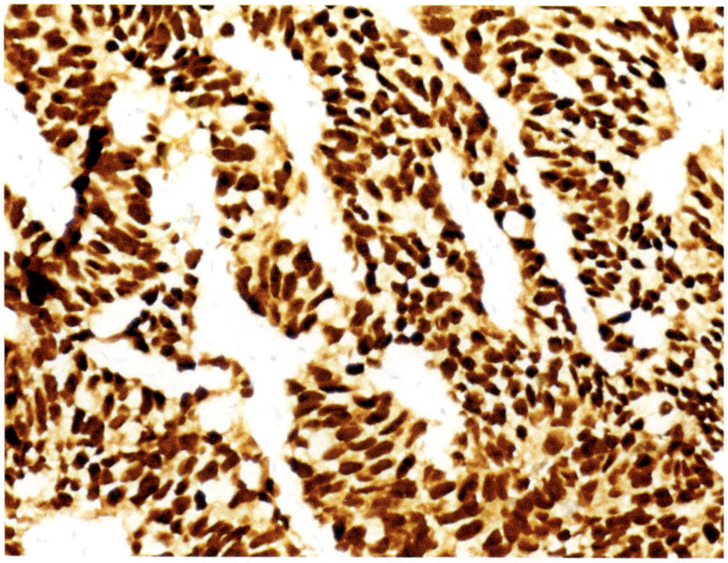

Figure 3.29e — Metastatic Breast Carcinoma, Spine Resection [GATA-binding protein 3 (GATA3) immunostain; low power]. GATA3 immunohistochemistry is a useful marker of breast cancer, particularly for metastatic tumors. Its sensitivity in detecting metastatic tumors is superior to that of mammaglobin and gross cystic disease fluid protein 15 (GCDFP15). This case of metastatic breast carcinoma shows diffuse strong immunoreactivity for GATA3. Of note, the mammaglobin immunostain was negative in this case.

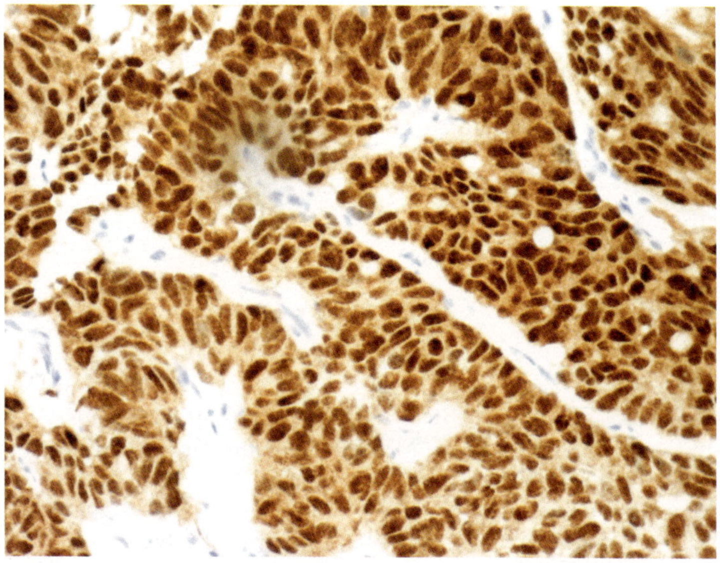

Figure 3.29f — Metastatic Breast Carcinoma, Spine Resection [Estrogen receptor (ER) immunostain; low power]. It is important to perform ER and progesterone receptor (PR) immunohistochemistry on metastatic breast cancer specimens because of potential therapeutic implications. As shown in the image, this case of metastatic breast carcinoma is diffusely positive for ER on the immunostain. The PR stain, on the other hand, was completely negative.

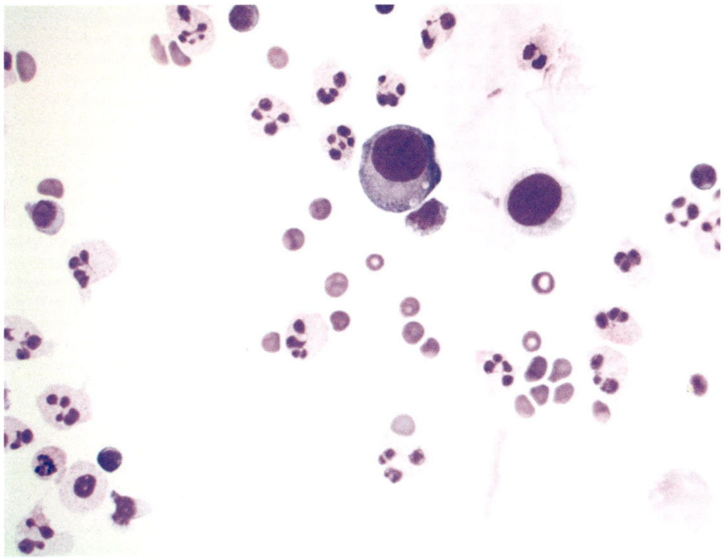

Figure 3.30a — Metastatic Prostatic Adenocarcinoma, Lumbar Puncture [Diff-Quik stain; high power]. Brain metastasis of prostatic carcinoma is rare. However, prostate carcinoma commonly metastasizes to bone, including the spine. An intrathecal metastasis can then spread to the cerebrospinal fluid (CSF). This CSF specimen is notable for two large, round cells with a large nucleus and ample delicate amphophilic cytoplasm. Small vacuoles can be seen in the cytoplasm. Nucleoli are visible.

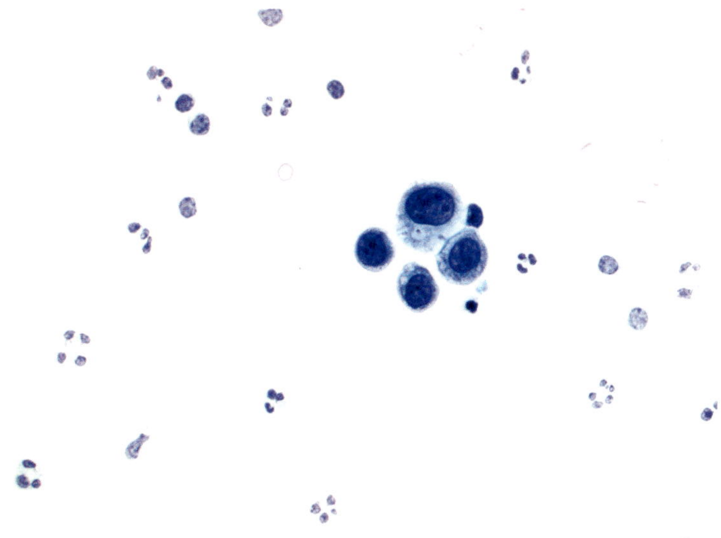

Figure 3.30b — Metastatic Prostatic Adenocarcinoma, Lumbar Puncture [Pap stain; high power]. A cluster of five cells of variable sizes is observed in the center of the field. The cells are characterized by distinct cell borders, ovoid nuclei, clumped chromatin, and prominent nucleoli. Metastatic prostatic carcinoma in the cerebrospinal fluid (CSF) is often associated with elevated prostate-specific antigen (PSA), which can be used with radiology and CSF cytology to improve the accuracy of the diagnosis.

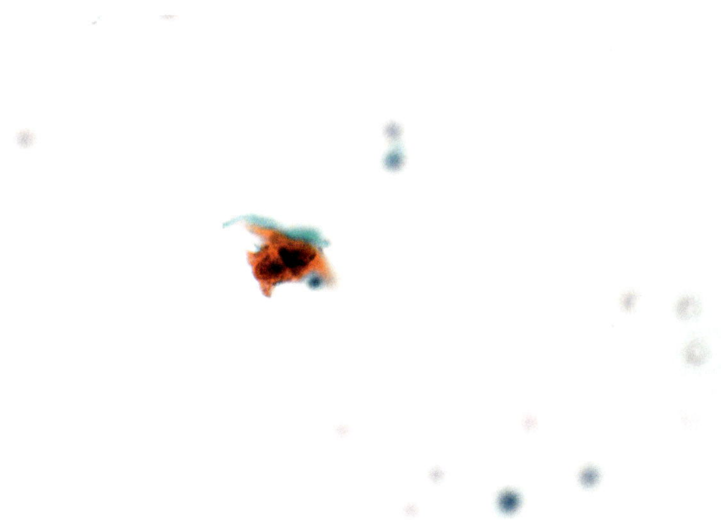

Figure 3.31 — Metastatic Head and Neck Squamous Cell Carcinoma, Lumbar Puncture [Pap stain; high power]. The Pap stain highlights one atypical epithelial cell with dense cytoplasmic orangeophilia, a classic cytologic finding of keratinizing squamous cell carcinoma. The cell is binucleated with hyperchromatic nuclei and a high nuclear-to-cytoplasmic ratio. The nuclear contour is irregular.

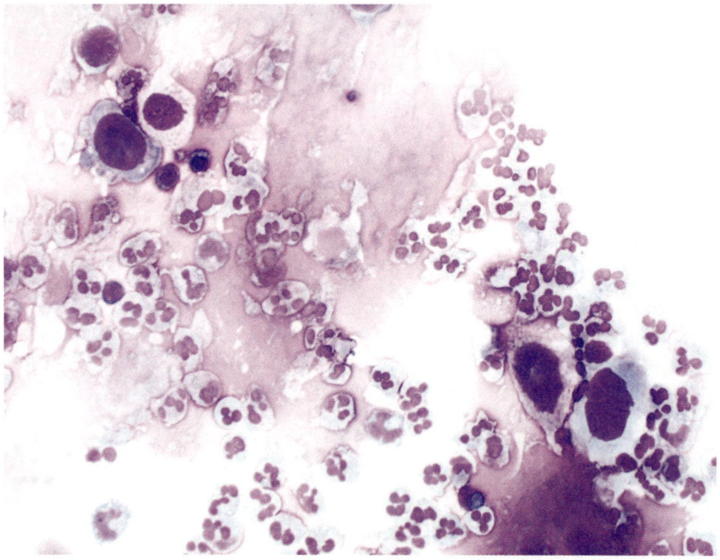

Figure 3.32 — Posttreatment Metastatic Squamous Cell Carcinoma, Lumbar Puncture [Diff-Quik stain; low power]. This CSF specimen shows numerous neutrophils and scattered large, highly atypical cells. The cells have large nuclei, irregular nuclear contour, smudgy chromatin, and moderate amounts of pale basophilic cytoplasm. Note that radiation can induce marked cytologic atypia, characterized by nuclear enlargement and irregularity.

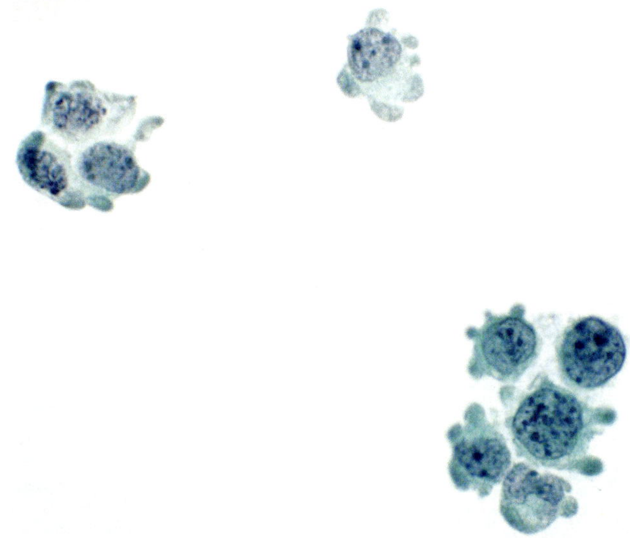

Figure 3.33a — Metastatic Lymphoepithelial Carcinoma, Lumbar Puncture [Pap stain; high power]. This high-power field shows a few large atypical cells with ruffled cytoplasm, vesicular nuclei, and prominent nucleoli. Unlike the surgical specimen, there are no lymphocytes in the background of this cerebrospinal fluid (CSF). Brain metastasis is an uncommon finding of lymphoepithelial carcinoma, which more frequently presents with nodal metastasis.

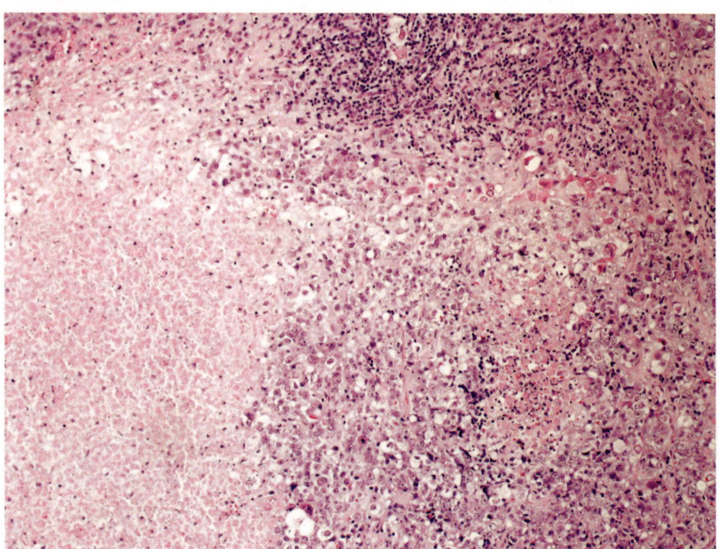

Figure 3.33b — Lymphoepithelial Carcinoma, Resection [H&E stain; low power]. Lymphoepithelial carcinoma is an undifferentiated squamous cell carcinoma that most commonly occurs in the nasopharynx. The histologic section demonstrates a malignant neoplasm with large areas of necrosis and sheets of pleomorphic malignant cells. Tumor cells have clear cytoplasm and prominent nucleoli. There are abundant lymphocytic infiltrates in the background.

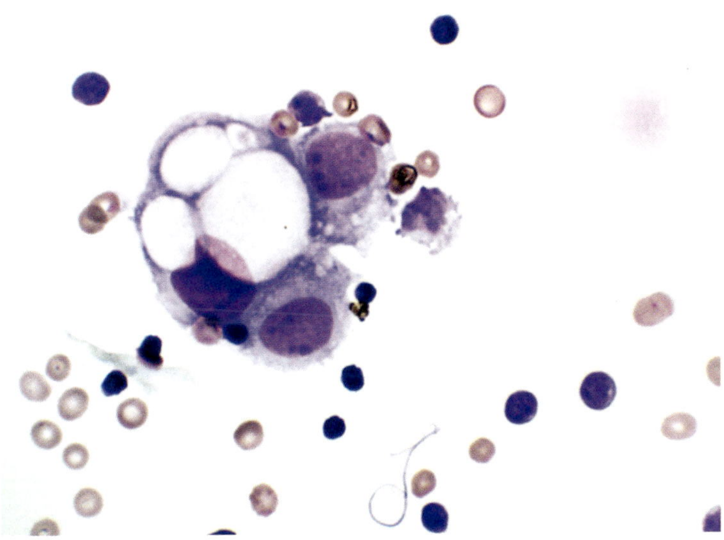

Figure 3.34 — Metastatic Ovarian Serous Carcinoma, Lumbar Puncture [Diff-Quik stain; high power]. A cluster of large, highly atypical cells is observed in this cerebrospinal fluid (CSF). The cells have an oval nucleus with abundant delicate cytoplasm and large mucin vacuoles. Three to five nucleoli are identified in the nuclei. Metastasis of ovarian cancer to the brain is rare.

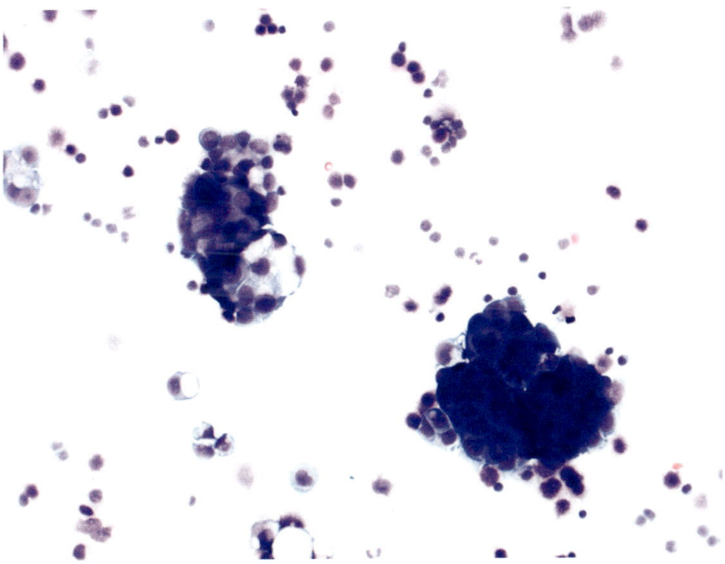

Figure 3.35a — Metastatic Uterine Serous Carcinoma, Lumbar Puncture [Diff-Quik stain; low power]. This field demonstrates two large fragments of highly pleomorphic cells surrounded by scattered single atypical cells. The cells have ample pale blue cytoplasm. Some contain a large intracytoplasmic mucin vacuole. Nucleoli are prominent.

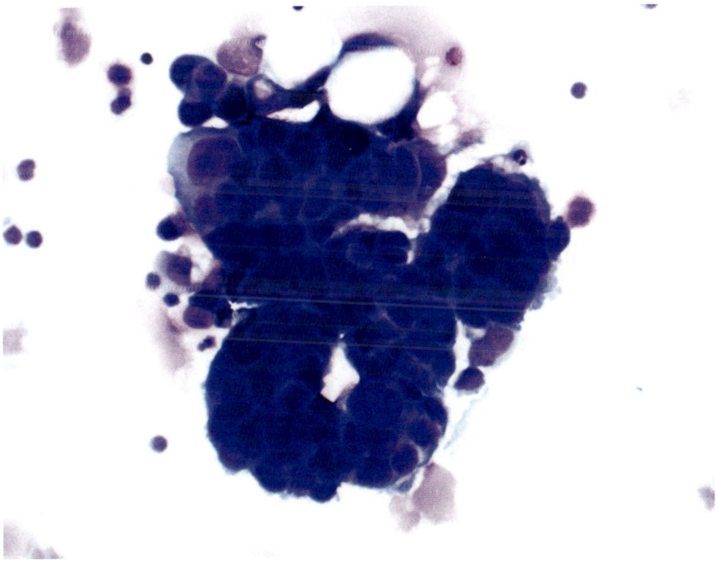

Figure 3.35b — Metastatic Uterine Serous Carcinoma, Lumbar Puncture [Diff-Quik stain; high power]. This tissue fragment contains many large tumor cells. The adjacent smaller lymphocytes can be used as a ruler to indicate how large the tumor cells are by comparison. Two cells at the top assume a signet-ring appearance as a result of the large clear mucin vacuole.

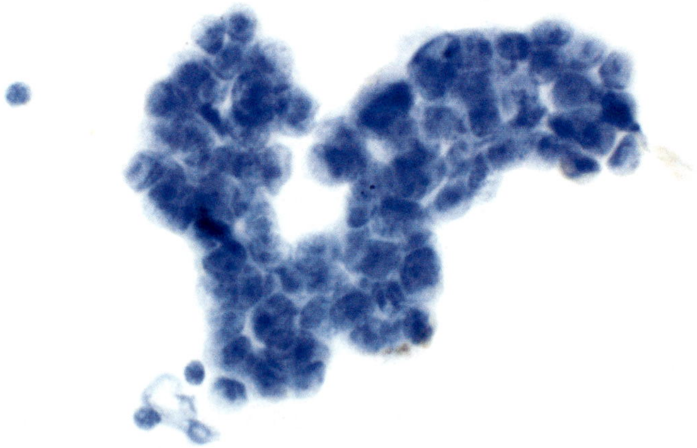

Figure 3.35c — Metastatic Uterine Serous Carcinoma, Lumbar Puncture [Pap stain; high power]. Uterine serous carcinoma is also known as *uterine papillary serous carcinoma*. Its metastasis is exceedingly rare in the central nervous system (CNS), with fewer than 10 cases described in the literature. The tumor fragment in this field is characterized by a papillary configuration. The tumor cells have hyperchromatic nuclei with coarse chromatin and visible nucleoli.

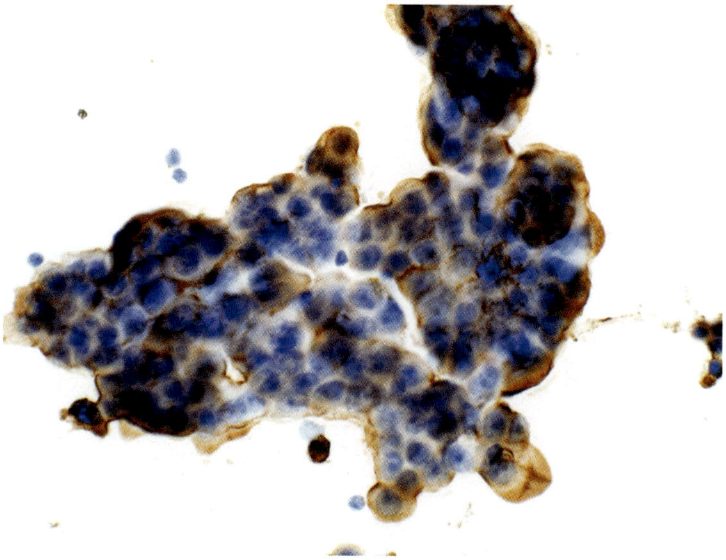

Figure 3.35d — Metastatic Uterine Serous Carcinoma, Lumbar Puncture [Ber-EP4 immunostain; high power]. Ber-EP4 is a cell membrane glycoprotein that is expressed in various carcinomas, especially adenocarcinomas. In this field, tumor cells demonstrate Ber-EP4 immunoreactivity in a membranous pattern. The stain does not distinguish uterine serous carcinoma from other adenocarcinomas metastasizing to the cerebrospinal fluid (CSF). Note the prominent nucleoli in the counter stain.

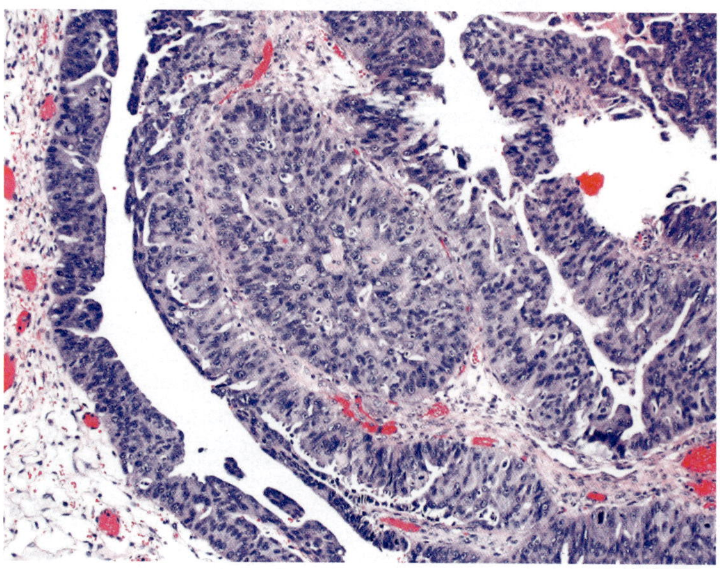

Figure 3.35e — Uterine Serous Carcinoma, Hysterectomy [H&E stain; low power]. This tumor exhibits a complex pattern of papillary, cribriform, and glandular architecture. Ragged luminal border and prominent nucleoli are important features that distinguish it from endometrioid carcinoma. Uterine serous carcinoma is the most aggressive histologic variant of endometrial cancer. Lymphovascular invasion and intraperitoneal spread are common.

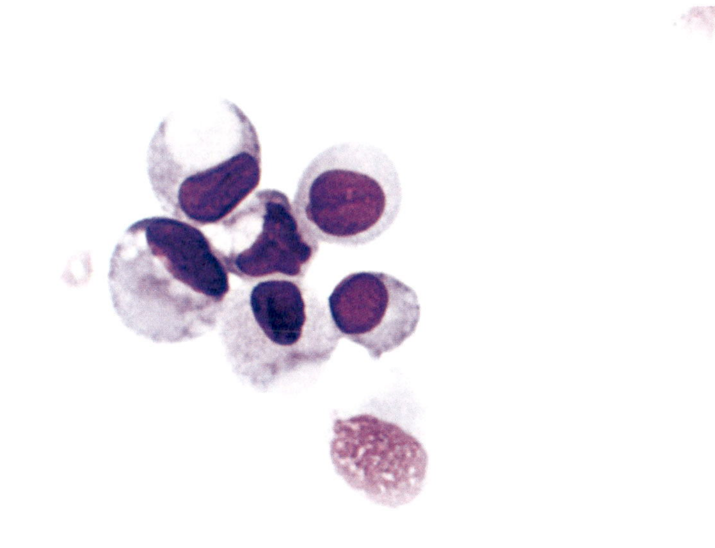

Figure 3.36a — Metastatic Gastric Adenocarcinoma, Lumbar Puncture [Diff-Quik stain; high power]. Metastatic gastric adenocarcinoma to the cerebrospinal fluid (CSF) is very rare. It usually occurs in the setting of leptomeningeal carcinomatosis. This field shows a cluster of large cells with ovoid nuclei and ample cytoplasm. The cytoplasmic mucin vacuoles are evident.

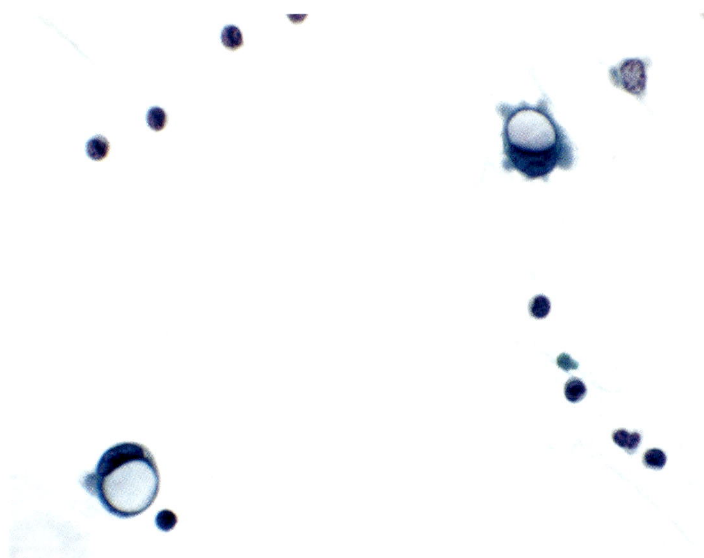

Figure 3.36b — Metastatic Gastric Adenocarcinoma, Lumbar Puncture [Pap stain; high power]. There are two metastatic gastric adenocarcinoma cells in this cerebrospinal fluid (CSF). The cells are large: They are 5 to 10 times the size of the background lymphocytes. The large mucin vacuole displaces the nuclei, giving the cells a signet-ring appearance.

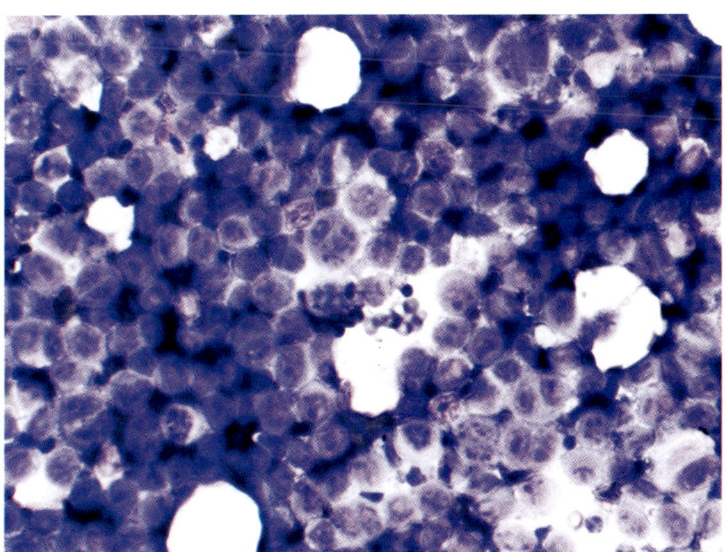

Figure 3.37a — Metastatic Melanoma, Lumbar Puncture [Diff-Quik stain; low power]. Melanoma is the third most common cancer that metastasizes to the central nervous system (CNS). Approximately, 10% to 40% of patients with melanoma experience CNS metastasis. Unlike the majority of the metastatic tumors that exhibit nonspecific morphology in the cerebrospinal fluid (CSF), metastatic melanoma cells have distinct cytologic features. This field shows numerous, overtly atypical cells characterized by frequent binucleation and prominent inclusion-like nucleoli.

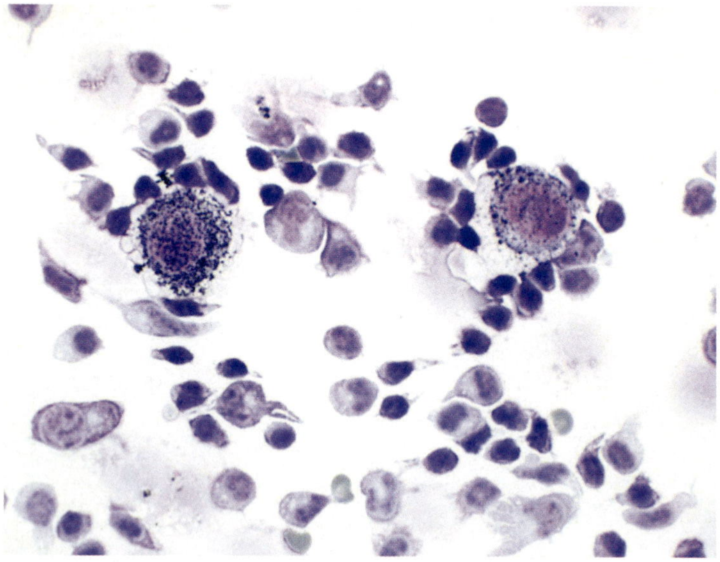

Figure 3.37b — Metastatic Melanoma, Lumbar Puncture [Diff-Quik stain; high power]. Tumor cells in this field demonstrate marked pleomorphism. There are small cells with hyperchromatic nuclei and scant cytoplasm; medium-sized cells with plasmacytoid morphology, binucleation, and prominent nucleoli; and large cells with abundant melanin pigment. Melanin pigment appears dark green on Diff-Quik stain.

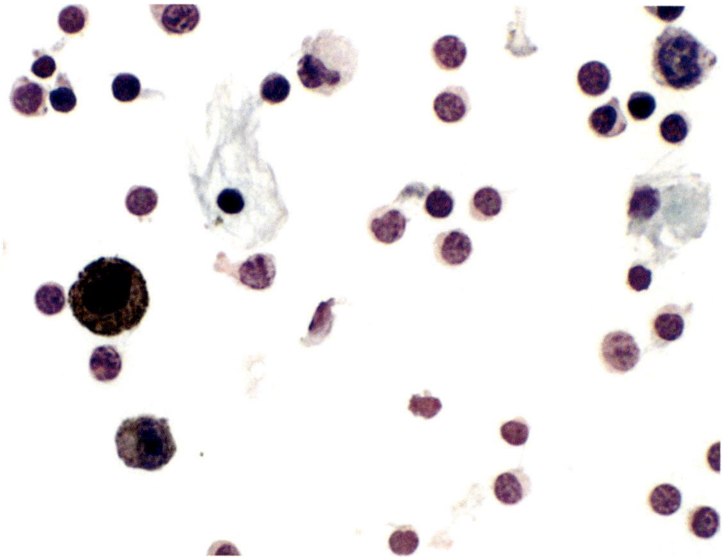

Figure 3.37c — Metastatic Melanoma, Lumbar Puncture [Pap stain; low power]. The majority of the cells in this field are small to medium-sized with oval nuclei, vesicular chromatin, and scant agranular cytoplasm. There are occasional large cells with abundant cytoplasm and conspicuous nucleoli. These cells also contain melanin pigment in the cytoplasm.

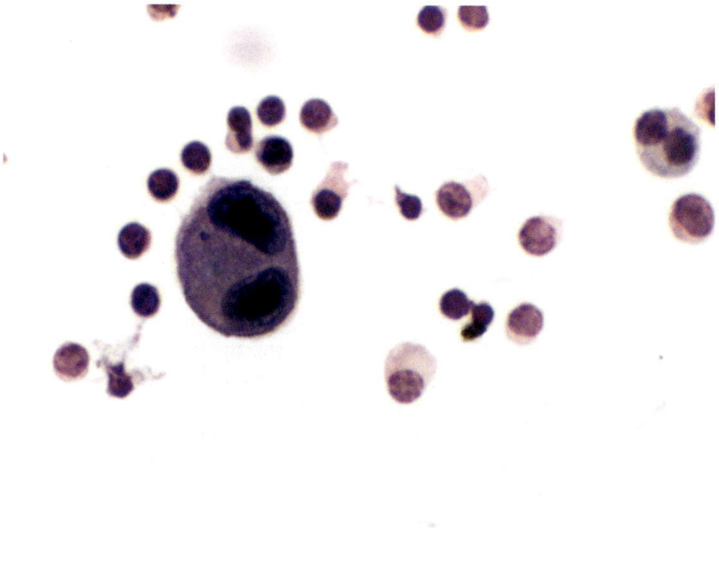

Figure 3.37d — Metastatic Melanoma, Lumbar Puncture [Pap stain; high power]. Metastatic melanoma cells are often more dispersed than clustered in the cerebrospinal fluid (CSF). They can be highly pleomorphic. The huge cell off-center demonstrates the classic morphologic features of melanoma: binucleation and prominent large nucleoli. The morphology resembles bug's eyes.

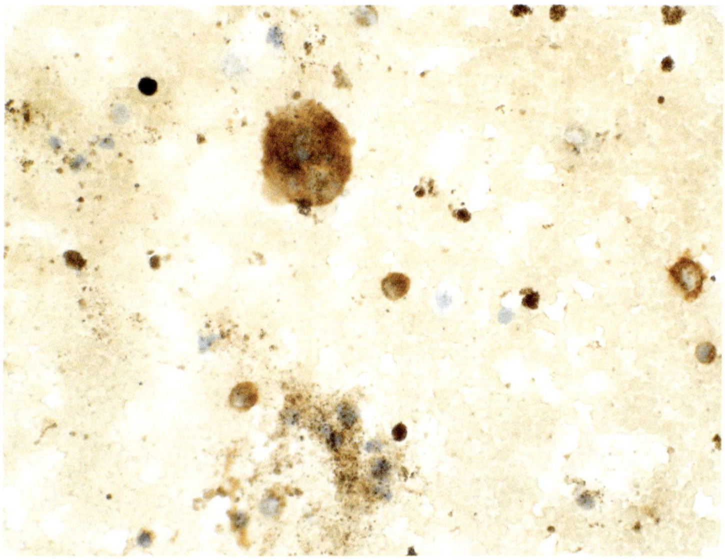

Figure 3.37e — Metastatic Melanoma, Lumbar Puncture [Melan-A stain; high power]. If the diagnosis is doubtful and there is additional material available, immunohistochemical studies can be performed. Melan-A is a commonly used marker for melanoma, demonstrating both high sensitivity and specificity. Its sensitivity is superior to that of HMB-45, and its specificity is better than that of S-100 protein.

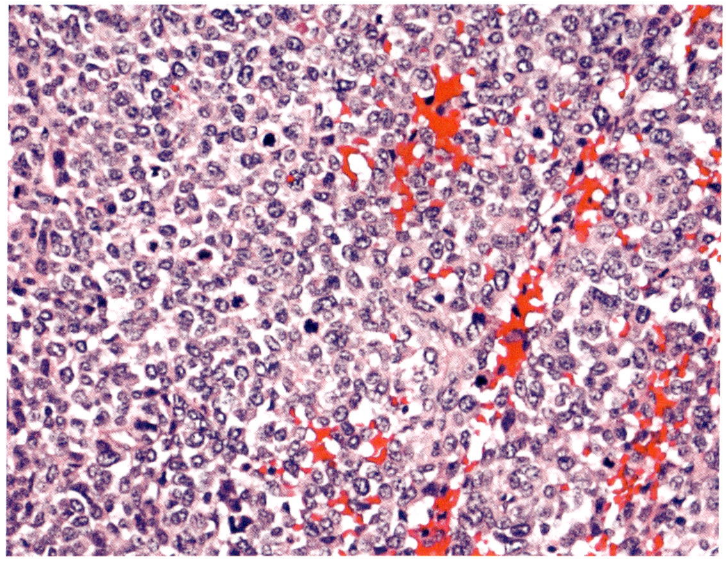

Figure 3.37f — Metastatic Melanoma, Brain Resection [H&E stain; low power]. Metastatic melanoma is a great mimic and can morphologically resemble other types of tumors. This case shows sheets of tumor cells that are relatively monomorphic. Nucleoli are visible but not prominent. The brisk mitotic activity is a clue to a diagnosis of malignancy. This dural-based tumor was initially mistaken as a malignant (anaplastic) meningioma because of the relatively bland cytologic features.

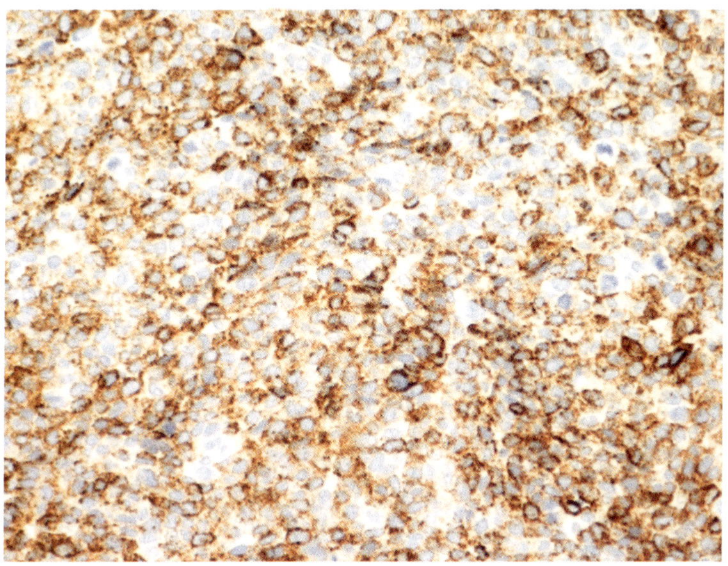

Figure 3.37g — Metastatic Melanoma, Brain Resection [HMB-45 immunostain; low power]. HMB-45 is a highly specific marker for melanoma (greater than 95% specificity). The sensitivity, in comparison, is not as good (50%–75%). This case of metastatic melanoma is strongly immunoreactive for HMB-45. It is important to avoid using S-100 as a melanoma marker on brain tissue because normal central nervous system (CNS) tissue and gliomas are also strongly positive for S-100 protein.

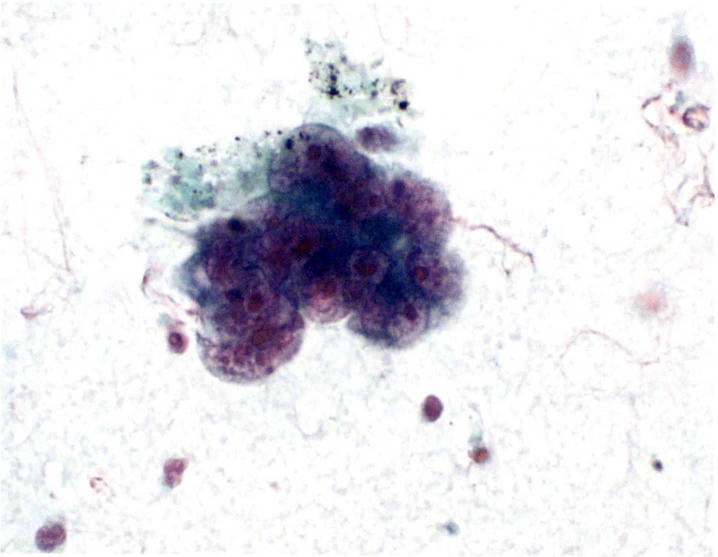

Figure 3.38a — Metastatic Embryonal Carcinoma, Lumbar Puncture [Pap stain; high power]. This high-power field demonstrates a three-dimensional fragment of large tumor cells. The cells. The cells have a high nuclear-to-cytoplasmic ratio, delicate chromatin, and large, red nucleoli. Compared with germinoma, metastatic embryonal carcinoma has scant cytoplasm and stays more cohesive in the cerebrospinal fluid (CSF).

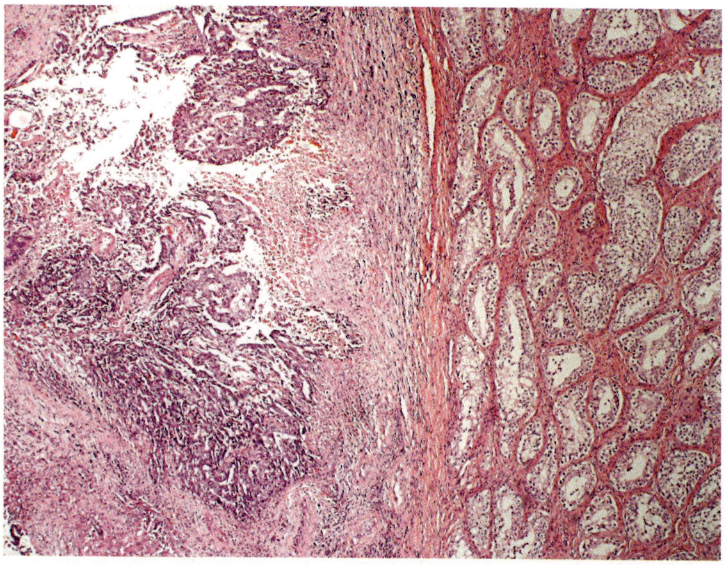

Figure 3.38b — Embryonal Carcinoma, Orchiectomy [H&E stain; low power]. Embryonal carcinoma comprises 2% to 10% of the pure germ cell tumors and is present in approximately 80% of the mixed germ cell tumors. It occurs in young adult patients aged 20 to 30 years. Unlike seminoma, the tumor usually does not replace the entire testis, as demonstrated in this image. This tumor shows tubular and papillary growth patterns with abundant coagulative necrosis.

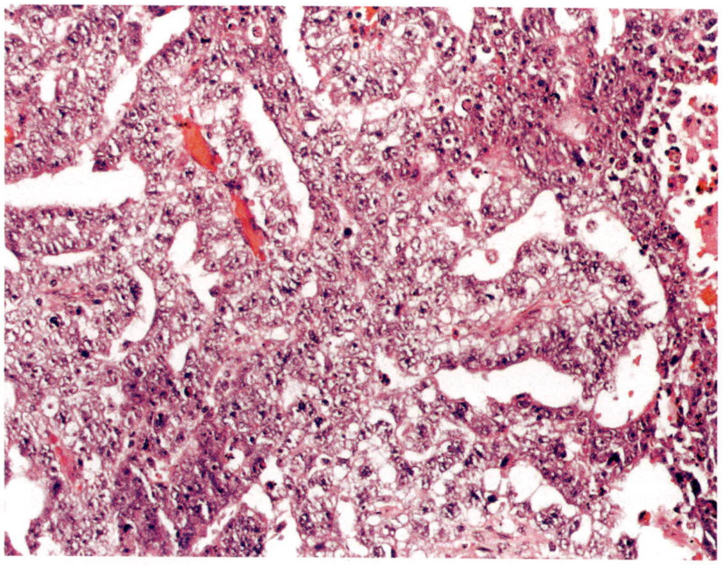

Figure 3.38c — Embryonal Carcinoma, Orchiectomy [H&E stain; high power]. At high magnification, tumor cells are remarkable for large vesicular nuclei with conspicuous nucleoli. Apoptosis is prominent on the right edge of the image. Immunohistochemically, embryonal carcinoma is positive for cytokeratin, CD30, and germ cell markers Oct3/4, SALL4, and placental alkaline phosphatase (PLAP).

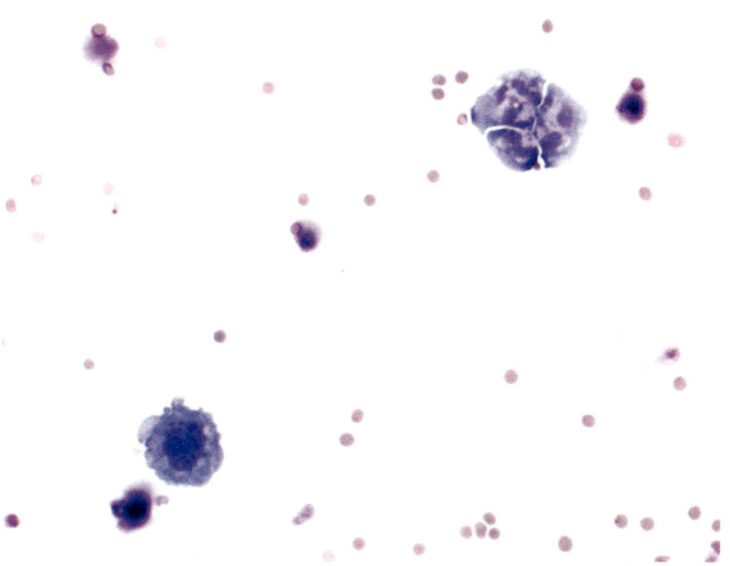

Figure 3.39a — Metastatic Embryonal Rhabdomyosarcoma, Lumbar Puncture [Diff-Quik stain; high power]. This field shows a loose aggregate of atypical cells in the right corner and a single atypical cell in the left corner. The cells are large and harbor abundant amphophilic cytoplasm. These tumor cells have fragmented nuclei, likely undergoing apoptosis triggered by treatment. The cytomorphologic features of metastatic embryonal rhabdomyosarcoma in the cerebrospinal fluid (CSF) are similar to those of other embryonal tumors such as medulloblastoma, central nervous system (CNS) embryonal tumor, and retinoblastoma.

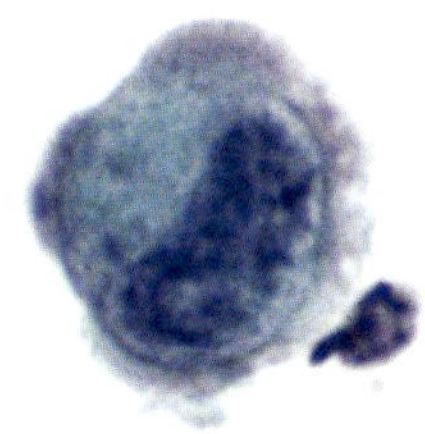

Figure 3.39b — Metastatic Embryonal Rhabdomyosarcoma, Lumbar Puncture [Pap stain; high power]. Embryonal rhabdomyosarcomas from the head and neck region are most likely to metastasize to the cerebrospinal fluid (CSF) due to proximity. The giant cell in this field has a rhabdoid morphology: ample cytoplasm and an eccentric nucleus. Chromatin is dense, nuclear membrane is smooth, and nucleoli are invisible. Unlike disseminated atypical teratoid/rhabdoid tumor (AT/RT), which also frequently shows rhabdoid morphology, intracytoplasmic inclusions are infrequent in metastatic embryonal rhabdomyosarcoma.

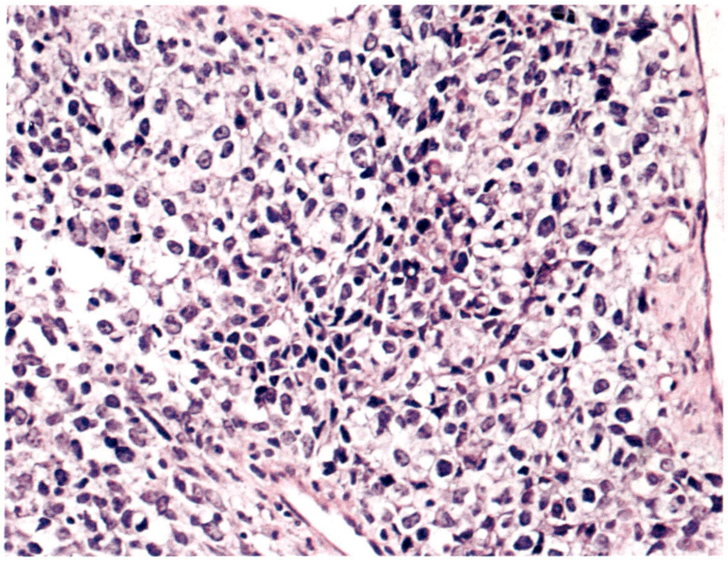

Figure 3.39c — Embryonal Rhabdomyosarcoma, Orbital Resection [H&E stain; low power]. Embryonal rhabdomyosarcoma is a poorly differentiated primitive malignant neoplasm that occurs in young children. The head and neck are common locations. This case demonstrates sheets of round cells with clear cytoplasm and a single large nucleus. Eosinophilic nucleoli are prominent. Necrosis can be seen in the center of the image. Embryonal rhabdomyosarcomas are positive for the muscle markers desmin and myogenin on immunohistochemistry.

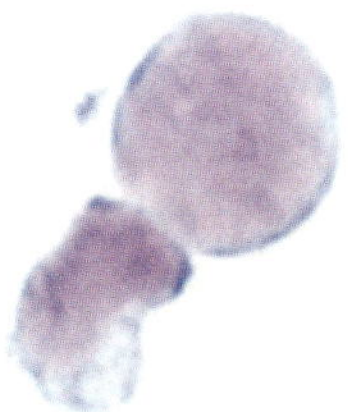

Figure 3.40 — Metastatic Alveolar Rhabdomyosarcoma, Lumbar Puncture [Diff-Quik stain; high power]. This field demonstrates two large tumor cells: One has a large, round nucleus and scant cytoplasm; the other contains a kidney-shaped nucleus with moderate amounts of basophilic cytoplasm. Nucleoli are indistinct. Although not present in this image, metastatic alveolar rhabdomyosarcoma in the cerebrospinal fluid (CSF) often shows rhabdoid morphology similar to that of metastatic embryonal rhabdomyosarcoma cells in Figure 3.39c.

4

Gastrointestinal and Pancreatobiliary Tract Exfoliative Cytopathology

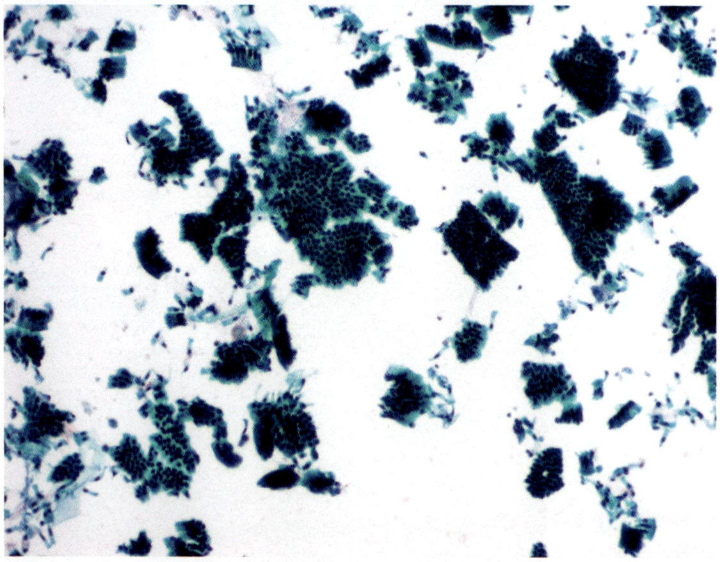

Figure 4.1 — Normal Epithelium, Biliary Brushing [Pap Stain; low power]. Low-power view showing multiple large and small fragments of columnar epithelium with preservation of the honeycomb architecture; each cell "respects" the borders of adjacent cells, and they are of equal size and shape. Depending on the view, one can appreciate well-polarized columnar epithelium with basally oriented nuclei and abundant apical cytoplasm. Biliary tract brushings are an excellent method of sampling small-caliber ducts, areas of stricture, and difficult-to-reach areas of fibrosis. As a result, biliary tract brushing often have a higher diagnostic yield than biopsy forceps and fine needle aspirations in this anatomic region. The main goal for biliary tract brushings is to distinguish a benign versus a malignant etiology for biliary strictures and to detect the development of high-grade dysplasia in at-risk patients, such as those with primary sclerosing cholangitis.

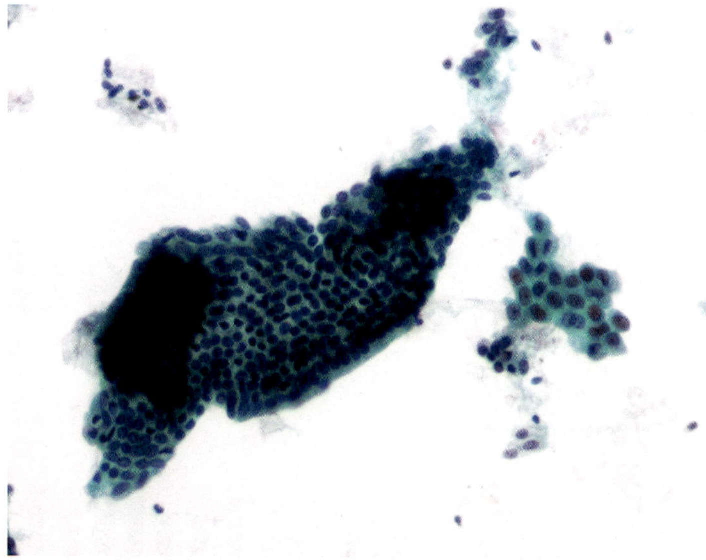

Figure 4.2a — Reactive Epithelium, Biliary Brushing [Pap stain; medium power]. The larger fragment of epithelial cells consists of cohesive cells with minimal nuclear overlap and preserved polarity. The luminal cells can be seen with basally oriented nuclei and an evenly arranged mucin cap. The adjacent smaller fragment shows mildly reactive epithelial cells. This fragment is flat and, although enlarged, contains cells with abundant cytoplasm and a normal corresponding nuclear-to-cytoplasmic ratio. The architecture is preserved with little to no nuclear overlap, smooth nuclear contours, and small nucleoli. A lack of necrosis, single atypical cells, high nuclear-to-cytoplasmic ratio, and atypical mitoses distinguish the reactive changes in this case from a diagnosis of adenocarcinoma.

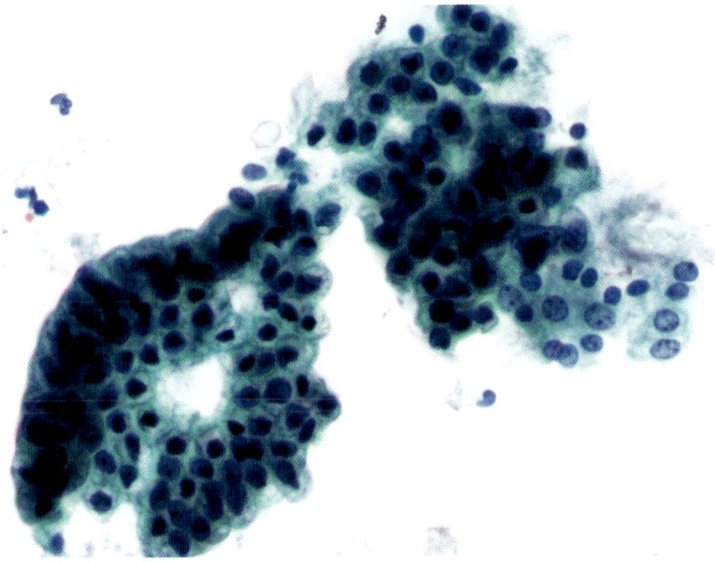

Figure 4.2b — Reactive Epithelium, Biliary Brushing [Pap stain; high power]. This field consists of two fragments of flat, ductal epithelial cells with well-defined cell borders and abundant vacuolated cytoplasm. The luminal epithelial cells can be seen at left with maintained polarity. At top right, there is a minimal degree of nuclear overlap and size variation (1–2 times); however, the nuclear contours are smooth, the chromatin is fine, and the nuclear-to-cytoplasmic ratio is normal. Occasional prominent nucleoli are observed. Atypia may be caused by inflammatory disorders such as sclerosing cholangitis, chronic pancreatitis, and biliary cirrhosis or by mechanical irritation from biliary stents, drains, or obstruction from a regional tumor.

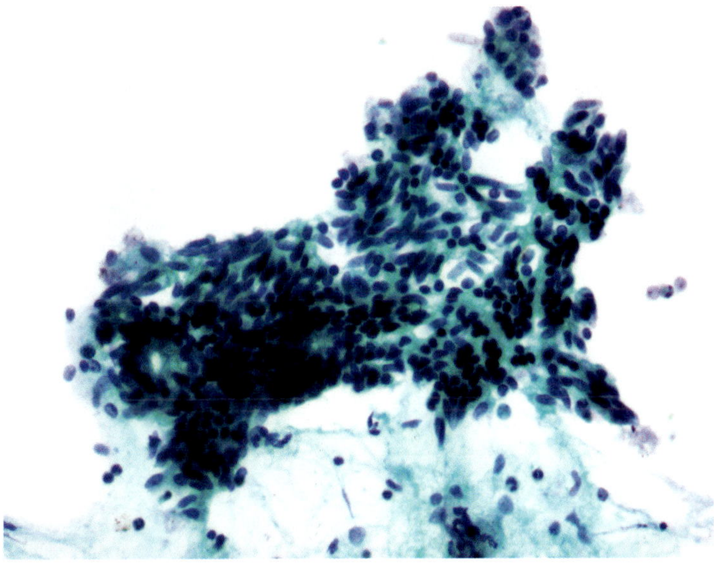

Figure 4.2c — Reactive Epithelium, Biliary Brushing [Pap stain; medium power]. This brushing reveals a large fragment of reactive biliary ductal epithelial cells. At first glance, the cells may appear to be disorganized and variable in size; however, note that the cells regionally maintain polarity and are lining up in different planes. This can be appreciated by the presence of a well-formed duct at the 1 and 9 o'clock designations. Furthermore, although some cells appear slightly elongated, when viewed in cross section, the same cell population is small and round. Overall, the epithelium has abundant granular cytoplasm, preserved nuclear-to-cytoplasmic ratios, smooth nuclear contours, and a relative lack of nucleoli. A few scattered acute inflammatory cells can be seen at the bottom of the image, perhaps suggesting an inflammatory etiology for the reactive changes identified in this specimen.

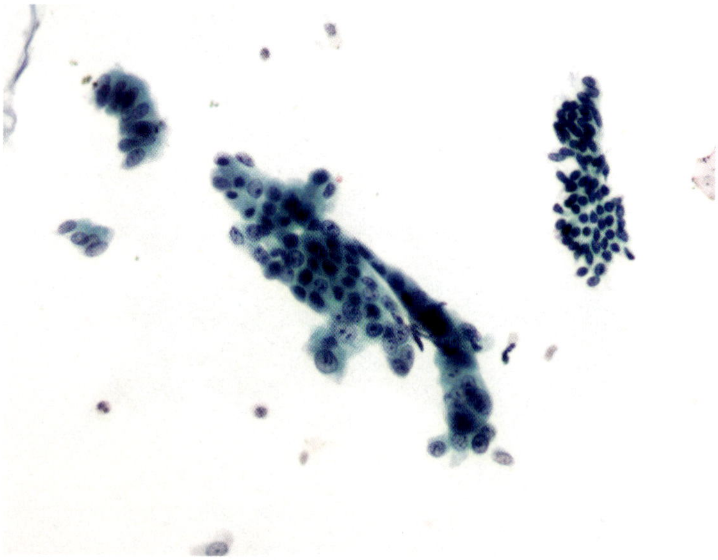

Figure 4.3a — Atypical Epithelium, Biliary Brushing [Pap stain; medium power]. In the center of the field, there is a flat fragment of enlarged epithelial cells with occasional overlapping and scattered prominent nucleoli. More important, the nuclear-to-cytoplasmic ratio is relatively preserved and the nuclear contours are smooth. These cytologic features can be juxtaposed with the adjacent normal epithelial fragment present at the far right. Although this group of cells are markedly different, the level of atypia may be consistent with that seen in reactive epithelial changes. Because there can often be significant overlap between reactive atypia and dysplastic changes/well-differentiated adenocarcinoma, this case warrants a descriptive diagnosis and categorization as atypical. Clinical correlation, as well as repeat sampling, may be helpful in this case.

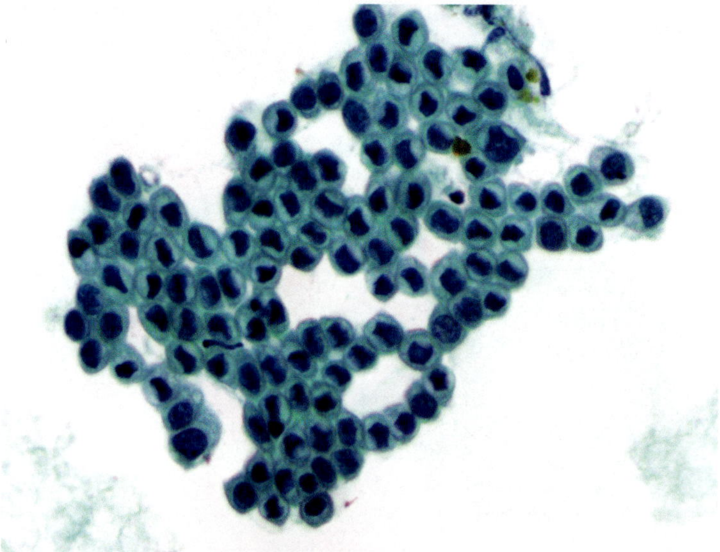

Figure 4.3b — Atypical Epithelium, Biliary Brushing [Pap stain; high power]. This field is composed of a monolayer of ductal epithelium notable for a lack of nuclear overlap and a relatively preserved normal architecture. The cytologic features, however, are more worrisome. There is scattered size variation and occasional nuclear enlargement reaching 2 times the size of the more normal-appearing epithelium. In most cases, however, the nuclear-to-cytoplasmic ratio is preserved. The nuclei appear hyperchromatic with indented and irregular contours, but there is a lack of prominent nucleoli. Taken together, this fragment of epithelial cells displays normal architecture but shows a degree of cytologic atypia warranting a descriptive diagnosis categorized as atypical. Although these changes most probably represent a reactive process, there is often a significant level of overlap with reactive atypia and well-differentiated carcinoma.

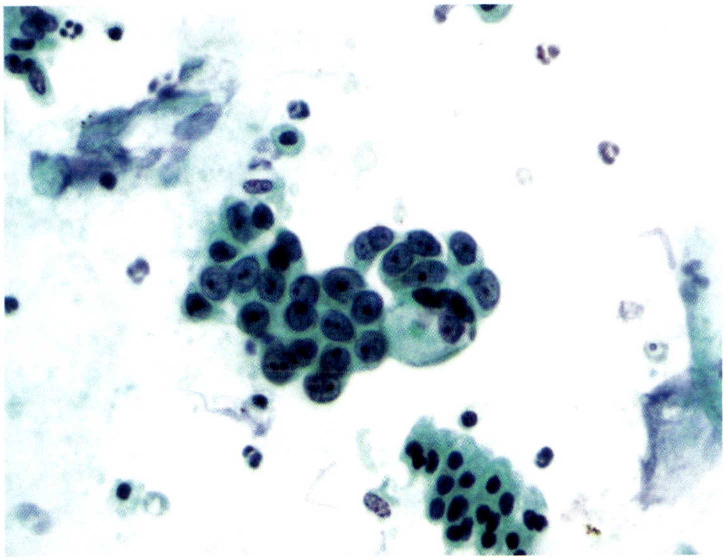

Figure 4.4a — Suspicious for Adenocarcinoma (ADC), Biliary Brushing [Pap stain; high power]. In the center of the field, there is a single, relatively flat fragment of atypical epithelial cells. These cells can be compared with the normal-appearing ductal epithelium at the bottom of the field. In contrast, the atypical epithelial cells show nuclear enlargement, increased nuclear-to-cytoplasmic ratio, occasionally cleaved nuclear membranes, and prominent nucleoli. Unlike the normal ductal epithelium which shows polarization, the atypical cells display a loss of normal architecture. In the background, however, there are scattered acute inflammatory cells, which may raise the possibility that the atypia present is reactive in nature. The overall degree of cytologic atypia in this case, though, is highly suggestive of ADC. Atypical mitoses or the presence of multiple atypical clusters may warrant an outright diagnosis of malignancy in the correct clinical context (i.e., the presence of a mass, a history of ADC).

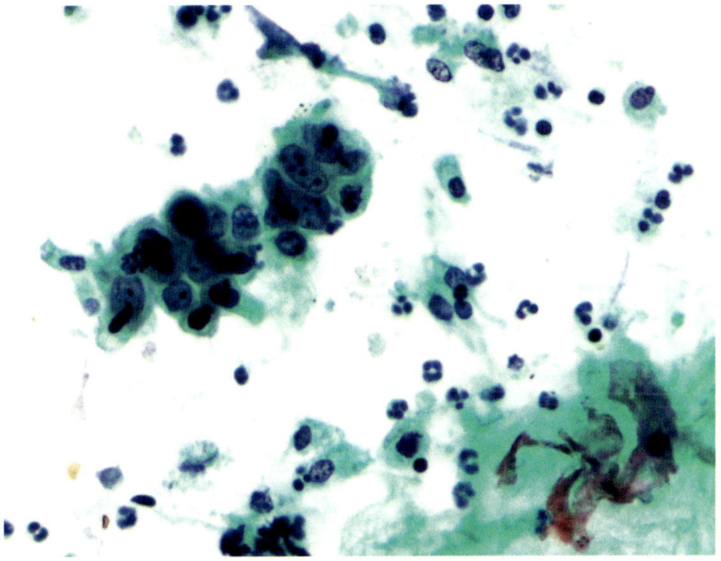

Figure 4.4b — Suspicious for Adenocarcinoma (ADC), Biliary Brushing [Pap stain; high power]. In this specimen, there is a single cluster of atypical epithelial cells showing size variation, nuclear enlargement, increased nuclear-to-cytoplasmic ratio, irregular nuclear membranes, prominent nucleoli, and chunky chromatin. Although, in isolation, these cells are nearly diagnostic for ADC, the presence of such a significant level of neutrophils should give the diagnostician pause. In acutely inflamed specimens, the degree of reactive epithelial atypia can be striking, often revealing what looks like tumor necrosis. In these cases, the best practice may be to give a descriptive diagnosis, stating that the lesion is suspicious for malignancy. If atypical mitoses, single and highly atypical cells, or atypia in other fields in an absence of inflammation are seen, a definitive diagnosis of malignancy may be made.

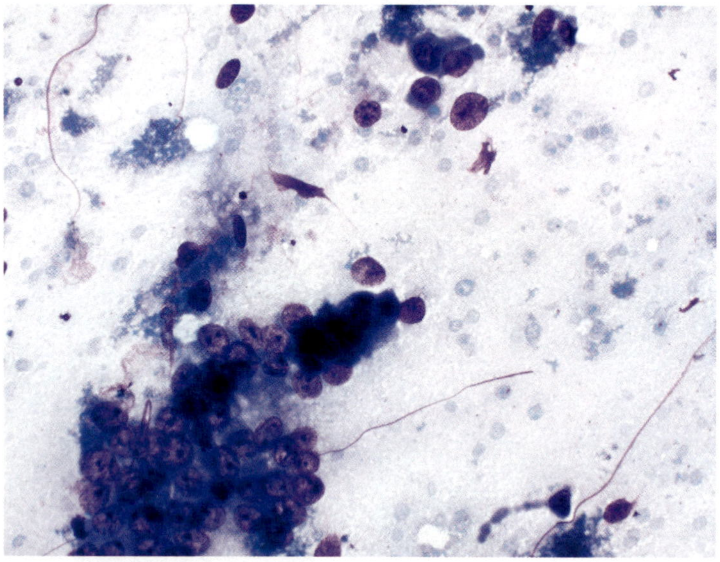

Figure 4.5a — Adenocarcinoma (ADC), Biliary Brushing [Pap stain; medium power]. This field contains a disorganized fragment and variably sized epithelial cells. The fragment shows nuclear crowding and overlap with a notable loss of normal spacing. In most cells, the nuclear-to-cytoplasmic ratio is increased and cytoplasm is scant. The nuclei are enlarged with prominent nucleoli. In this case, single, intact atypical cells are seen and are highly suggestive of a malignant process.

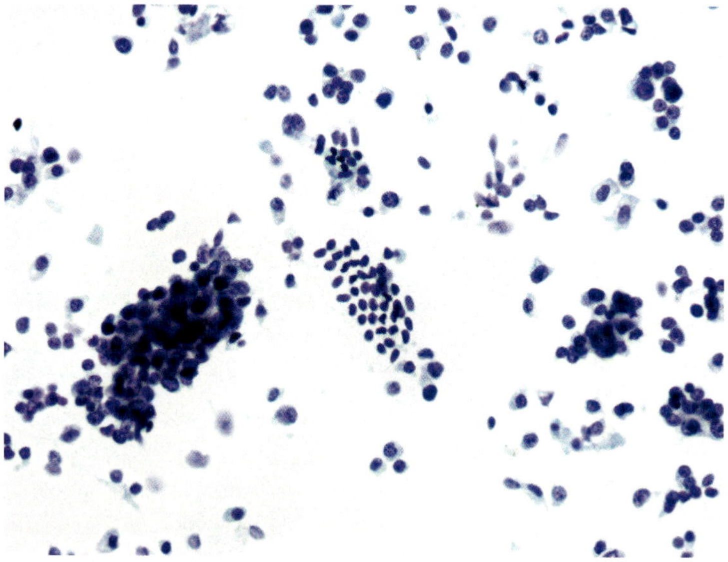

Figure 4.5b — Adenocarcinoma (ADC), Biliary Brushing [Pap stain; medium power]. In the center of the field, a fragment of relatively normal biliary epithelium is composed of evenly spaced, small, uniform cells. In contrast, the neighboring fragments of atypical epithelial cells are disorganized, with a significant degree of nuclear overlap. These cells are enlarged, with hyperchromatic nuclei, and show prominent nucleoli. In addition, single atypical cells are scattered throughout. The main differential diagnosis is with reactive atypia, which can be distinguished from malignancy by a notable normal nuclear-to-cytoplasmic ratio, minimal nuclear overlap, smooth nuclear contours, small nucleoli (when present), and rare/lack of single atypical cells.

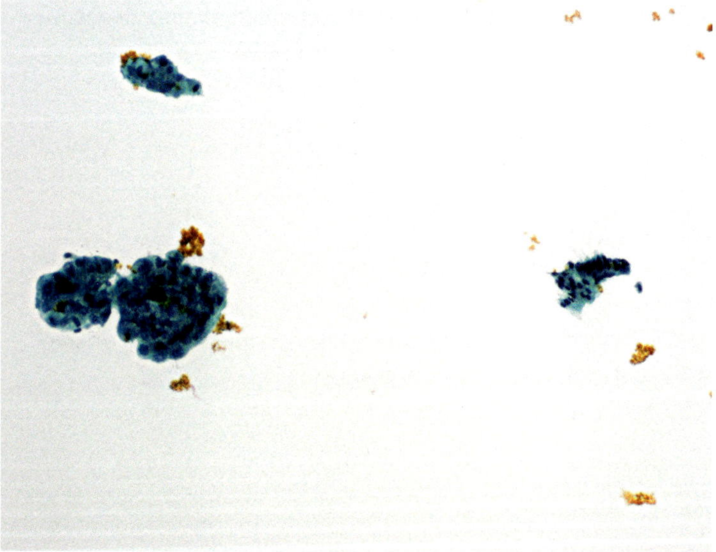

Figure 4.5c — Adenocarcinoma (ADC), Biliary Brushing [Pap stain; low power]. In this low-power view, two distinct cell populations are evident. On the right, there is a normal-appearing small fragment of polarized columnar epithelial cells with basally oriented nuclei and abundant cytoplasm. In juxtaposition, there are three-dimensional atypical fragments of epithelial cells with significant overlap and loss of normal architecture. The nuclear-to-cytoplasmic ratio is increased, and the nuclei are enlarged with irregular borders. These features are diagnostic of ADC; the clinical, radiographic, and surgical material confirmed a diagnosis of cholangiocarcinoma. Risk factors for cholangiocarcinoma include primary sclerosing cholangitis, fibropolycystic liver disease, hepatolithiasis, and chronic liver disease, to name a few. Cholangiocarcinoma can arise in the liver (intrahepatic), at the hilum (hilar), or from the hilum to the ampulla of Vater (extrahepatic). Overall survival is poor; localized intrahepatic stage I disease is associated with a 5-year survival of 15%.

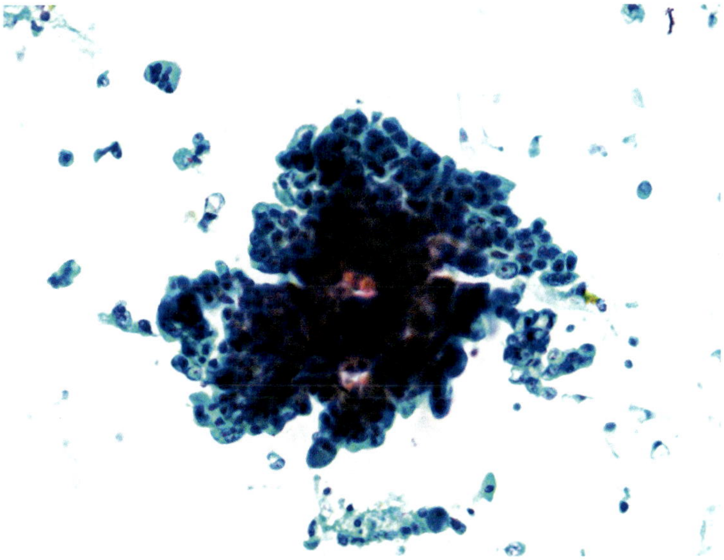

Figure 4.5d — Poorly Differentiated Adenocarcinoma (ADC), Biliary Brushing [Pap stain; medium power]. This specimen consists of an irregular, three-dimensional fragment of atypical epithelial cells. Although a few rare cells have vacuolated, possibly mucinous, cytoplasm, the majority contain little cytoplasm with an increased nuclear-to-cytoplasmic ratio. The enlarged nuclei are pleomorphic and hyperchromatic, with significant nuclear overlapping. Prominent nucleoli are observed in a subset of cells. Single atypical cells, necrotic debris, and degenerating cells can be seen scattered in the background. A minimal amount of yellow to green staining bile can be seen at 3 o'clock, a common finding in biliary tree brushing specimens.

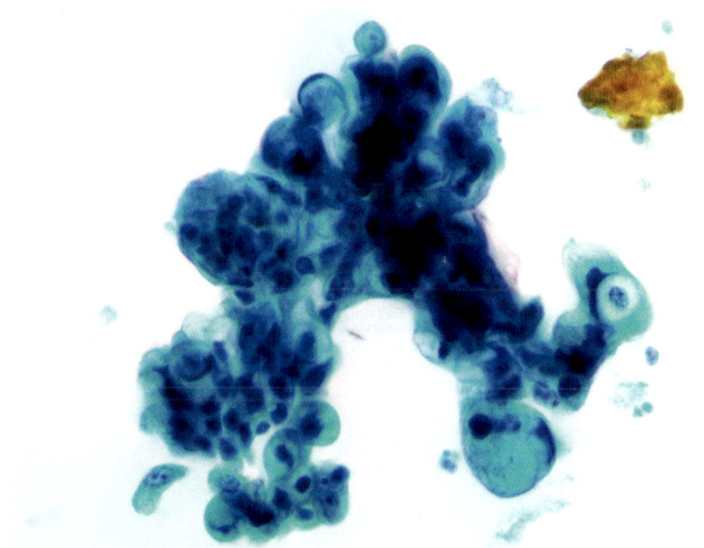

Figure 4.5e — Poorly Differentiated Adenocarcinoma (ADC), Biliary Brushing [Pap stain; high power]. This high-power view shows an irregular, three-dimensional fragment of markedly atypical epithelial cells with varying amounts of mucinous, vacuolated cytoplasm; hyperchromatic and irregular nuclei; and conspicuous nucleoli. Note the presence of a signet ring cell between 11 and 12 o'clock, as well as at 7 o'clock. In addition, a multinucleated cell can be seen at 9 o'clock. This finding is nonspecific but is more commonly observed in highly anaplastic pancreatic ADCs with osteoclastic and pleomorphic giant cells versus cholangiocarcinomas. Finally, chunky yellow to brown bile pigment can be seen at the top right corner. Ultimately, the distinction between cholangiocarcinoma and pancreatic ADC is often made based on clinical features and imaging.

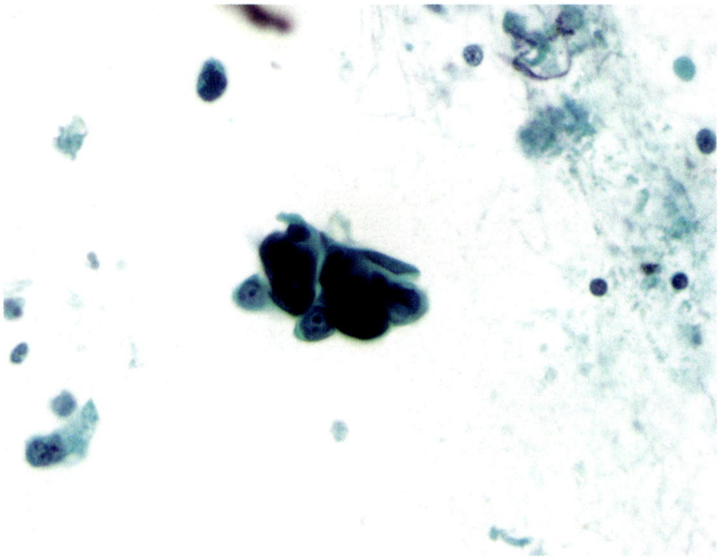

Figure 4.5f — Poorly Differentiated Adenocarcinoma (ADC), Biliary Brushing [Pap stain; high power]. This high-power view shows a small, irregular cluster of overlapping atypical epithelial cells, as well as a few scattered highly atypical single cells with nuclear enlargement, an increased nuclear-to-cytoplasmic ratio, irregular nuclear membranes, and prominent macronucleoli. The single cell at the bottom left corner of the image contains abundant, vacuolated cytoplasm suggestive of mucin. Furthermore, note the presence of necrotic debris and a few scattered lymphocytes. Taken together, these findings favor an invasive, poorly differentiated ADC.

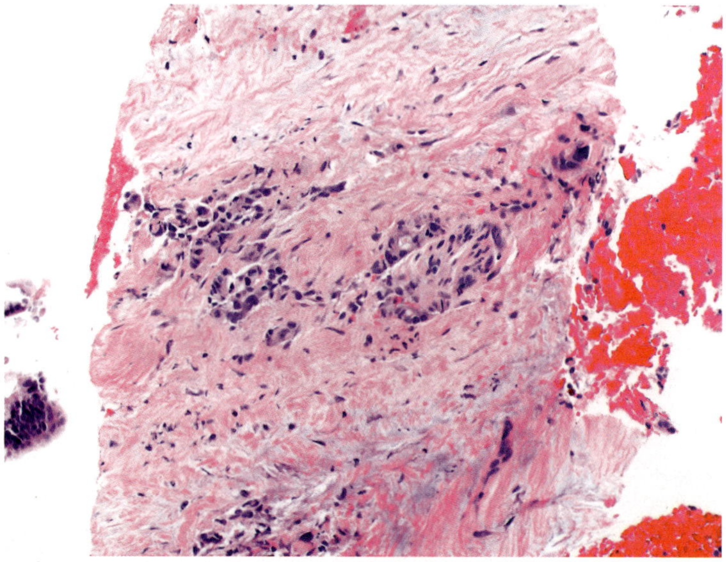

Figure 4.5g — Adenocarcinoma (ADC), Common Bile Duct Biopsy [H&E stain; intermediate power]. Core needle biopsy of this distal common bile duct stricture consists primarily of dense fibrous tissue and blood, but there are small embedded foci of irregular glands. Accurate diagnosis can be challenging, first to even locate the atypical glands within the specimen and second to correctly classify them. The differential diagnosis includes benign biliary ducts and glands, foci of chronic pancreatitis, and infiltrating ADC. At times, it may not be possible to render a definitive diagnosis of ADC, but rather only identify glands that are atypical.

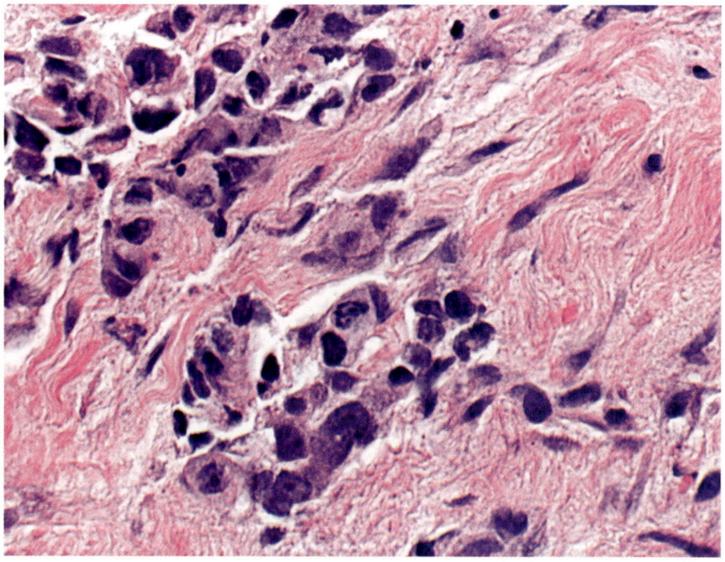

Figure 4.5h — Adenocarcinoma (ADC), Common Bile Duct Biopsy [H&E stain; high power]. Several histologic features aid in the diagnosis of ADC, despite the limited material available. These features include cytologic atypia consisting of a high nuclear-to-cytoplasmic ratio with at least a three-fold nuclear variation between cells, a haphazard growth pattern, the presence of incomplete glandular lumina and intraluminal necrosis, atypical mitotic figures, and the presence of perineural or vascular invasion. This field demonstrates irregular glands with nuclear size variation, nuclear hyperchromasia with prominent nucleoli, and scattered mitotic figures. It is generally not possible to differentiate between pancreatic, biliary, and upper gastrointestinal tract ADCs on the limited material of a core biopsy, but the clinicoradiographic presentation in conjunction with the biopsy results guide patient management.

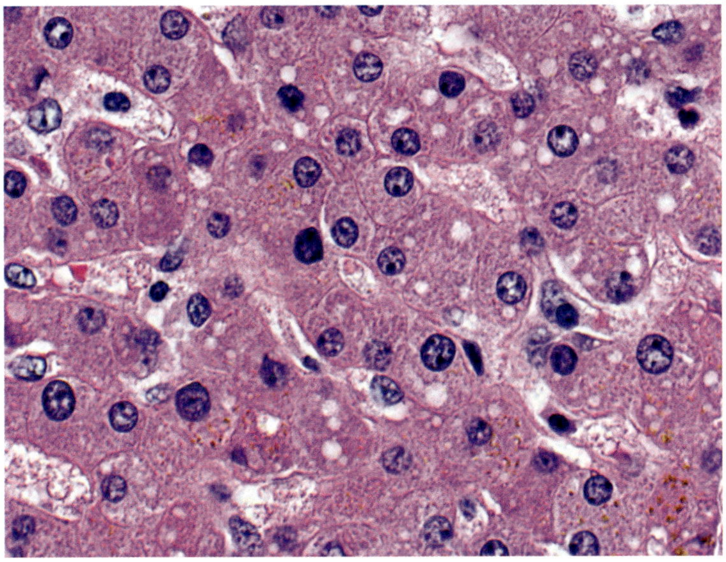

Figure 4.6 — Benign Hepatic Parenchyma, Core Biopsy [H&E stain; intermediate power]. Benign hepatocytes are large polygonal cells arranged in plates that are two to three cells thick and are lined by hepatic sinusoids. Hepatocytes contain fairly uniform, round nuclei with coarse chromatin and punctate nucleoli. Hepatocytes make bile and can display golden-brown cytoplasmic bile pigment, as illustrated here. This example of benign hepatic parenchyma also displays mild microvesicular steatosis (i.e., intracytoplasmic accumulation of lipid). When a liver mass is examined, evaluation of the adjacent and background hepatic parenchyma can provide useful clues to the etiology of the liver mass; for instance, the presence of cirrhosis suggests an underlying diseased liver state and predisposition to primary hepatic malignancies.

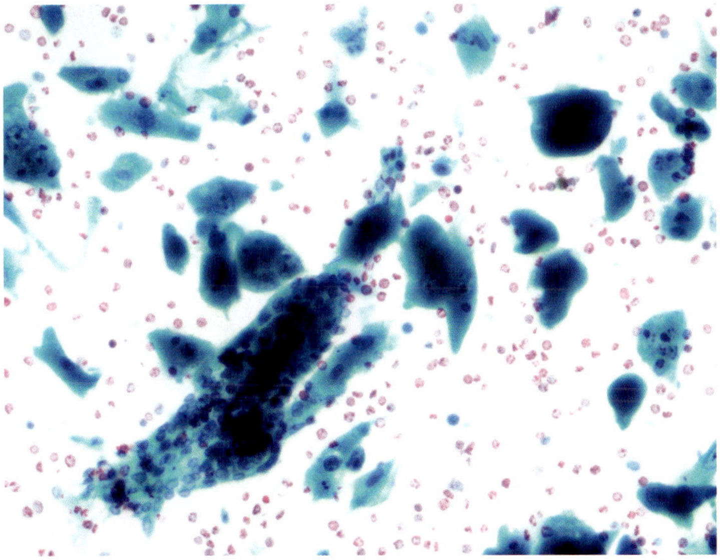

Figure 4.7a — Hepatocellular Carcinoma (HCC), Biliary Brushing [Pap stain; medium power]. Large, scattered, single, atypical hepatocytes are seen with abundant granular cytoplasm and prominent macronucleoli. Several cells show binucleation and the presence of green bile pigment in the cytoplasm, a finding that is indicative of hepatocyte differentiation. It is often difficult to distinguish well-differentiated HCC from normal hepatocytes on cytology. The most helpful features include enlarged nuclei and the presence of prominent macronucleoli. Poorly differentiated HCCs typically consist of malignant cells with little to no resemblance to a normal hepatocyte and often contain numerous enlarged, bare nuclei in the background. Biliary brushings are rarely used to diagnose HCC, with fine needle aspiration (FNA) being the modality of choice. The follow-up in this case revealed a fibrolamellar variant of HCC.

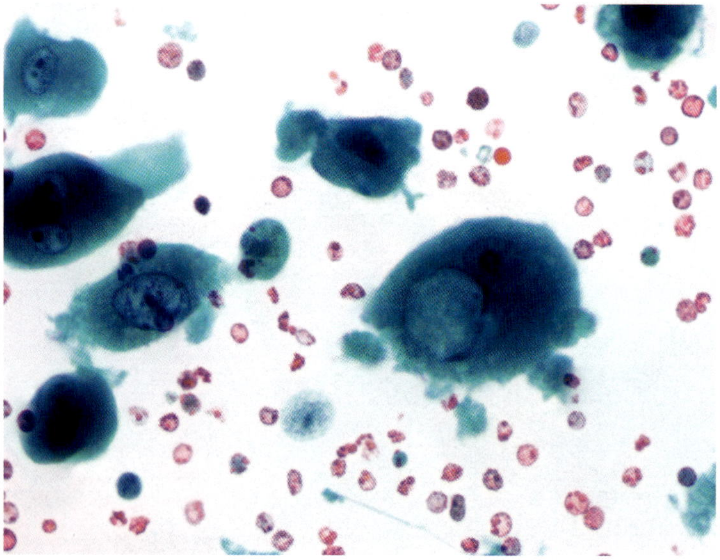

Figure 4.7b — Hepatocellular Carcinoma (HCC), Biliary Brushing [Pap stain; high power]. This image shows a high-power view of several individual large, malignant hepatocytes with vacuolated cytoplasm, increased nuclear-to-cytoplasmic ratio, peripherally condensed chromatin, and prominent and irregular macronucleoli. A binucleated cell is seen at the far left edge of the image. Reactive hepatocytes may show significant size variation and prominent nucleoli; however, they typically retain a normal nuclear-to-cytoplasmic ratio with smooth nuclear membranes and rarely exhibit mitosis. The cell in the middle displays a cytoplasmic pale body, a feature often seen in the fibrolamellar variant. It would also be rare to find reactive and/or normal hepatocytes in a biliary tree brushing; however, ultimately, clinical correlation is required. As many as 40% to 60% of patients experience metastasis, most commonly to the porta hepatis lymph nodes, biliary system, pancreas, and celiac axis.

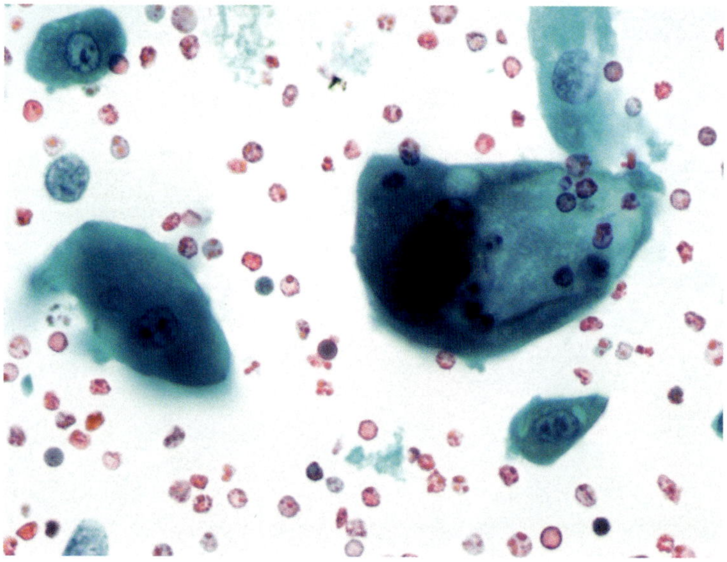

Figure 4.7c — Hepatocellular Carcinoma (HCC), Biliary Brushing [Pap stain; high power]. This image reveals large, malignant epithelial cells with hepatocytic differentiation, including pleomorphism, increased nuclear-to-cytoplasmic ratio, irregular nuclear membranes, and macronucleoli. Furthermore, a cytoplasmic pale body can be seen, as can the presence of intracytoplasmic green bile at 11 o'clock. As many as 70% to 90% of HCCs arise in the setting of cirrhosis, which can make the diagnosis challenging. Clinical correlation, such as the presence of a large mass or a significantly elevated alpha-fetoprotein (AFP) level may be helpful in the diagnosis.

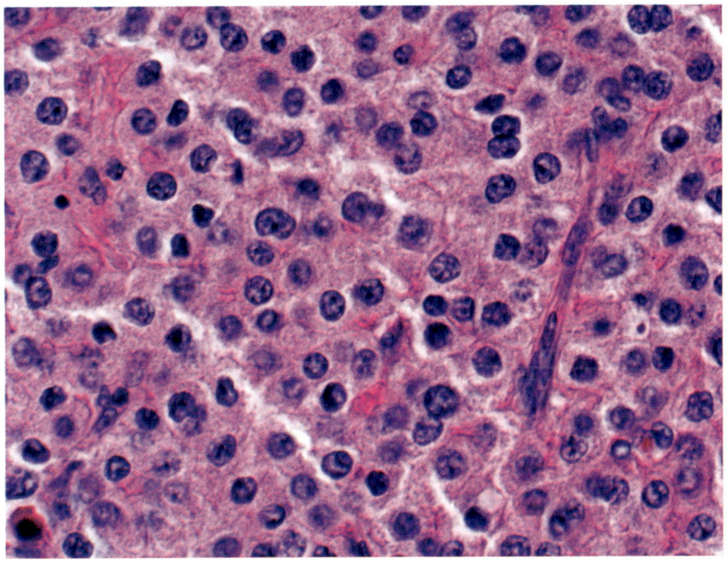

Figure 4.7d — Well-Differentiated Hepatocellular Carcinoma (HCC), Resection [H&E stain; intermediate power]. This field of an HCC is taken at the same magnification as the field of benign hepatic parenchyma in Figure 4.6. The cells in this well-differentiated HCC are *smaller* than the cells in the benign hepatic parenchyma, and indeed they show minimal to no nuclear atypia. However, the cells are forming solid nests and no longer conform to the two-cell-thick hepatic plate architecture of normal liver. Risk factors for the development of hepatocellular carcinoma include hepatitis B or C virus infection, cirrhosis of any cause (including hemochromatosis, alpha 1 antitrypsin deficiency, nonalcoholic fatty liver disease, and alcoholism), and aflatoxin exposure. HCC is in the differential diagnosis of "pink cell tumors" of unknown origin, along with melanoma, adrenocortical carcinoma, renal cell carcinoma, and other rare oncocytic tumors.

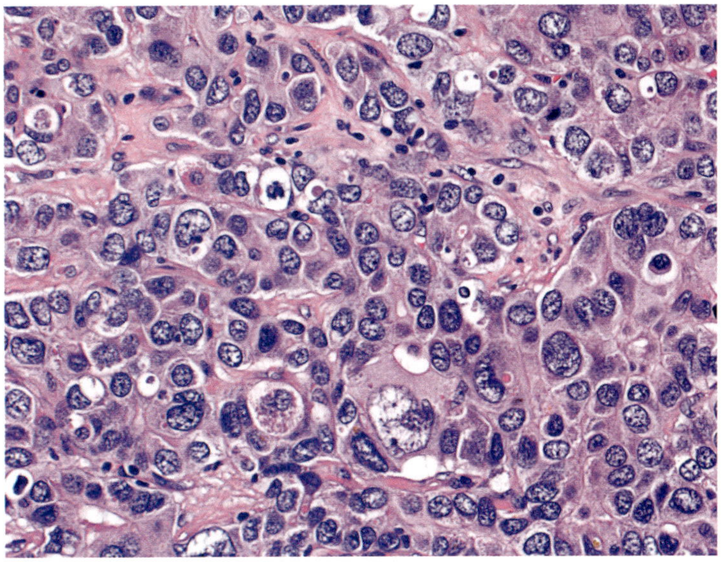

Figure 4.7e — Poorly Differentiated Hepatocellular Carcinoma (HCC), Resection [H&E stain; intermediate power]. This poorly differentiated hepatocellular carcinoma (HCC) displays marked cytologic atypia, with nuclear pleomorphism, high nuclear-to-cytoplasmic ratio, and irregular nuclear membranes. The growth pattern no longer resembles the normal hepatic plates. By immunohistochemistry, HCC cells (and benign hepatocytes) label for glypican 3, arginase-1, Hep Par1, and CAM5.2 (specifically, cytokeratin 18). Polyclonal carcinoembryonic antigen (CEA) labels HCC in a canalicular pattern. Other neoplasms that have hepatoid differentiation, such as hepatoid gastric adenocarcinoma or hepatoid yolk sac carcinoma, can also label for hepatic cell markers, including glypican 3, Hep Par1, and arginase-1; hence caution must be used in interpreting these markers in a given lesion.

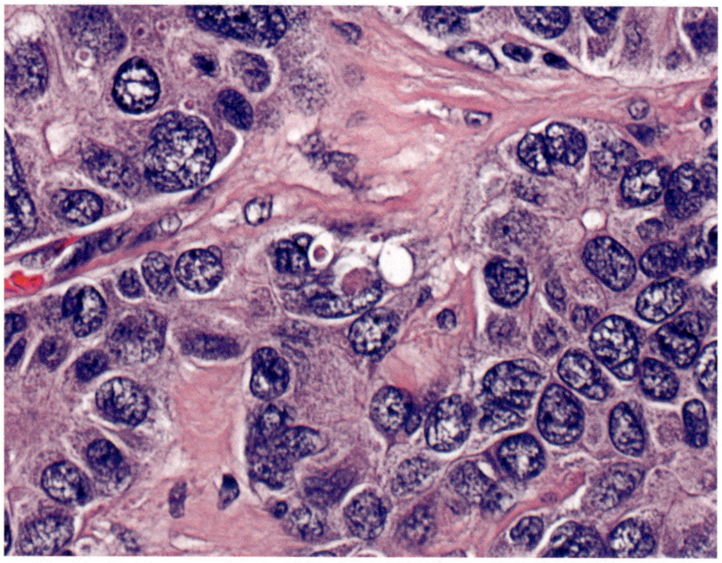

Figure 4.7f — Poorly Differentiated Hepatocellular Carcinoma (HCC), Resection [H&E stain; high power]. High-power examination of this poorly differentiated HCC reveals small, round, pink cytoplasmic inclusions that have a clear halo around them. These intracellular hyaline bodies are accumulations of protein seen in a subset of HCC. HCC can display other intranuclear and cytoplasmic inclusions or pigment. These include cytoplasmic Mallory's hyaline, hepatitis B virus surface antigen, pale bodies of fibrinogen, bile pigment, and fat. In addition to variable cytologic features, HCC can display a range of architectural patterns, including clear cell, giant cell, pelioid (vascular), pseudoglandular, sarcomatoid, solid, and trabecular.

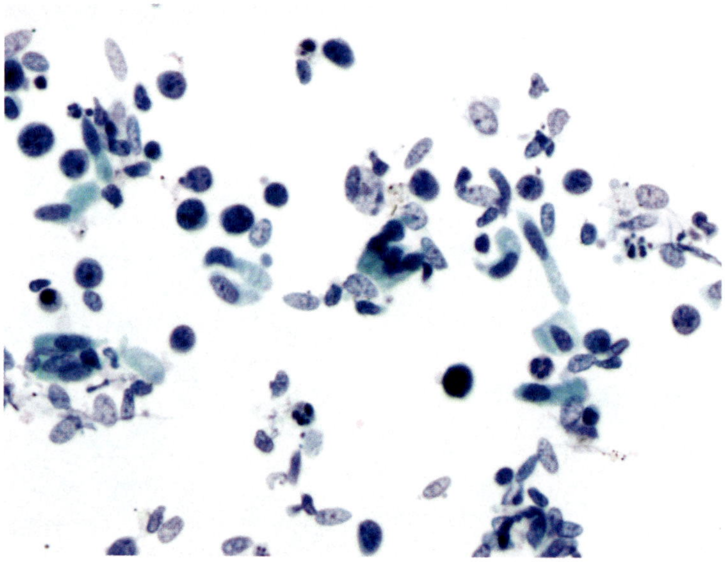

Figure 4.8a — B-Cell Lymphoma, Biliary Brushing [Pap stain; high power]. In this field, a variety of cells can be visualized, including scattered, bland-appearing columnar epithelium and inflammatory cells such as neutrophils, lymphocytes, and plasma cells. In addition, there are large, round atypical cells 2 to 4 times the size of the normal-appearing lymphocytes. These atypical lymphocytes have scant cytoplasm, an increased nuclear-to-cytoplasmic ratio, irregular nuclear membranes, clumpy chromatin, and multiple prominent nucleoli. Cell block studies in this case showed this abnormal population of cells to be strongly positive for CD20 and completely negative for CD3, establishing the diagnosis of a B-cell lymphoma.

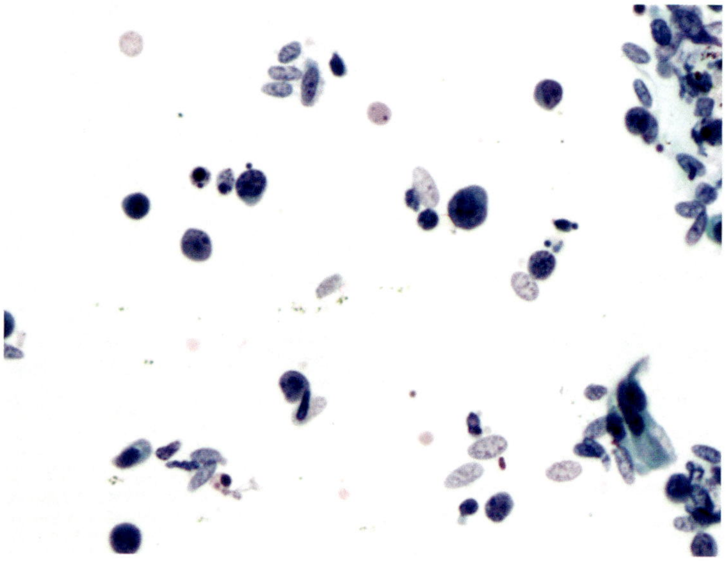

Figure 4.8b — B-Cell Lymphoma, Biliary Brushing [Pap stain; high power]. This high-power view shows scattered, bland-appearing columnar epithelium at the bottom right corner and a prominent population of large, round, atypical lymphocytes with an increased nuclear-to-cytoplasmic ratio, irregular nuclear membranes, clumpy chromatin, and multiple prominent nucleoli. Apoptotic bodies can be seen at 10 o'clock. This level of atypia is suggestive of a high-grade malignant lymphoma. The main differential for lymphoma in the gastrointestinal tract is between an extranodal marginal zone lymphoma (MALToma) versus a diffuse large B-cell lymphoma (DLBCL). MALTomas are typically composed of small B-lymphocytes, are low grade, and are usually not seen eroding into distant sites.

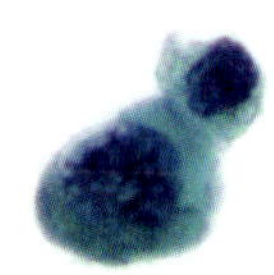

Figure 4.9 — Metastatic Melanoma, Biliary Brushing [Pap stain; high power]. In this high-power image, two highly atypical cells with an increased nuclear-to-cytoplasmic ratio, irregular nuclear membranes, and hyperchromatic and irregular chromatin are seen. These two cells in isolation are concerning for a poorly differentiated malignant neoplasm; however, no distinctive cytologic features suggesting a tissue of origin are present. It is best to approach such as case with a broad differential, including lymphoma, carcinoma, and melanoma, to name a few. Melanomas may show epithelioid, small cell, or even spindle cell morphologies. In addition, melanomas may contain cytoplasmic pigment and can show a variety of nuclear patterns. As a result, making a definitive diagnosis of metastatic melanoma is often challenging without a cell block to perform immunohistochemical stains, such as S-100 protein, SOX-10, Melan-A, and HMB-45.

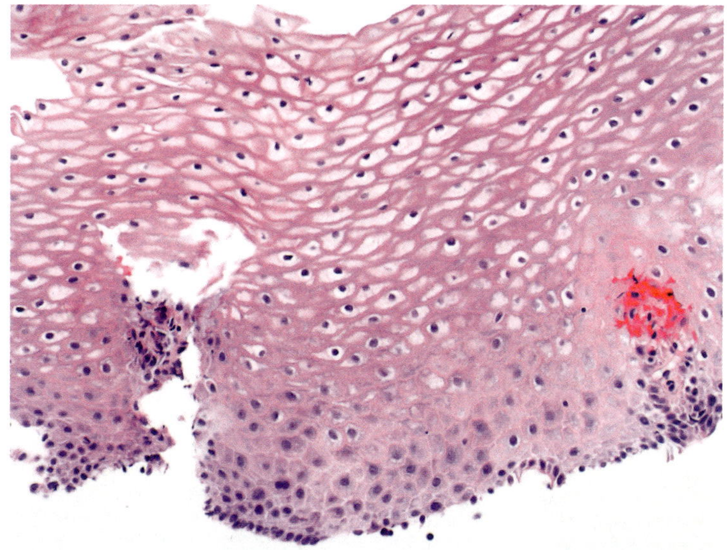

Figure 4.10 — Normal Mucosa, Esophageal Biopsy [H&E stain; intermediate power]. Normal esophageal mucosa consists of stratified squamous epithelium in which the squamous cells mature from a basal cell layer through a spiny cell layer. Normal esophageal mucosa is nonkeratinizing but can keratinize under conditions of irritation. The papillae are typically less than a third of the thickness of the squamous epithelium, but extension into the upper third, along with prominent eosinophils, are common features in gastrointestinal reflux disease.

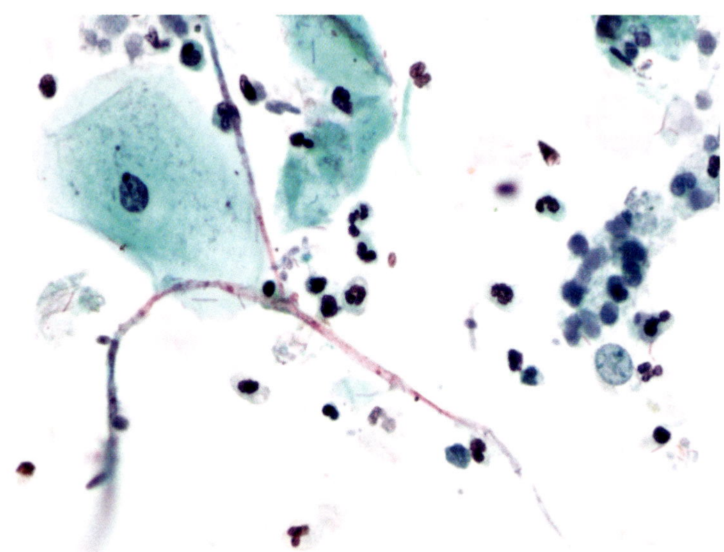

Figure 4.11a — Candidiasis, Esophageal Brushing [Pap stain; high power]. This high-power view shows two squamous epithelial cells and multiple inflammatory cells composed predominately of neutrophils. In the center of field, there is a fungal organism forming hyphae, along with several small adjacent yeast forms. This bimorphic fungal organism is most consistent with *Candida*. Candidal esophagitis is most commonly due to *Candida albicans* and is typically associated with a neutrophilic inflammatory response. The most common endoscopic findings include characteristic white plaques and ulcerated mucosa where the plaques have been disrupted or scraped away. Most patients are significantly immunocompromised and require antifungal therapy.

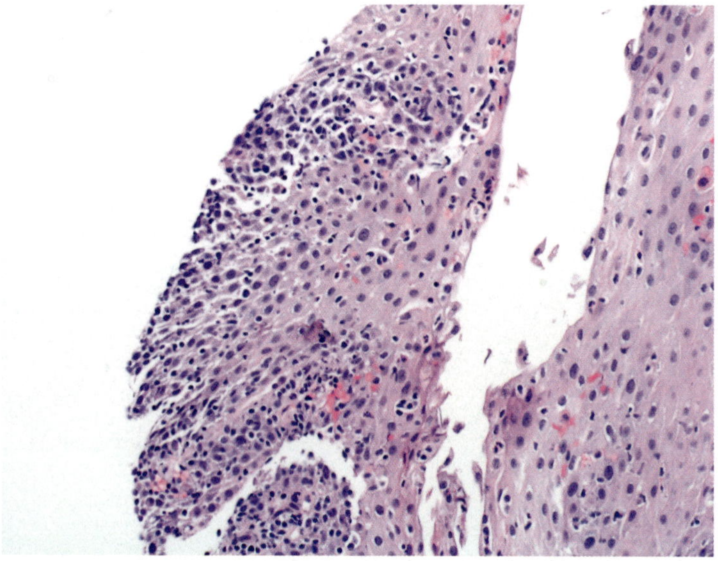

Figure 4.11b — Candidiasis, Esophageal Biopsy [H&E stain; low-intermediate power]. This esophageal biopsy shows erosive esophagitis with intraepithelial acute and chronic inflammatory and reactive squamous epithelial changes; elsewhere in the biopsy is evidence of erosion. The presence of acute inflammation in the esophagus should raise suspicion for an infectious process such as fungal or viral infection. *Candida* frequently colonizes the oral cavity, and yeast forms alone are not indicative of an infection. The presence of superficial, mature squamous cells with bacterial organisms or *Candida* yeast forms in an upper gastrointestinal tract endoscopic biopsy likely represents oral cavity contamination. However, the presence of fungal hyphal forms suggests infection.

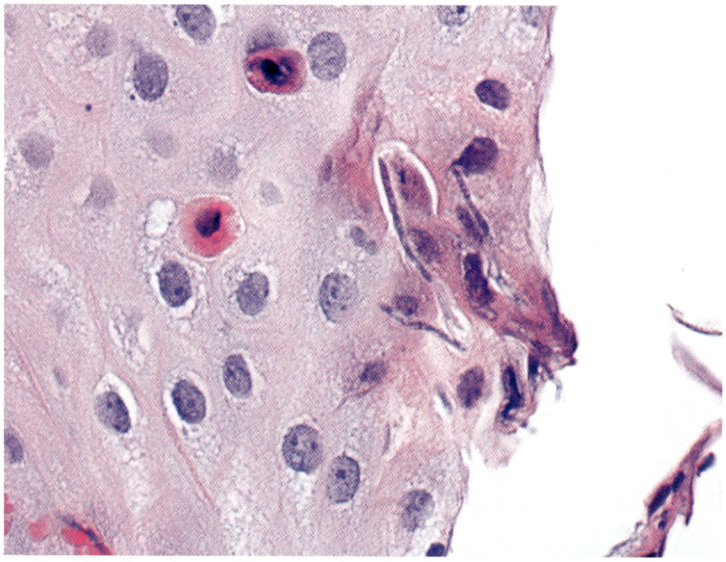

Figure 4.11c — Candidiasis, Esophageal Biopsy [H&E stain; high power]. High-power examination of the superficial layer of squamous epithelium reveals hyphal forms diagnostic of candidal esophagitis. The hyphal forms are basophilic, slender, and nonbranching. The presence of keratin flakes at the superficial aspect of the epithelium is often seen in association with candidiasis and should prompt examination for the hyphal forms. The fungal hyphae are not always readily visible on H&E stain, and a low index of suspicion for adding a special stain such as periodic acid–Schiff (PAS) is warranted in this setting.

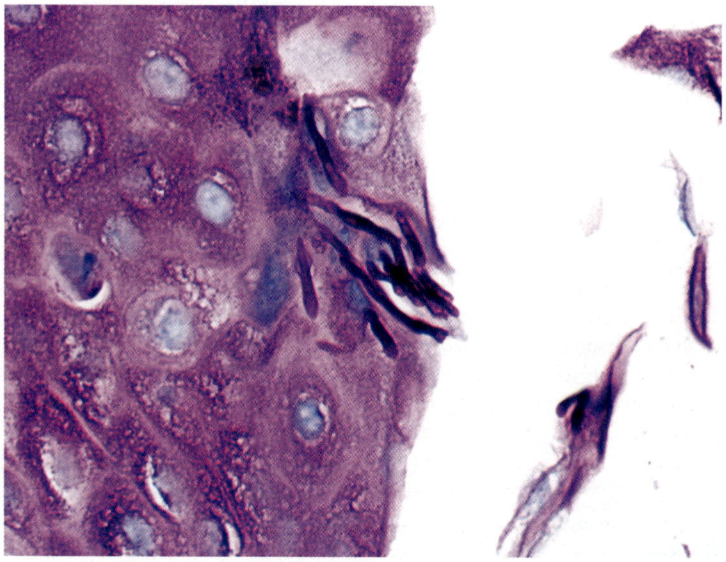

Figure 4.11d — Candidiasis, Esophageal Biopsy [Periodic acid–Schiff (PAS) stain; high power]. The PAS special stain highlights *Candida* hyphal forms as bright magenta. The use of special stains can be very helpful in identifying focal microorganisms. Although both the yeast forms (small ovoid structures) and the hyphal forms can be highlighted by PAS, only the hyphal forms are indicative of an invasive pathogenic process. Candidiasis can affect both immunocompromised and immunocompetent individuals.

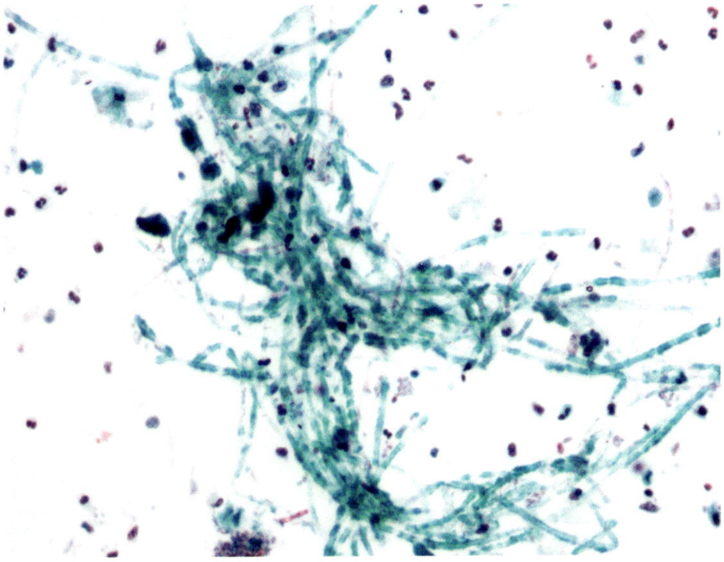

Figure 4.12 — *Aspergillus* Esophagitis, Esophageal Brushing [Pap stain; medium power]. This specimen consists of acutely branched, septate hyphae in a background of scattered neutrophils and rare lymphocytes, a finding that favors an invasive fungal infection. Based on the 45° angle branching and the septate hyphae that are 3 to 6 µm in diameter, these findings are compatible with *Aspergillus. Aspergillus* infection is rare in the esophagus and gastrointestinal tract. When it occurs, it is usually seen in the context of neutropenia, high-dose steroids, or prior mucosal injury. Although not well developed in this case, terminal swelling can be observed in oxygenated locations and mimics yeast forms. In addition, fruiting bodies (vesicles with conidia) may be seen in highly oxygenated areas, typically the paranasal sinuses and sometimes the lung.

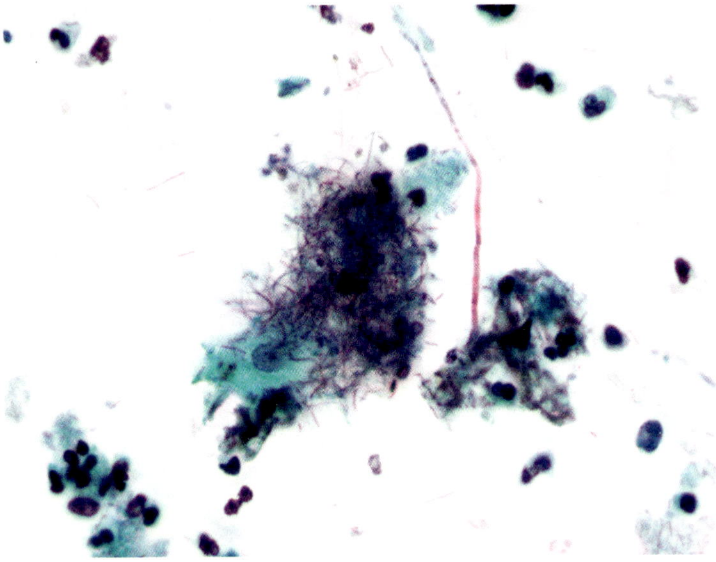

Figure 4.13 — *Actinomyces*, Esophageal Brushing [Pap stain; high power]. In the center of the field, a few squamous epithelial cells are seen with overlying clumps of thin, filamentous bacterial rods. Neutrophils are seen overlying the squamous cells, as well as scattered in the background. *Actinomyces* thrives in anaerobic and microaerophilic environments and is often a commensal organism in the oropharynx and gastrointestinal and genitourinary tracts. As a result, this organism is incidentally seen in multiple sites and preparations. In immunocompromised hosts, *Actinomcyes* can cause acute and chronic inflammation and form granulomas. Infections can be serious, with tissue necrosis and sinus tract formation. Therefore it is important recognize this organism and determine whether it is pathologic.

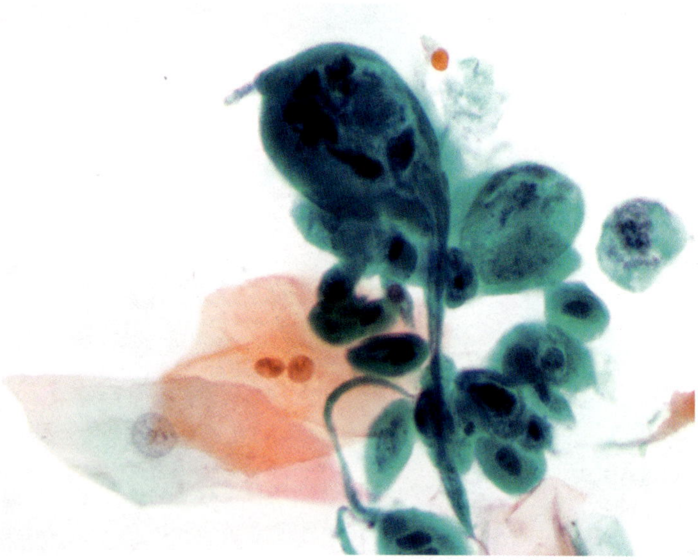

Figure 4.14a — Herpes Esophagitis, Esophageal Brushing [Pap stain; high power]. This reveals several squamous epithelial cells displaying significant viral cytopathic effect. For example, the large cell at top displays multinucleation and a hint of nuclear molding. At bottom right, there is a single atypical cell with a dense nucleus and well-developed perinuclear halo, as seen with a Cowdry type A inclusion. Although these cells are large, there is abundant cytoplasm. In this case, the cytopathic effect is consistent with a herpesvirus family. Microbiologic studies confirmed the virus to be herpes simplex virus (HSV). Clinically, most patients are immunocompromised and present with odynophagia or dysphagia and nausea and vomiting. Endoscopy typically reveals small, exudative ulcers in the middle and distal esophagus.

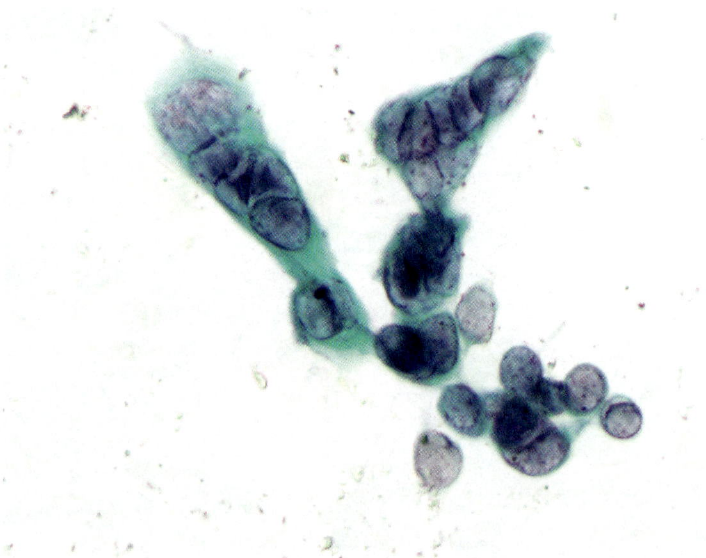

Figure 4.14b — Herpes Esophagitis, Esophageal Brushing [Pap stain; high power]. In this specimen, the classic features of herpes simplex virus (HSV) viral cytopathic effect are seen. These findings include the three Ms: **M**ultinucleation, nuclear **M**olding, and **M**argination of the chromatin. In addition, individual single cells are seen with ground-glass nuclear change and margination—sometimes referred to as Mononuclear cells (the fourth M). The presence of these viral cytopathic features is diagnostic of HSV. Viral culture, in this case, confirmed the diagnosis. In the modern era, molecular testing with polymerase chain reaction (PCR) has largely supplanted viral culture as the gold standard of diagnosis. In tissue or on cell block material, HSV immunohistochemistry is available and can aid in the diagnosis.

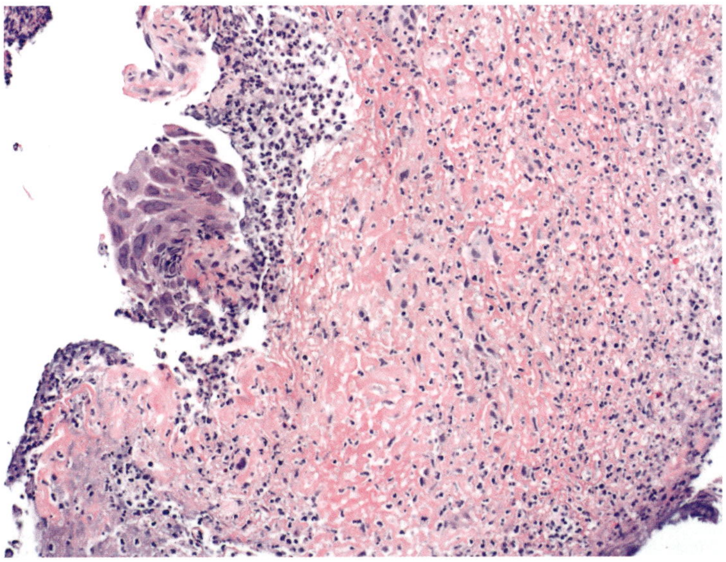

Figure 4.14c — Herpes Esophagitis, Esophageal Biopsy [H&E stain; intermediate power]. This endoscopic biopsy of the distal esophagus reveals extensive erosion and ulceration with fibrinopurulent exudate. There is only focal residual intact squamous mucosa, and from this power, the nuclei appear enlarged. The differential diagnosis of ulceration in an endoscopic biopsy includes infection for a variety of microorganisms (viral, bacterial, or fungal), neoplasia, and reactive ulceration, such as in gastroesophageal reflux disease or due to medication such as iron pills. Higher-power examination with targeted special stains or immunostains is necessary to differentiate these processes.

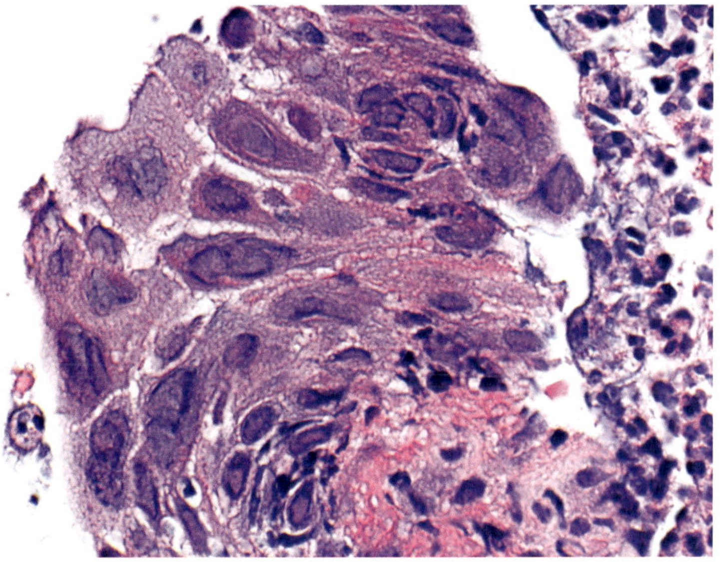

Figure 4.14d — Herpes Esophagitis, Esophageal Biopsy [H&E stain; high power]. The squamous epithelium in this field displays nuclear enlargement, nuclear molding, and a glassy appearance of the nuclei. The findings are characteristic of herpesvirus infection. The most common causes of infectious esophagitis are *Candida*, herpes simplex virus (HSV), and cytomegalovirus (CMV), and herpes esophagitis most often affects immunocompromised individuals. The ground-glass inclusions, nuclear molding, and multinucleation are characteristic features. Cowdry A inclusions, which are intranuclear and cytoplasmic inclusions of dense eosinophilic material with a peripheral halo, may be observed. The viral cytopathic changes and inclusions are more likely to be present in biopsies taken from the periphery of the gross ulcer, rather than from the center. An immunostain for HSV can confirm the diagnosis.

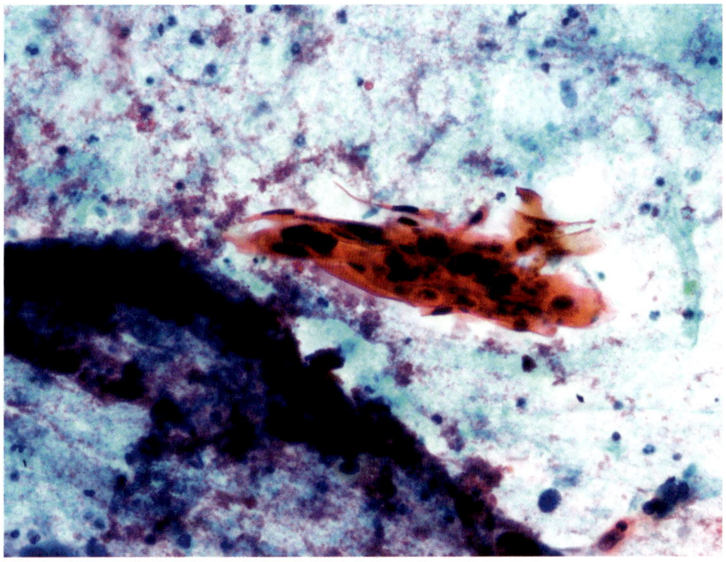

Figure 4.15a — Keratinizing Squamous Cell Carcinoma (SqCC), Esophageal Brushing [Pap stain; medium power]. This specimen consists of malignant epithelial cells with dense orangeophilic cytoplasm, occasional refractile cytoplasmic perinuclear rings, and a tadpole cell at 12 o'clock. The nuclei are hyperchromatic, angular, enlarged, and variable in shape (pleomorphic) and size (anisonucleosis). This fragment of cells displays a level of atypia far beyond that seen in squamous metaplasia and is present in the context of a dirty background, a feature that favors an invasive process. Patients with esophageal SqCC are generally treated with a combination of neoadjuvant chemotherapy and radiation therapy after esophagectomy.

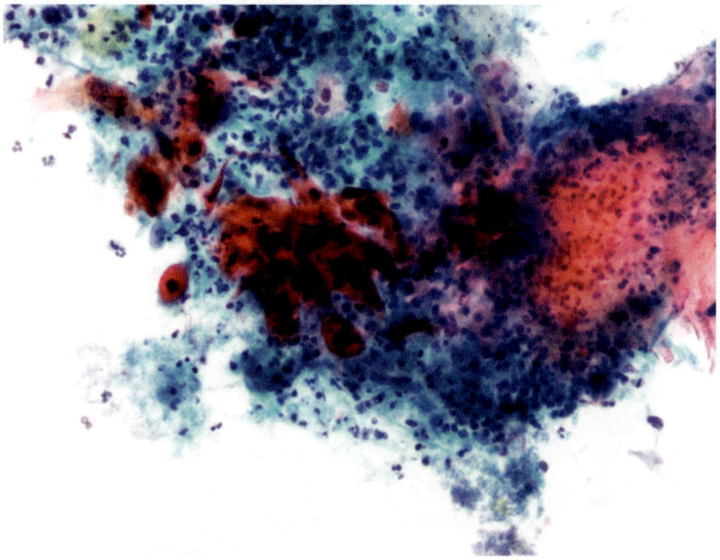

Figure 4.15b — Keratinizing Squamous Cell Carcinoma (SqCC), Esophageal Brushing [Pap stain; medium power]. This specimen shows scattered bizarre keratinizing epithelial cells with hyperchromatic and enlarged nuclei with irregular nuclear contours. The fragment of atypical squamous cells with orangeophilic cytoplasm in the center of the field contains a significant amount of nuclear debris. In addition, the background is highly necrotic with a robust acute and chronic inflammatory infiltrate. If a patient has a history of prior radiation therapy, necrosis may be encountered; however, with radiation change, any nuclear enlargement should be accompanied by an increase in cytoplasm (normal nuclear-to-cytoplasmic ratio). In addition, the level of atypia and abnormal keratinization seen in this case indicate a malignant process.

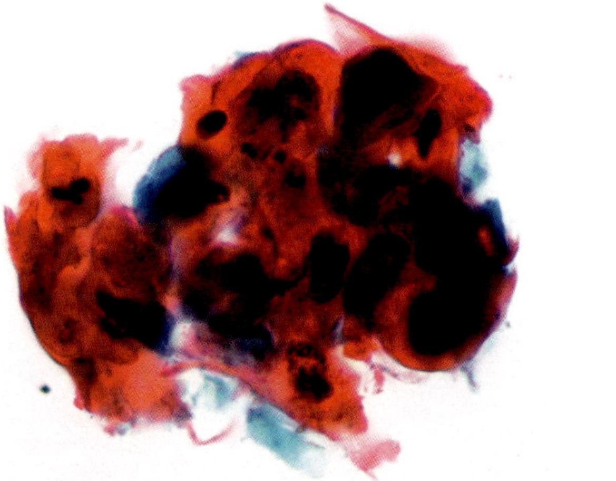

Figure 4.15c — Keratinizing Squamous Cell Carcinoma (SqCC), Esophageal Brushing [Pap stain; high power]. This field contains a disorganized cluster of brightly orangeophilic, keratinizing squamous epithelial cells with enlarged and bizarre hyperchromatic nuclei with irregular nuclear contours and abnormal chromatin. Refractile perinuclear cytoplasmic rings can be seen in several cells. Although not entirely diagnostic of invasion, the severity of the atypia and pleomorphism present in this case favors an invasive process over high-grade dysplasia. Major risk factors for esophageal SqCC include tobacco use and heavy alcohol consumption. In contrast, the major risk factors for esophageal adenocarcinoma (ADC) include esophageal reflux and corresponding obesity. As a result, in the United States, the incidence of esophageal SqCC has been declining, whereas the incidence of esophageal ADC has been climbing.

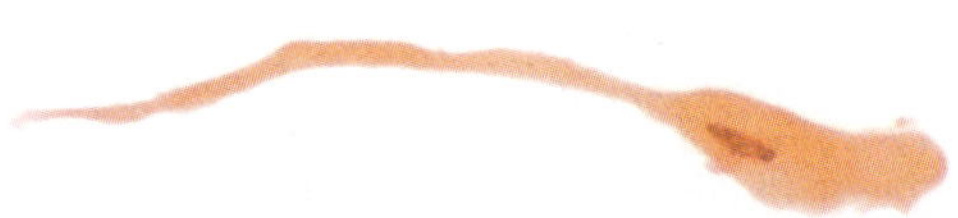

Figure 4.16a — Squamous Cell Carcinoma (SqCC), Esophageal Brushing [Pap stain; high power]. This high-power view shows a single, highly atypical keratinizing squamous cell with an irregular, elongated tail of orangeophilic cytoplasm. The nucleus is eccentrically placed at the broad-based end. Although this nucleus lacks hyperchromasia and appears relatively normal in size, the bizarre cytoplasmic tail and atypia is consistent with a cell type called a *tadpole cell,* which is indicative of SqCC in Pap-stained cytologic preparations. These bizarre cells are often seen in the context of other atypical cells and tumor diathesis.

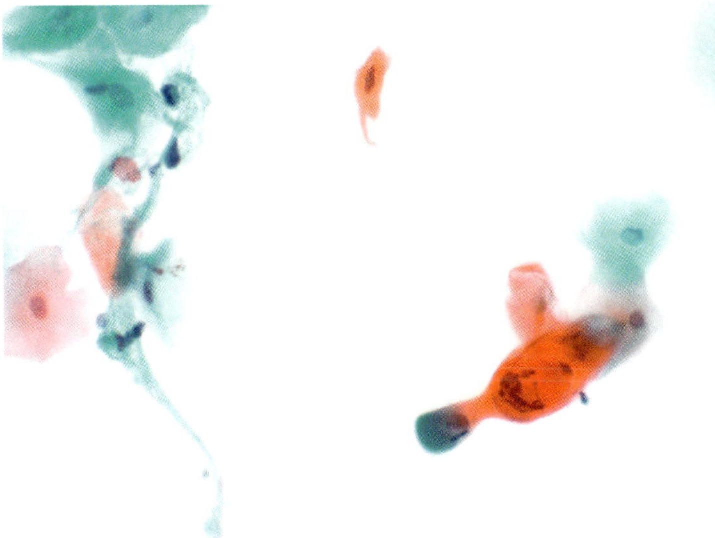

Figure 4.16b — Squamous Cell Carcinoma (SqCC), Esophageal Brushing [Pap stain; high power]. In this image, two atypical squamous cells are present at 12 and 4 o'clock. Both cells have dense, orangeophilic cytoplasm with irregular features. The cell at 12 o'clock has a small cytoplasmic tail and an increased nuclear-to-cytoplasmic ratio with irregular nuclear borders and hyperchromasia. The cell at 4 o'clock is enlarged with polygonal cytoplasm and highly irregular chromatin. Although not present in this case, well-differentiated SqCCs sometimes have Herxheimer spirals, which are strands of twisted keratin that indicate atypical squamous differentiation.

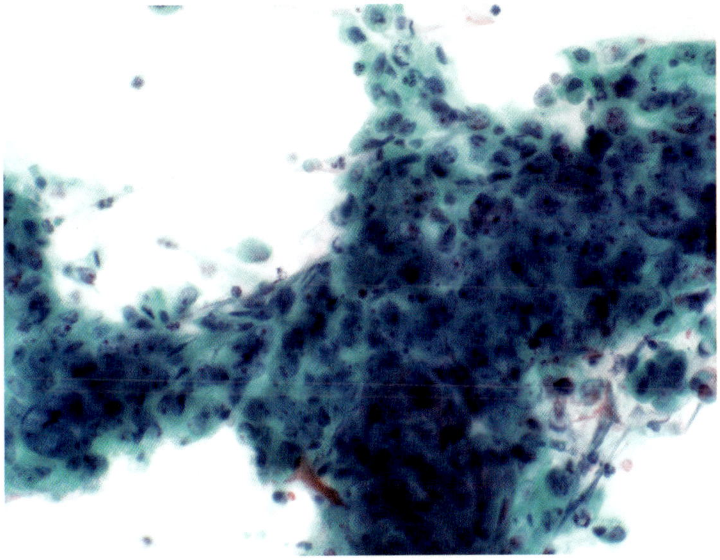

Figure 4.16c — Poorly Differentiated Squamous Cell Carcinoma (SqCC), Esophageal Brushing [Pap stain; high power]. This high-power view shows a large, three-dimensional fragment of highly atypical cells with considerable nuclear overlap and occasional keratin debris. The cells have enlarged nuclei with significant nuclear size variation. The nuclear-to-cytoplasmic ratio is increased, the nuclear membranes are irregular, and the chromatin is coarse and chunky. Prominent nucleoli are seen, and a significant amount of nuclear debris and apoptotic bodies are scattered throughout. In comparison to overtly keratinizing SqCC, the nuclear-to-cytoplasmic ratio of minimally keratinizing and nonkeratinizing SqCC is significantly greater. Reaching a diagnosis of carcinoma is not difficult in this case; however, distinguishing this specimen from an adenocarcinoma may not be possible. The keratin fragments in this case, however, support a diagnosis of SqCC.

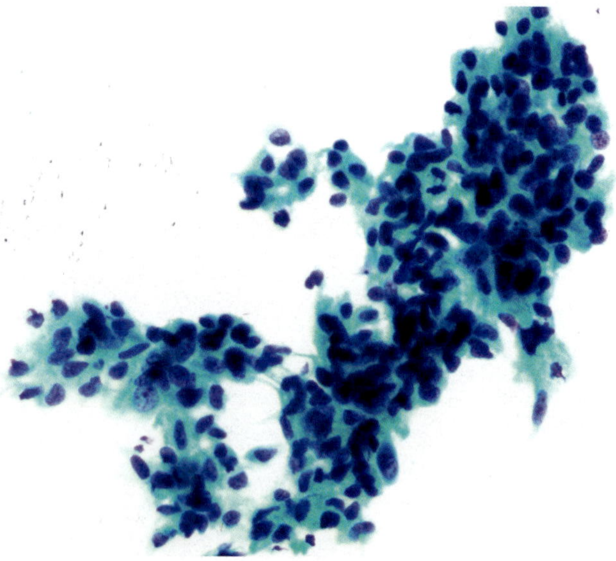

Figure 4.16d — Poorly Differentiated Squamous Cell Carcinoma (SqCC), Esophageal Brushing [Pap stain; medium power]. This view shows a three-dimensional disorganized sheet of mitotically active pleomorphic epithelial cells with a high nuclear-to-cytoplasmic ratio, irregular nuclear contours, and coarse to hyperchromatic chromatin. Although the cytoplasm is dense in some areas, there is a notable lack of overt orangeophilic keratinization, as well as a notable lack of glandular/mucinous differentiation. The main differential diagnosis for cases with this degree of atypia and lack of clear differentiation is between a poorly differentiated SqCC and a poorly differentiated adenocarcinoma (ADC). As a result, it is often necessary to perform immunostains on cell block material; p40 is positive in SqCC and negative in ADC. Furthermore, the location of the sampled area can be helpful, as SqCC is typically found in the middle and upper third of the esophagus versus ADC, which is typically at the lower third of the esophagus, including the gastroesophageal junction.

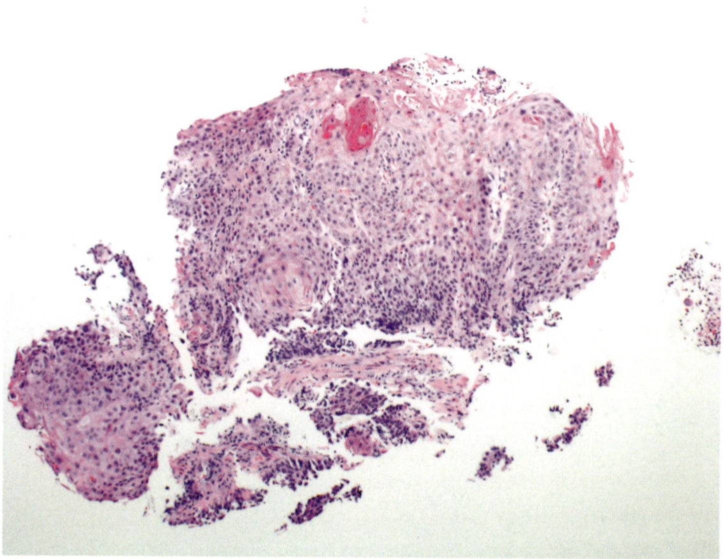

Figure 4.17a — Squamous Cell Carcinoma (SqCC), Esophageal Biopsy [H&E stain; low power]. Low-power examination of this midesophageal endoscopic biopsy shows a thickened squamous epithelium with atypia. The nuclei are hyperchromatic and enlarged, there is paradoxical maturation at the base, and there is superficial ulceration. From this low-power view, the differential diagnosis includes reactive changes and SqCC in situ. Esophageal SqCCs typically involve the middle third of the esophagus and, unlike esophageal adenocarcinoma, are unrelated to Barrett's esophagus.

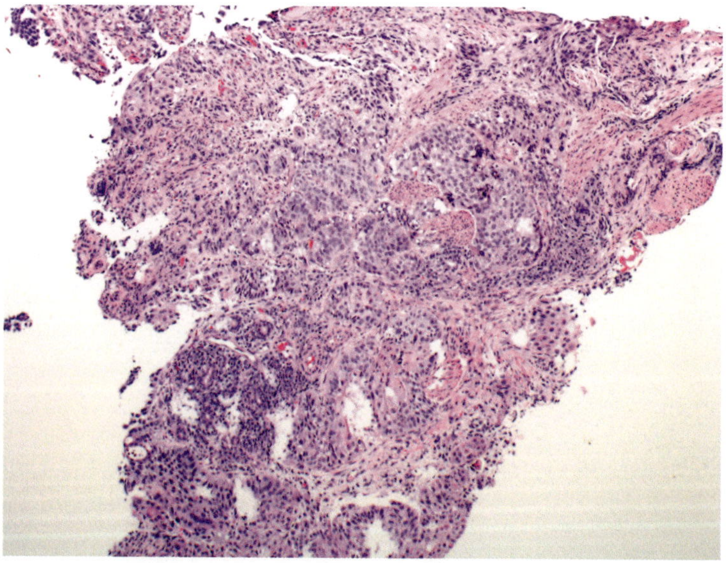

Figure 4.17b — Squamous Cell Carcinoma (SqCC), Esophageal Biopsy [H&E stain; low power]. A second fragment of tissue from the same endoscopic esophageal biopsy reveals nests of squamous cells with irregular, infiltrative growth into smooth muscle bundles. The findings are diagnostic of an invasive SqCC. Esophageal carcinomas can access the lymphatic system even with invasion at the level of only the lamina propria (intramucosal). Risk factors for the development of SqCC include tobacco use, alcohol use, consuming scalding beverages, and ingestion of nitrates and nitrosamines in fermented foods. In the United States, esophageal adenocarcinoma is more common that esophageal SqCC; however, world-wide, SqCC predominates.

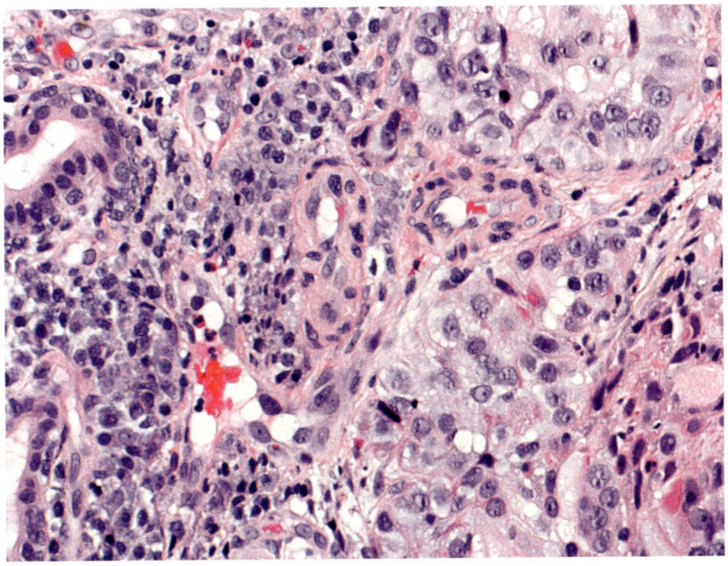

Figure 4.17c — Squamous Cell Carcinoma (SqCC), Esophageal Biopsy [H&E stain; intermediate power]. This field demonstrates infiltrates nests of malignant cells (right half of field) adjacent to benign capillaries, benign inflammatory cells, and benign esophageal submucosal glands and ducts (left half of field). The malignant cells display a high nuclear-to-cytoplasmic ratio, conspicuous nucleoli, pale pink cytoplasm with foci of bright pink keratinization, and readily apparent mitotic figures. In contrast, the benign ducts display uniform cuboidal cells with bland, uniform nuclei. The presence of the submucosal glands should not be misinterpreted as a component of adenocarcinoma or adenosquamous carcinoma.

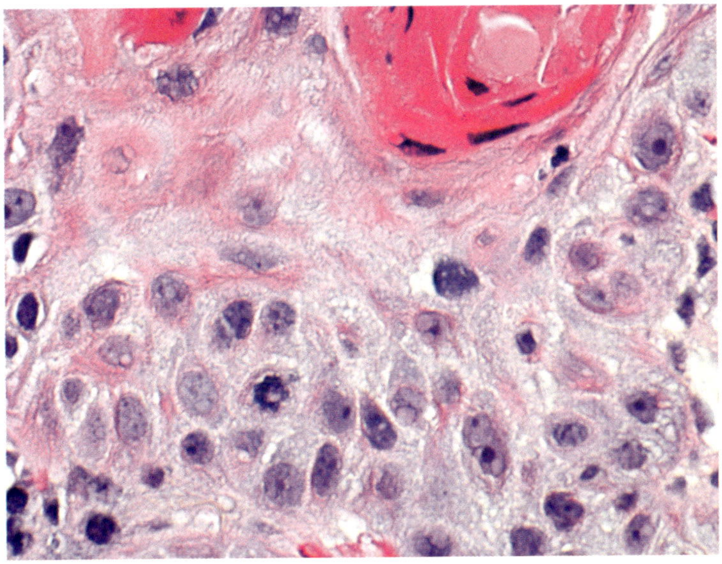

Figure 4.17d — Squamous Cell Carcinoma (SqCC), Esophageal Biopsy [H&E stain; high power]. SqCCs of any site, including this SqCC in situ component in the esophagus, display similar morphologic features. As seen here, the cells display intercellular bridges (desmosomes), glassy pink cytoplasm, and foci of keratinization. SqCCs are immunoreactive for high-molecular-weight cytokeratins, p63, and p40. In the absence of an association with high-risk human papillomavirus (HPV), as seen in head and neck, anal, or gynecologic SqCCs, no immunostains or special stains can differentiate the site of origin of an SqCC.

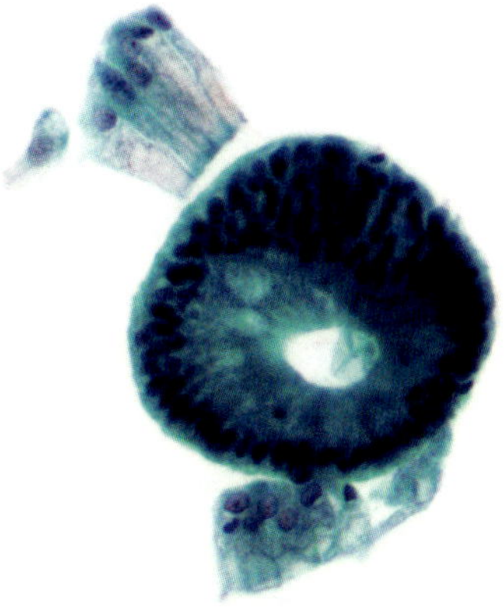

Figure 4.18a — Barrett's Esophagus, Esophageal Brushing [Pap stain; medium power]. In this field, an intact, well-formed glandular structure and two smaller fragments of intact columnar epithelium are present. The nuclei are basally oriented, and the cytoplasm is vacuolated. Several mucin-filled goblet cells are easily identified from 8 to 10 o'clock in the gland. These findings represent intestinal metaplasia without atypia. Barrett's esophagus is a clinicopathologic diagnosis. The biopsy must be taken near the gastroesophageal junction in the tubular esophagus in an area of salmon-colored mucosa. In addition, the biopsy must show intestinal metaplasia with goblet cells present. To be a true goblet cell, the mucin vacuoles must be at least three times the size of the nucleus of the cell.

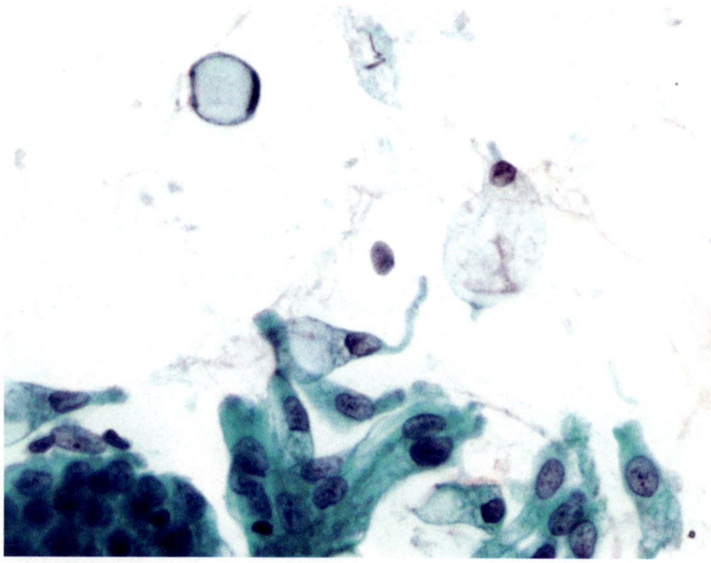

Figure 4.18b — Barrett's Esophagus, Esophageal Brushing [Pap stain; high power]. This high-power view shows columnar epithelium with basally oriented nuclei and foamy, apical mucin vacuoles. The nuclei are round to oval, and the nuclear membranes are smooth. The chromatin is open without conspicuous nucleoli. A single goblet cell, identified at the top of the image, has abundant mucin and a compressed nucleus, resembling a signet ring cell. Note the absence of nuclear overlap, loss of polarity, and irregular or enlarged nuclei. Although the presence of this goblet cell may raise concern for a signet ring cell carcinoma, the overall features of this specimen are more consistent with a benign, metaplastic process. This specimen was taken from the tubular esophagus and was clinically consistent with Barrett's esophagus. The risk of progression of Barrett's esophagus without dysplasia or with low-grade dysplasia to esophageal adenocarcinoma (ADC) is increased but remains low; however, for patients with high-grade dysplasia, the risk of ADC is approximately 4% to 8% per year and approximately 30% overall.

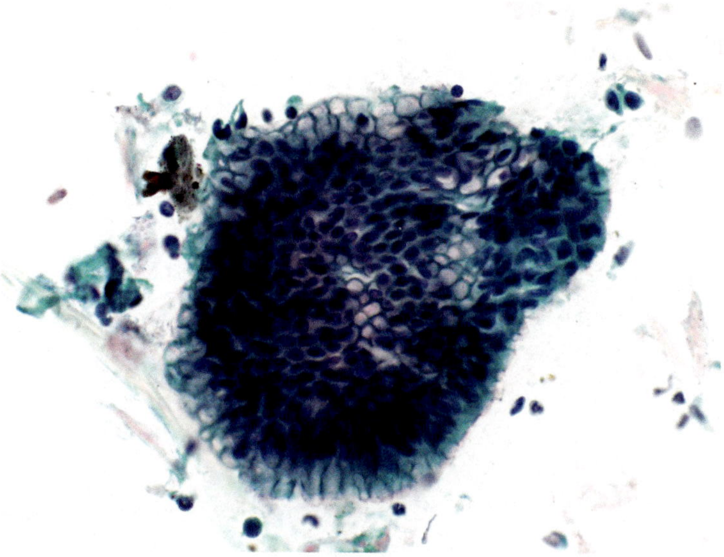

Figure 4.18c — Barrett's Esophagus, Esophageal Brushing [Pap stain; medium power]. This image reveals a large, flat, cohesive sheet of columnar cells with a typical honeycomb architecture. Seen best at the edges of the cluster, the nuclei are basally oriented, are uniform, contain delicate mucinous cytoplasm, and are reminiscent of intestinal-type epithelium. Occasionally, goblet cells are seen scattered throughout this fragment with mucin vacuoles larger than three times the size of the nucleus. Distinguishing reactive atypia from low-grade dysplasia in cytology brushings is not possible in Barrett's esophagus. High-grade dysplasia, however, shows a loss of normal polarity, decreased cytoplasmic mucin, elongated irregular nuclei, and an increased nuclear-to-cytoplasmic ratio.

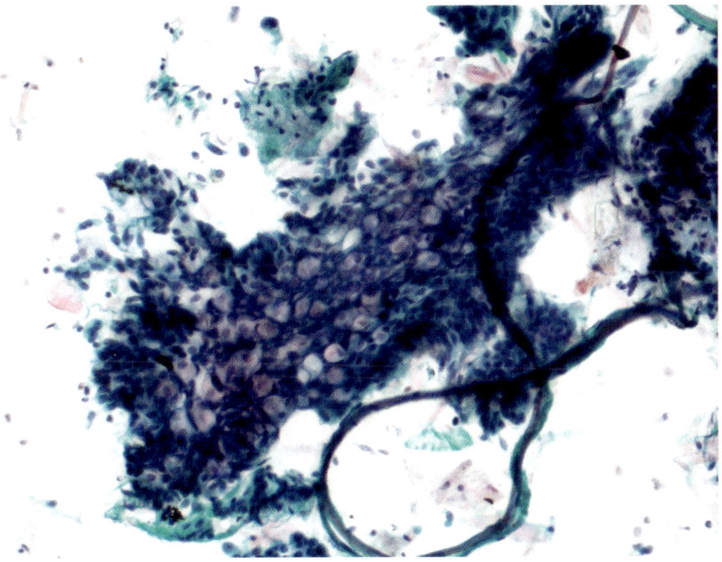

Figure 4.18d — Barrett's Esophagus, Esophageal Brushing [Pap stain; medium power]. This image shows a large sheet of flat columnar epithelium punctuated by goblet cells with large cytoplasmic vacuoles of mucin. Overall, the architecture is intact, and at this power, the nuclei appear regular without significant enlargement, hyperchromasia, or atypia. This specimen is highly cellular and shows several fragments of cells with reactive changes. Furthermore, there is a large atypical cell at the top of the image. Although it is difficult to evaluate the cell at this power, a single, atypical, and hyperchromatic cell such as this one may raise the possibility of high-grade dysplasia.

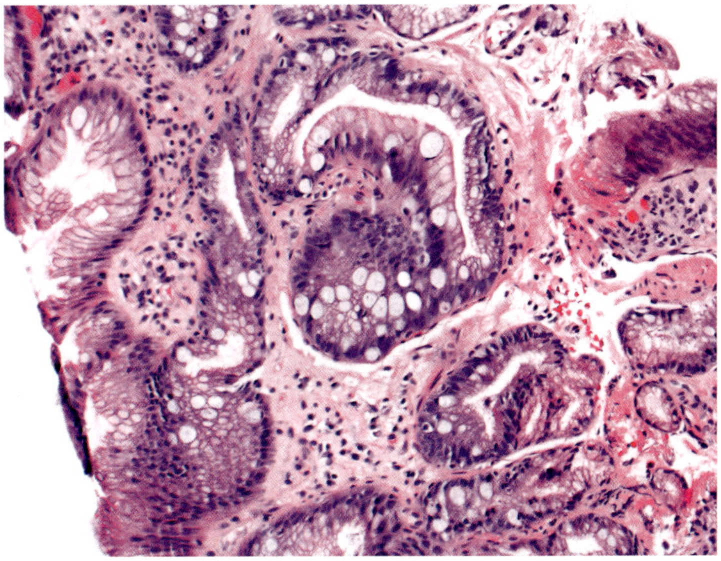

Figure 4.19 — Barrett's Esophagus, Esophageal Biopsy [H&E stain; intermediate power]. This endoscopic biopsy is taken from the distal third of the esophagus; however, there is no recognizable squamous epithelium in the sample. Rendering a diagnosis of Barrett's esophagus requires knowledge of the anatomic site of the biopsy. Instead of the normal squamous mucosa, the surface epithelium is cardiac-type gastric mucosa, and intestinal metaplasia with mucin-filled goblet cells is evident in the deeper glands.

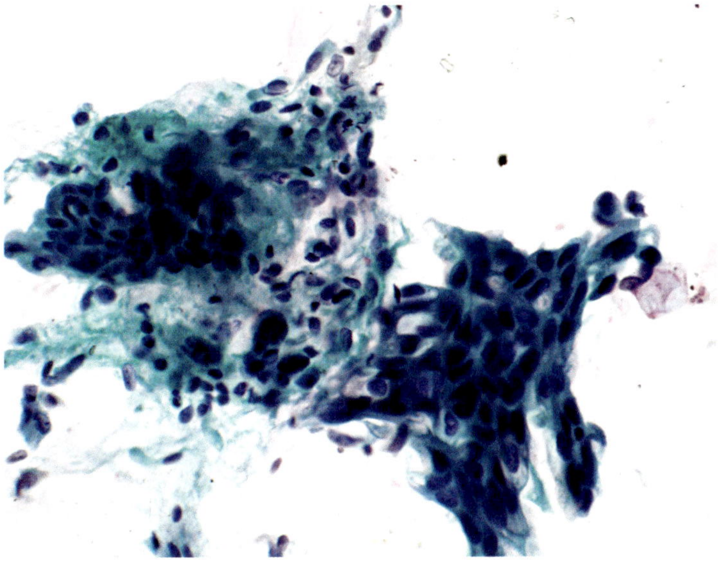

Figure 4.20a — Adenocarcinoma (ADC) Arising in a Background of Barrett's Esophagus, Esophageal Brushing [Pap stain; medium power]. This image consists of fragments of glandular epithelium with several distinct features. The fragment on the far right contains a disorganized sheet of glandular epithelial cells with significant overlap, nuclear enlargement and elongation, a high nuclear-to-cytoplasmic ratio, hyperchromasia, and occasional nucleoli. Furthermore, the fragment is punctuated by goblet cells with abundant mucin vacuoles. In the middle of the field, there are several single large, atypical cells with hyperchromatic, irregular nuclei three to four times the size of a normal nucleus. The presence of goblet cells indicates Barrett's esophagus; however, the architectural and cytologic atypia indicates at least high-grade dysplasia. The presence of single atypical cells and a dirty background are findings concerning for invasive ADC, which was confirmed on surgical follow-up. Barrett's esophagus, particularly with high-grade dysplasia, is a well-established risk factor for esophageal ADC.

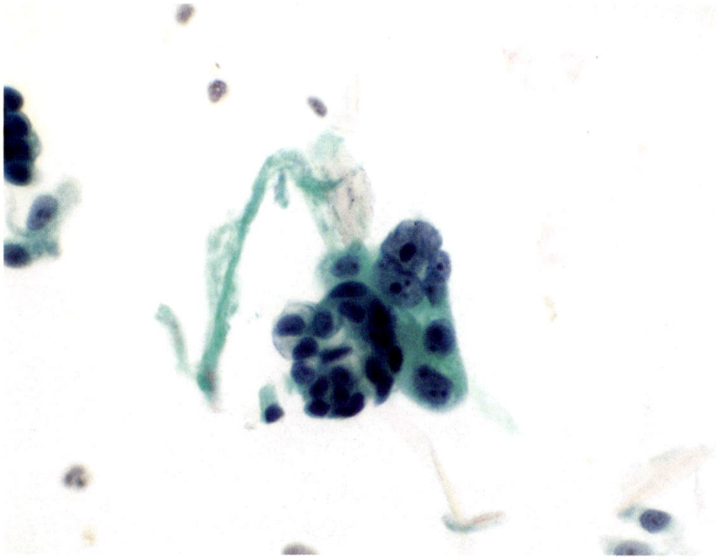

Figure 4.20b — Adenocarcinoma (ADC) Arising in a Background of Barret's Esophagus, Esophageal Brushing [Pap stain; high power]. In this high-power specimen, two distinctive cell populations are present. A round, glandular structure consists of relatively small columnar cells with vacuolated cytoplasm, a normal nuclear-to-cytoplasmic ratio, round and regular nuclei, and small nucleoli. Immediately adjacent to these cells, there are several extremely large pleomorphic cells with finely vacuolated cytoplasm, an increased nuclear-to-cytoplasmic ratio, and enlarged and bizarre nuclei, and prominent nucleoli. This level of atypia is seen in carcinoma. The most likely explanation of these two cell populations in an esophageal brushing specimen taken near the gastroesophageal junction is this: The glandular structure represents Barrett's esophagus, and the bizarre, atypical cells represents an arising ADC.

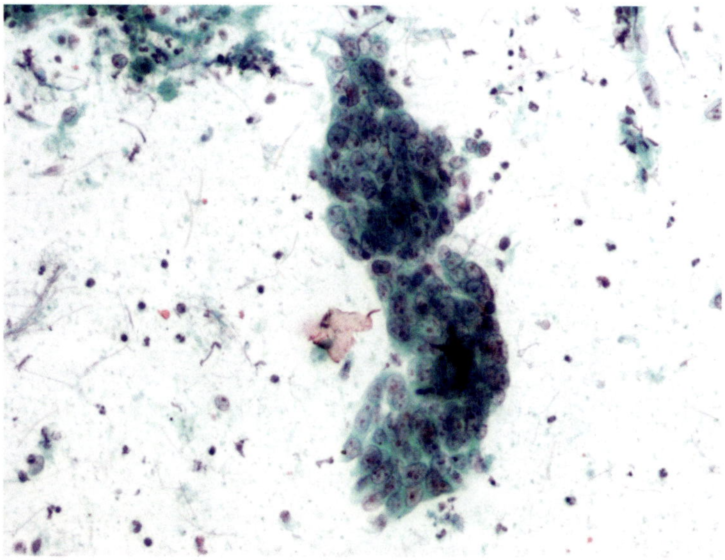

Figure 4.21a — Adenocarcinoma (ADC), Esophageal Brushing [Pap stain; medium power]. In this medium power view, a single large fragment of atypical epithelial cells shows a loss of polarity. There are significant cell overlap and size variation. The nuclear-to-cytoplasmic ratio is increased, the nuclear contours are irregular, and prominent nucleoli are evident. The background, especially at the top left corner of this view is necrotic. These features suggest a diagnosis of ADC. Esophageal ADCs typically arise in a background of Barrett's esophagus with high-grade dysplasia. Thus these cancers arise in the lower third of the esophagus. In cytologic brushings, the distinction between Barrett's esophagus with high-grade dysplasia and ADC can be challenged; however, in Barrett's esophagus with high-grade dysplasia, cells are less cohesive and contain crowded strips of pseudostratified elongated cells. The typical three-dimensional clusters, prominent nucleoli, and necrosis should be absent.

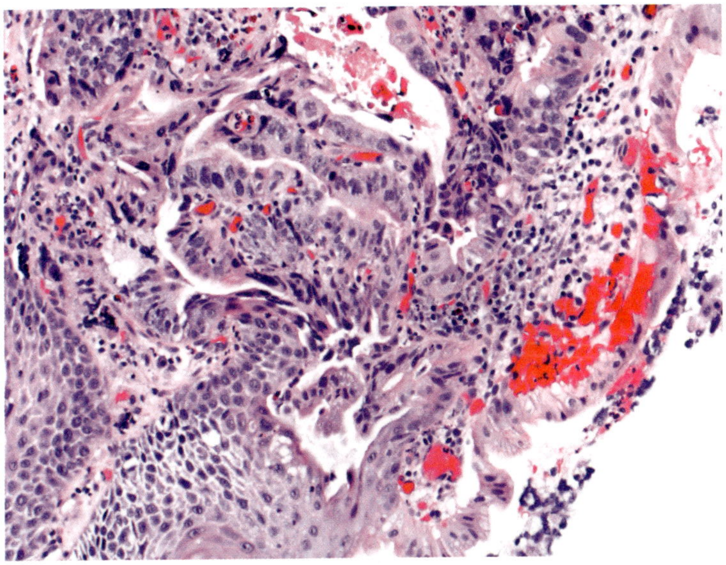

Figure 4.21b — Adenocarcinoma (ADC), Esophageal Biopsy [H&E stain; intermediate power]. Endoscopic biopsy of this distal esophageal lesion reveals surface epithelium that is both squamous (left half of field) as well as glandular (right half of field). The surface glandular epithelium resembles cardiac-type gastric mucosa. Beneath the surface lining, within the lamina propria, there is a complex arrangement of glands with atypical nuclei and necrosis. These features are in keeping with at least intramucosal ADC arising in Barrett's esophagus. Intramucosal ADC in the esophagus and stomach has the potential to access the lymphatics and metastasize, in contrast to colonic intramucosal ADC, which does not.

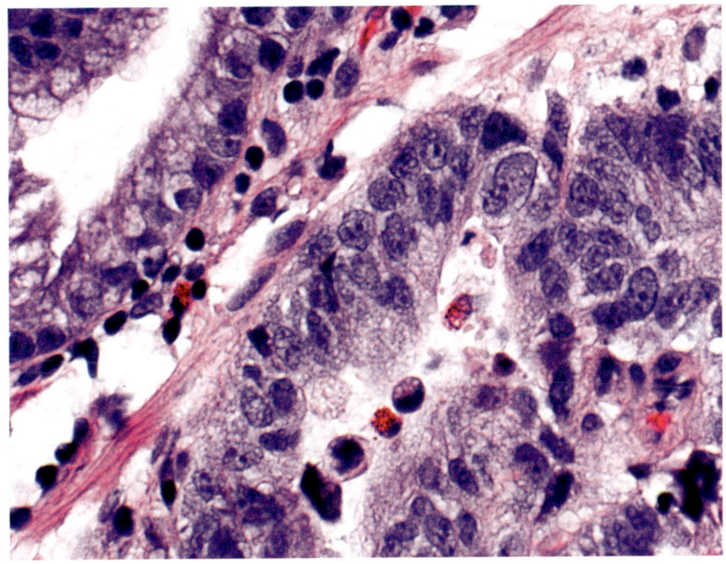

Figure 4.21c — Adenocarcinoma (ADC), Esophageal Biopsy [H&E stain; high power]. This field provides a high-power comparison of an ADC gland (right half of field) with a benign gland (left half of field). The malignant gland displays a high nuclear-to-cytoplasmic ratio, punctate nucleoli, nuclear crowding with loss of nuclear polarity, mitotic figures, and cellular necrosis. In contrast, the benign gland displays nuclear polarization with basally oriented nuclei and abundant cytoplasm, uniform nuclei, and inconspicuous nuclei. By immunohistochemistry, esophageal ADCs are typically immunoreactive for cytokeratin 7 (CK7) and CDX2 and negative for cytokeratin 20 (CK20). Of note, a subset of esophageal ADCs can show aberrant labeling for TTF-1, which is a diagnostic pitfall with pulmonary ADC.

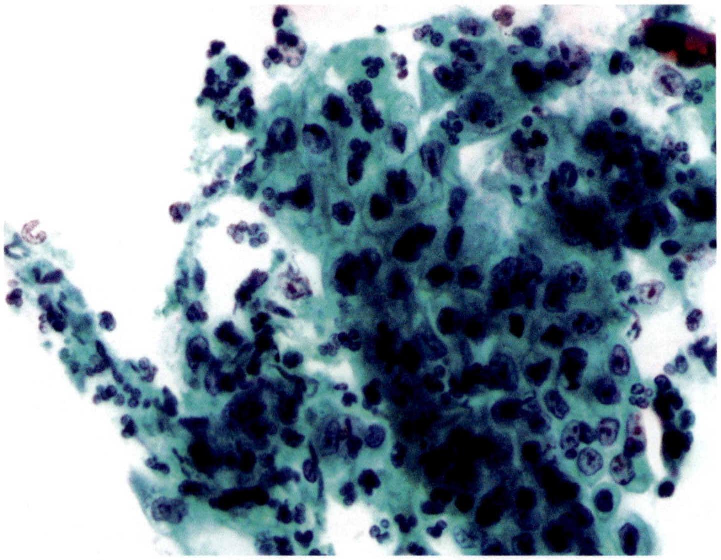

Figure 4.22 — Reactive Changes, Gastric Brushing [Pap stain; high power]. This field shows a large, flat, cohesive sheet of epithelial cells with well-defined cytoplasmic borders and a significant neutrophilic infiltrate. The nuclei show many atypical features, including moderate size variation, nuclear enlargement, and prominent nucleoli. However, the overall architecture is intact with no nuclear overlap, and the nuclear-to-cytoplasmic ratio is preserved. In the context of such brisk acute inflammation, these findings are consistent with reactive epithelial changes and should not be overinterpreted as dysplasia or neoplasia. *Helicobacter pylori* is a common cause of acute gastritis. Interestingly, gastric brush cytology has been shown to be more sensitive for detecting *H. pylori* than biopsy.

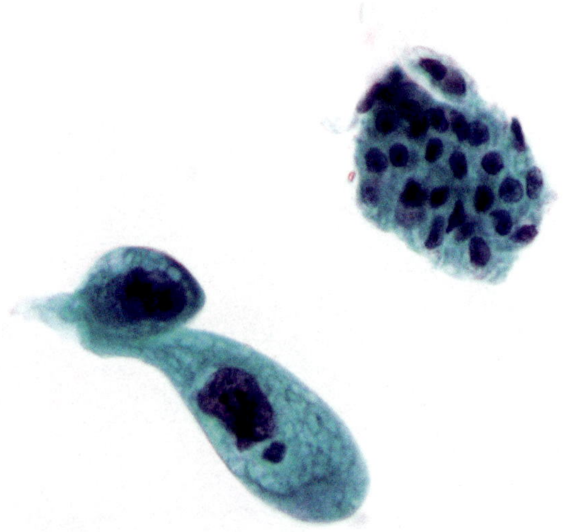

Figure 4.23a — Mucinous Adenocarcinoma (ADC), Gastric Brushing [Pap stain; high power]. There is a single fragment of reactive columnar epithelial cells at the top right portion of the image. The architecture is relatively preserved and the nuclear-to-cytoplasmic ratio is normal. The degree of nuclear membrane irregularity and the presence of occasional nucleoli favor these cells to be reactive gastric epithelium. In juxtaposition, the two cells in the center of the field are nearly the size of the entire reactive tissue fragment. The nuclei are greater than four to five times the size of the reactive nuclei and show significantly more nuclear membrane irregularity and coarse chromatin. Note the presence of numerous mucin vacuoles that give the cytoplasm a vacuolated appearance in this preparation. Although the overall nuclear-to-cytoplasmic ratio may not be too elevated because of the mucin, the degree of nuclear enlargement and atypia is diagnostic for ADC; surgical resection confirmed a diagnosis of mucinous ADC.

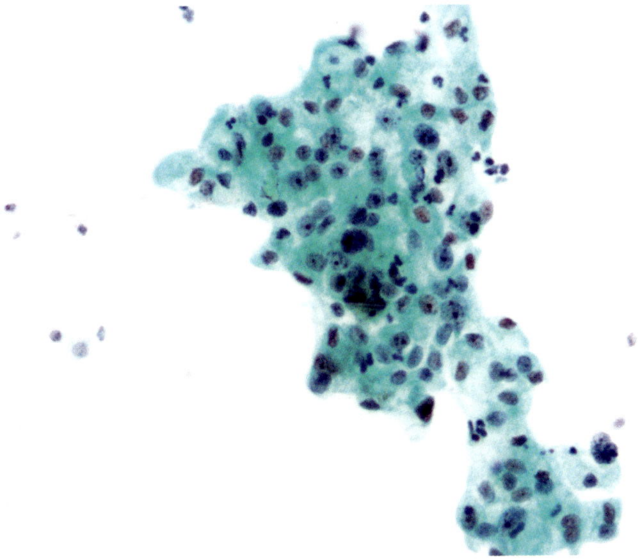

Figure 4.23b — Adenocarcinoma (ADC), Gastric Brushing [Pap stain; high power]. This image shows a large cluster of atypical epithelial cells with delicate, finely vacuolated cytoplasm. There is a loss of polarity with significant cellular overlap, leading to a disorganized architecture lacking the honeycomb appearance of benign glandular epithelium. Nuclear size variation, nuclear enlargement, and prominent nucleoli are visible throughout. These findings support a diagnosis of ADC. Risk factors for gastric ADC include a diet rich in salt-preserved foods and those containing nitroso compounds, obesity, smoking, chronic *Helicobacter pylori* infection, and excessive alcohol intake. The prognosis is dependent mainly on the stage of the tumor; stage T1 disease is associated with a greater than 90% survival, whereas stage T4 disease is associated with a 15% to 20% survival.

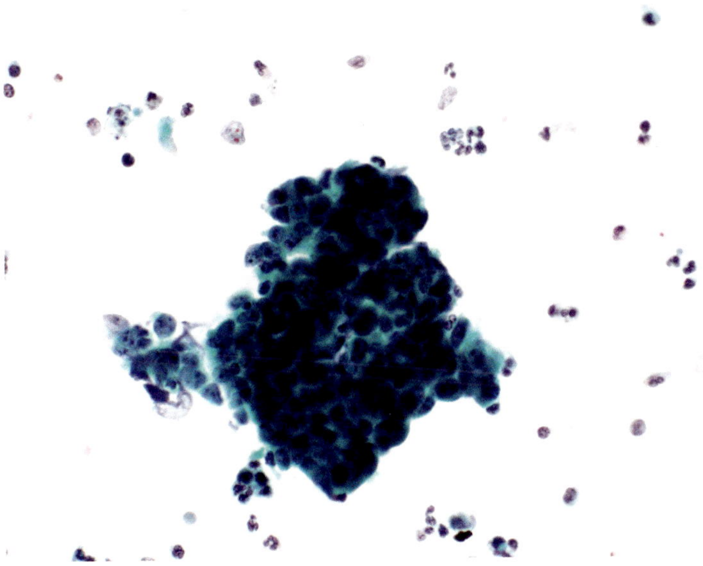

Figure 4.23c — Adenocarcinoma (ADC), Gastric Brushing [Pap stain; high power]. This high-power view shows a cluster of atypical epithelial cells with significant loss of the typical honeycomb architecture. There are nuclear enlargement, size variation, and overlap. The nuclear-to-cytoplasmic ratio is markedly increased, the chromatin is hyperchromatic and coarse, and nucleoli are present. The background contains neutrophils; however, this degree of architectural and nuclear atypia is far above that accepted for reactive changes. In addition, reactive and ulcerative lesions tend to be less cellular, but when fragment of epithelial cells are intact, they are flat and organized. Furthermore, a background of necrosis and inflammation can be present in both ulcerative and malignant lesions. Therefore special attention should be given to cytologic details when evaluating these cases.

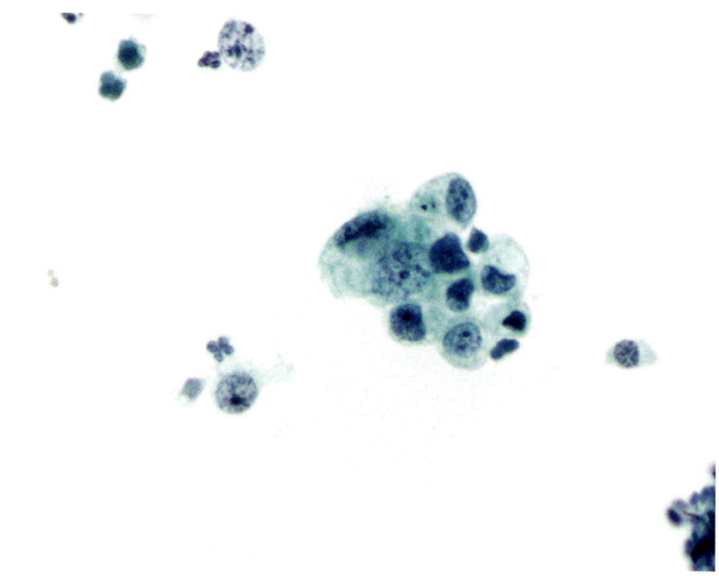

Figure 4.23d — Adenocarcinoma (ADC), Gastric Brushing [Pap stain; high power]. In this view, there is a small cluster of highly atypical cells with significant pleomorphism and a lack of normal architecture. The larger cells have abundant vacuolated cytoplasm and enlarged nuclei with prominent nucleoli. The smaller cells, such as the cell on the edge at 3 o'clock, contain delicate, mucinous cytoplasm, with a mucin cap causing nuclear indentation. Several other small cells contain more hyperchromatic, indented nuclei. Finally, several single high, atypical, naked nuclei with prominent nucleoli are seen in the background. Taken together, these features favor a diagnosis of ADC. The main differential diagnosis in this case would be high-grade dysplasia; however, the level of atypia and the presence of single atypical cells favor a carcinoma diagnosis in this case.

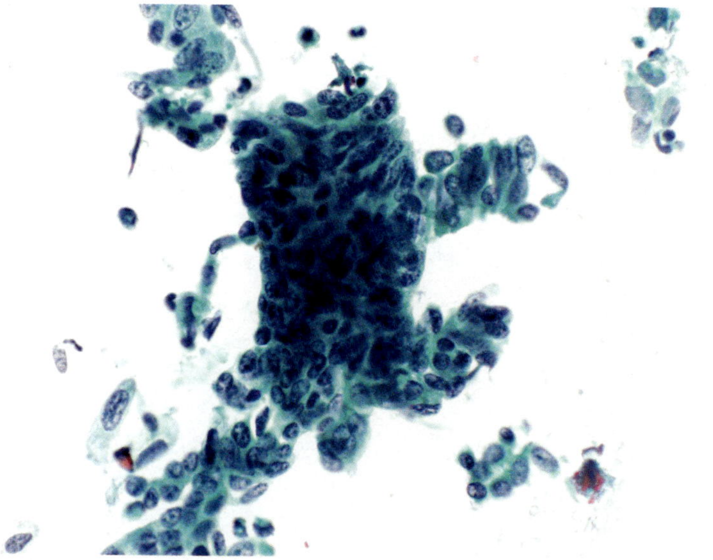

Figure 4.24 — Metastatic Colonic Adenocarcinoma (ADC), Gastric Brushing [Pap stain; high power]. This high-power view shows a population of atypical epithelial cells with several malignant features. First, there is loss of an organized, normal honeycomb architecture. There are significant nuclear enlargement and overlap. The nuclei are oval with irregular nuclear contours. The chromatin is coarse, and occasionally, apoptotic nuclear debris is seen in the center of the large fragment. In this view, no obvious mucin production is present. Although these features are consistent with a diagnosis of ADC, the origin is not automatically gastric. This patient had a history of colonic ADC and was suspected of having metastasis from a colonic primary tumor. Because gastric carcinomas can be of intestinal-type differentiation, distinguishing between these two primaries of cytopathologic features is impossible. Even if cell block material is available, there can be immunohistochemical overlap.

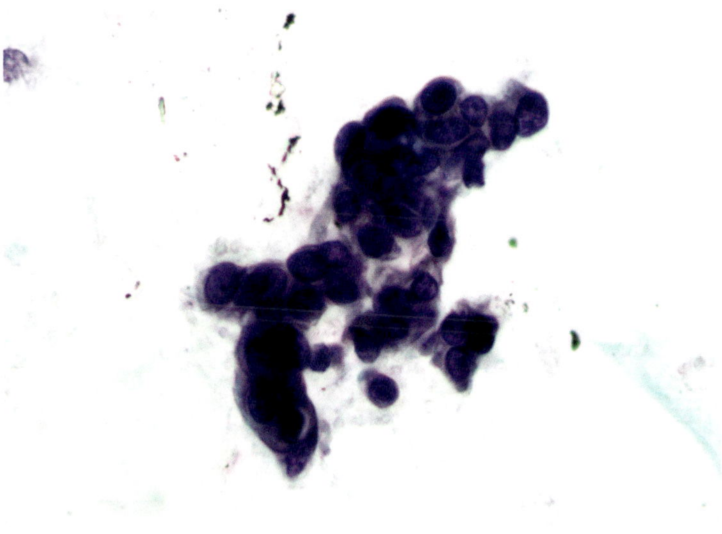

Figure 4.25 — Atypical Glandular Epithelium, Duodenal Brushing [Pap stain; high power]. This view shows an atypical fragment of glandular epithelium with nuclear overlap and architectural disorganization. Although the nuclear-to-cytoplasmic ratio is increased, the nuclei are only slightly enlarged, are round to oval, and contain smooth nuclear membranes. In addition, a few scattered small nucleoli are seen. Reactive epithelium can show minimal nuclear overlap and nucleoli; however, the degree of architectural disorganization and increased nuclear-to-cytoplasmic ratio should raise concern for high-grade dysplasia or carcinoma in this specimen. In the absence of severely malignant features, such as single atypical cells, atypical mitotic figures, and tumor necrosis, the distinction between high-grade dysplasia and carcinoma cannot be made. Clinical correlation and/or biopsy is required.

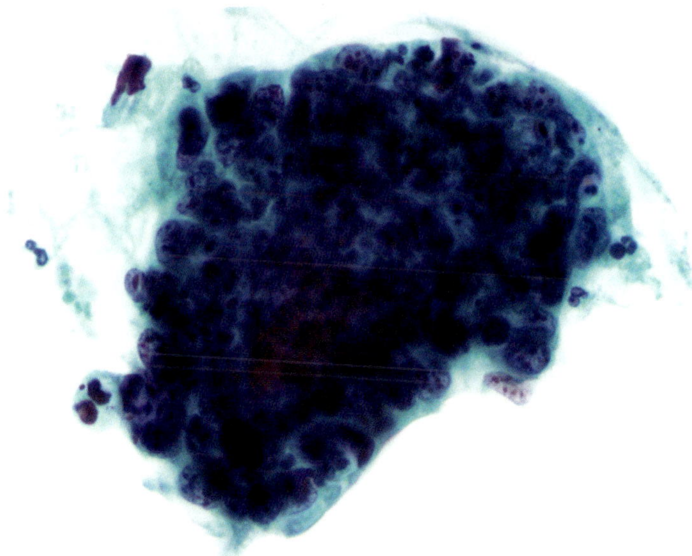

Figure 4.26a — Adenocarcinoma (ADC), Colon Brushing [Pap stain; high power]. This high-power view depicts a single three-dimensional fragment of highly atypical epithelial cells. There is significant nuclear overlap with a loss of polarity and disruption of the normal architecture. The more easily visualized cells at the periphery show nuclear membrane irregularities and nuclear pleomorphism—no two cells look exactly the same. The chromatin is clumpy with prominent nucleoli. It is often difficult to make a distinction between an adenoma and ADC in these specimens. For example, adenomas with low-grade dysplasia show increased cellularity; however, they lack the architectural and cytologic atypia described previously. In contrast, adenomas with high-grade dysplasia can have many of the described features. As a result, a diagnosis of suspicious for ADC may be warranted if all the features of malignancy are not well developed.

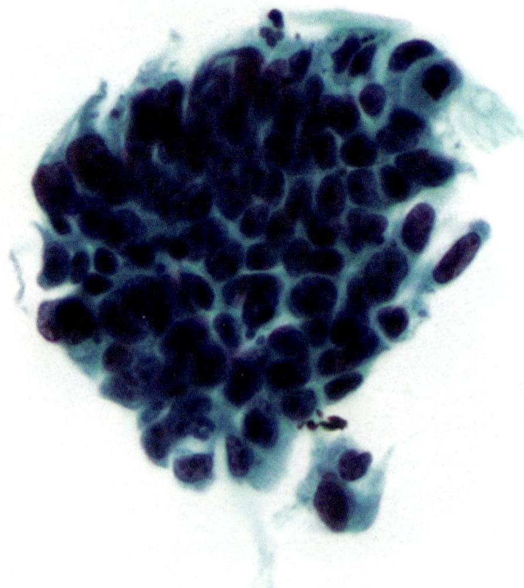

Figure 4.26b — Adenocarcinoma (ADC), Colon Brushing [Pap stain; high power]. This image shows an atypical fragment of ductal epithelium with scant mucinous cytoplasm. There is nuclear overlapping and a loss of polarity. The nuclei are enlarged, are hyperchromatic, and contain prominent nucleoli. The level of atypia seen in this case is suggestive of high-grade dysplasia or ADC. On biopsy, the patient was found to have an invasive ADC. Colorectal cancer is the third leading cause of cancer death in the United States. The prognosis depends largely on the depth of invasion, which cannot be determined on colonic brush specimens, and the presence or absence of lymph node metastasis. Positive cytology on a colonic brushing should trigger additional sampling.

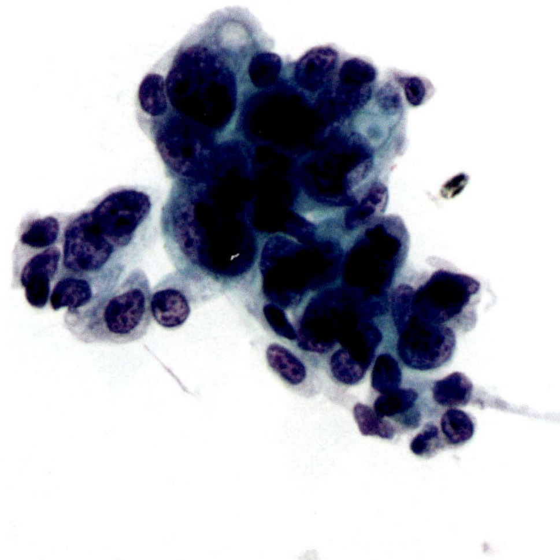

Figure 4.26c — Adenocarcinoma (ADC), Colon Brushing [Pap stain; high power]. This high-power view shows an irregular fragment of overlapping and disorganized epithelial cells. There are nuclear enlargement and size variation, in addition to an increased nuclear-to-cytoplasmic ratio, irregular nuclear membranes, hyperchromasia, and prominent nucleoli. At 12 and 2 o'clock, large mucin vacuoles can be seen. At 8 o'clock, smaller multiple cytoplasmic vacuoles, most likely also representing mucin, are present. Near 3 o'clock, there is a cell with highly atypical chromatin, possibly representing a mitotic figure. This degree of cytologic atypia warrants a diagnosis of carcinoma, and the presence of mucin indicates a glandular origin. Thus a diagnosis of ADC is justifiable. Although colonic brushings having largely fell out of favor, recognizing the appearance of colonic ADC is important. Metastatic colon ADC may be seen in other locations where exfoliative and fluid cytology is a common practice and is the standard of care.

5

The Anal Pap Test

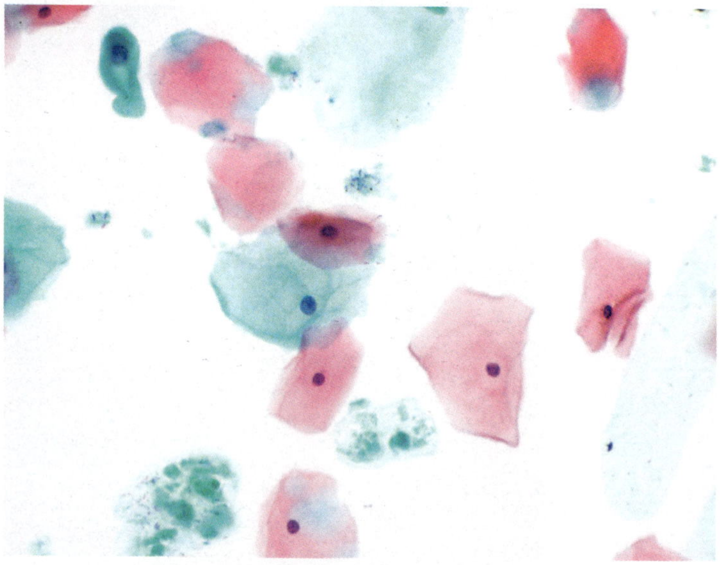

Figure 5.1a — Benign Squamous Cells, Anal Pap [Pap stain; high power]. This field consists of unremarkable squamous epithelial cells. The majority of the cells are superficial, mature nucleated and anucleated squamous epithelial cells with pink cytoplasm, indicative of keratinization, and small nuclei. Anal Paps are collected by inserting the swab approximately 2 to 3 in. into the anus to reach the anorectal junction, rotating the swab 360° while applying lateral pressure and then slowly withdrawing it. The sample must then be quickly fixed in preservative within 15 seconds to avoid drying artifact. Liquid-based preparations have improved adequacy over the conventional method and can be used for human papillomavirus (HPV) testing.

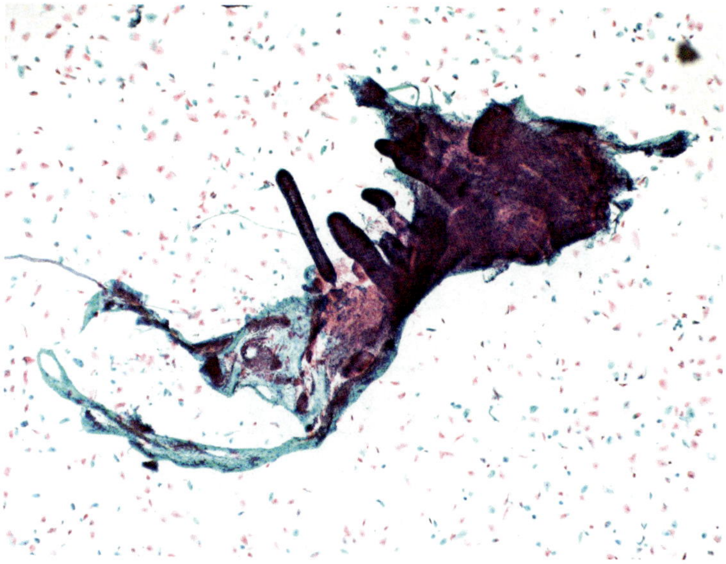

Figure 5.1b — Benign Glandular Cells, Anal Pap [Pap stain; low power]. In this low-power view, one can appreciate a large fragment of columnar epithelium in this specimen. Furthermore, multiple squamous epithelial cells are seen evenly spaced throughout the field. Examination at higher magnification is required to appreciate the cytologic features and to render a diagnosis.

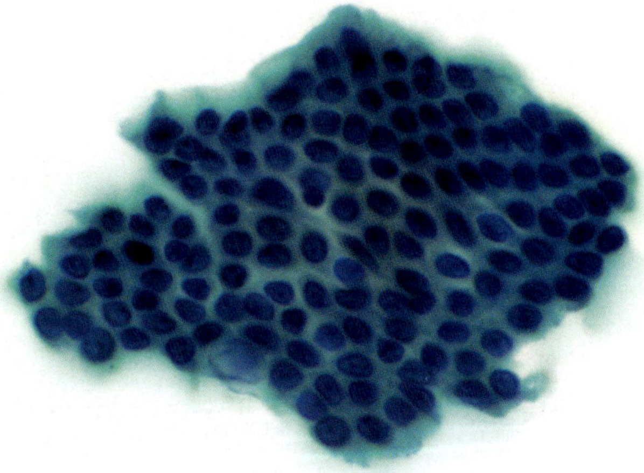

Figure 5.1c — Benign Glandular Cells, Anal Pap [Pap stain; high power]. This image consists of rectal mucosa, which is indicative of anorectal transformation zone sampling. Note the honeycomb architecture indicative of benign glandular epithelium, the basally oriented nuclei with apical cytoplasm (best shown between 10 and 1 o'clock), and a goblet cell at 7 o'clock. The nuclei are round and regular with rare nucleoli. Nuclear overlap, hyperchromasia, and size variation are not seen. It is important to assess the cytomorphologic features of glandular epithelium and not simply note its presence for adequacy; human papillomavirus (HPV) infection can involve the glandular epithelium and is a known cause of adenocarcinoma of the anus.

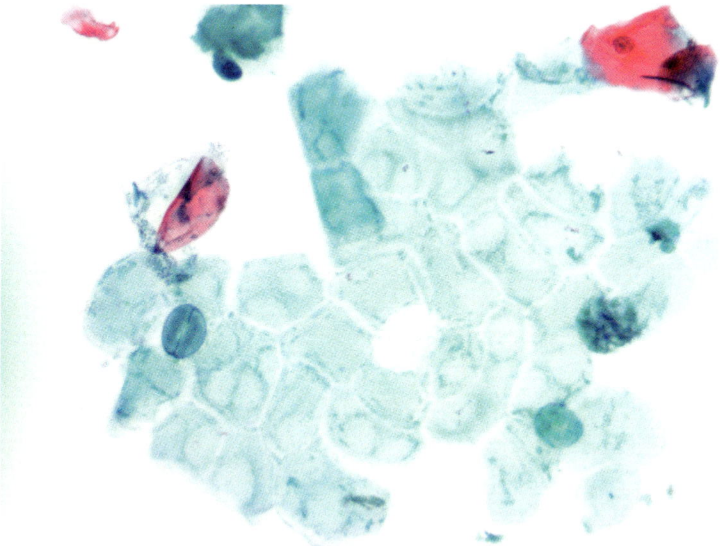

Figure 5.2a — Vegetable Material, Anal Pap [Pap stain; high power]. This image contains a couple of unremarkable, mature, and squamous epithelial cells. Another population consists of an intact monolayer of organized cells with distinct cell borders, consistent with vegetable material from fecal matter. Fecal matter and bacteria are normal and common findings in these specimens. Liquid-based preparations typically show a cleaner background and reduce the amount of obscuring material. Larger fragments of fecal matter, such as intact, undigested vegetable material, however, are difficult to remove. As long as the material does not obscure the cytopathologic examination of the squamous epithelium, its presence does not affect adequacy.

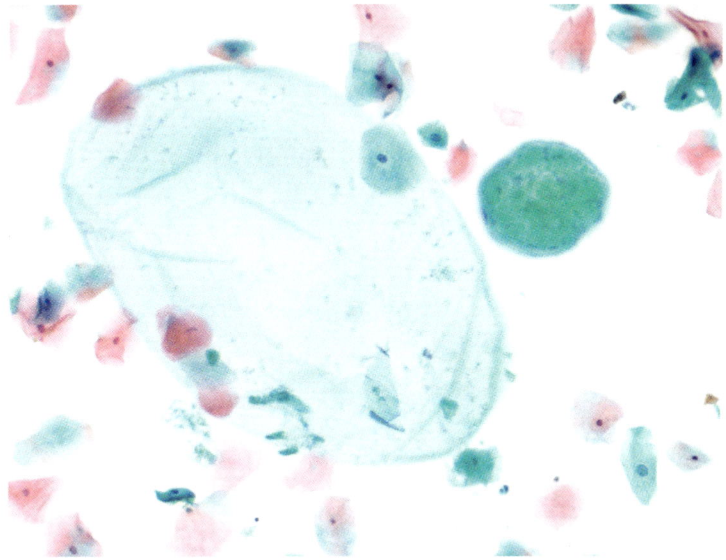

Figure 5.2b — Vegetable Material, Anal Pap [Pap stain; low power]. Two large fragments of anucleated foreign material are seen in the center of the field, most consistent with vegetable material. The intermediate cell overlying the vegetable material at 12 o'clock displays perinuclear clearing; however, it is not the thick perinuclear halo characteristic of koilocytic change. Furthermore, the nucleus is not enlarged and the nuclear membrane is smooth. The remaining specimen shows several unremarkable intermediate and superficial squamous epithelial cells, as well as anucleated squamous debris. Foreign material, including cotton and fecal matter, are often seen in anal Pap smears.

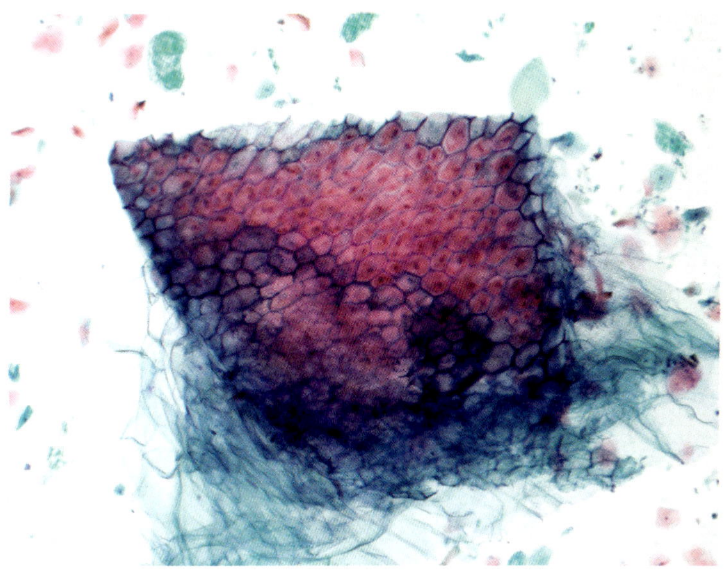

Figure 5.2c — Vegetable Material, Anal Pap [Pap stain; low power]. This low-power image shows a large tissue fragment containing polygonal cells with well-defined borders, ample cytoplasm, and one or two small nuclei per cell. The cell borders, in this case, represent plant cell walls, and the nuclei indicate that the vegetable has not been cooked. In anal Pap smears, fecal matter is often present and typically consists of high-fiber vegetables that have been incompletely digested. Occasionally, strands of striated skeletal muscle fibers are seen. It is important to recognize foreign material and to not mistake it for an abnormal population of cells arising from the patient.

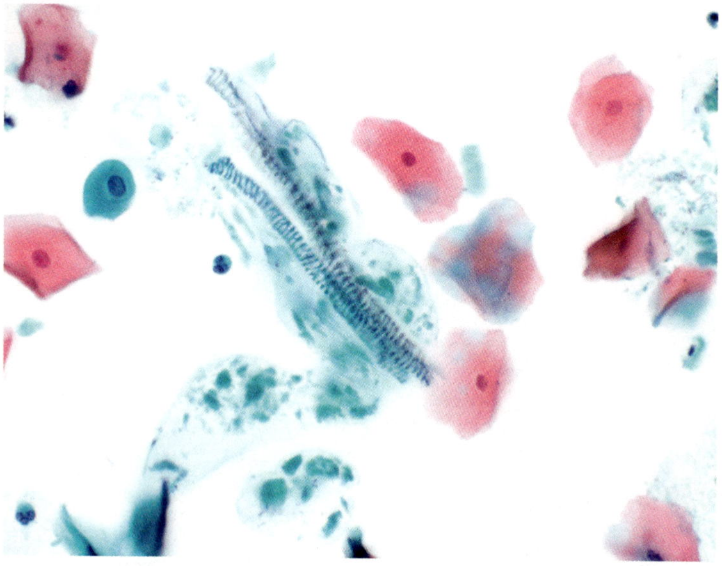

Figure 5.2d — Vegetable Material, Anal Pap [Pap stain; high power]. Superficial and anucleated mature squamous cells are seen scattered throughout this field. In addition, a parabasal cell is present at 10 o'clock and two neutrophils are present at 7 and 9 o'clock. Notice that several cells have a small rim of perinuclear clearing; however, there is a notable lack of nuclear enlargement and nuclear membrane irregularity. This thin, blurry rim of cytoplasmic clearing is most consistent with reactive changes. In the center of the field, there is a large, fragmenting foreign structure with a zipper-like configuration. These individual cells are much too small to represent human columnar epithelial cells. Furthermore, the arrangement of cells does not resemble or recapitulate a human structure.

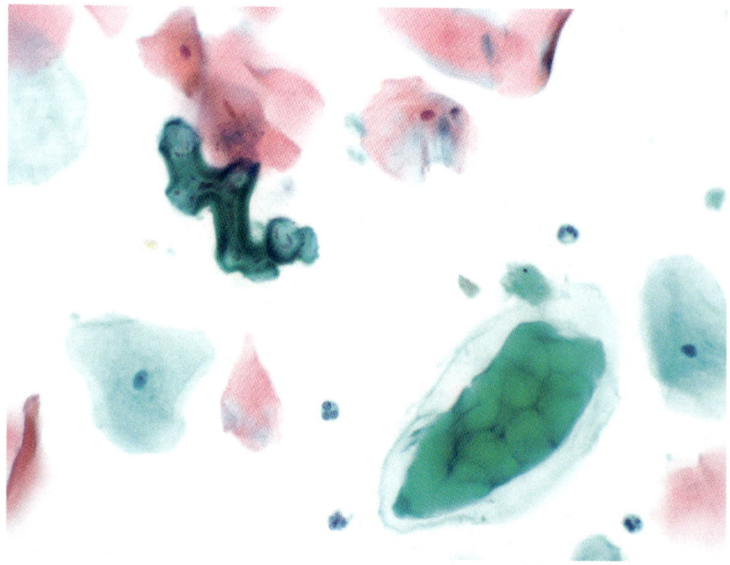

Figure 5.2e — Vegetable Material, Anal Pap [Pap stain; high power]. This high-power view shows two fragments of foreign material, most likely representing vegetable matter. The fragment at the bottom right does bear some resemblance to herpes simplex virus (HSV) cytopathologic effect. For example, it almost appears as though there is multinucleation with a ground-glass appearance and nuclear margination. The overall shape and comparison to normal-appearing background squamous epithelial cells, however, reveal that this material is not a human cell: infected, noninfected, or malignant. Finally, the fragment at the top left could be mistaken for a fungal organism; however, it is much too large and does not resemble the morphology of a known fungal organism.

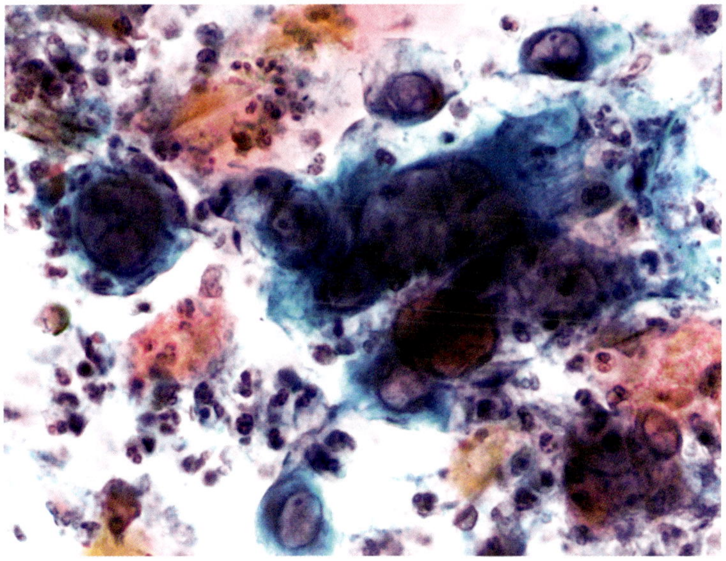

Figure 5.3a — Herpes Simplex Virus, Anal Pap [Pap stain; high power]. This is a conventional preparation of a specimen taken from the anal canal. Note the dirty background and less-preserved nuclear detail versus liquid-based preparations. This specimen, however, shows several interesting findings. First, numerous neutrophils are seen in the background and overlie atypical squamous epithelial cells. Second, few normal, mature squamous epithelial cells are present. Third, the predominant squamous population displays striking viral cytopathic effect consistent with infection by the herpes simplex virus (HSV). The cells have enlarged, glassy-appearing nuclei with peripheral margination of the chromatin. In the center of the field and at 9 o'clock, multinucleation is present. Given the poor nuclear resolution in this preparation, it is difficult to discern well-developed nuclear molding; however, single atypical mononuclear cells are seen scattered at the top and bottom of the image. Taken together, these findings are diagnostic for HSV infection and should not be interpreted as a high-grade squamous intraepithelial lesion (HSIL) or an invasive carcinoma.

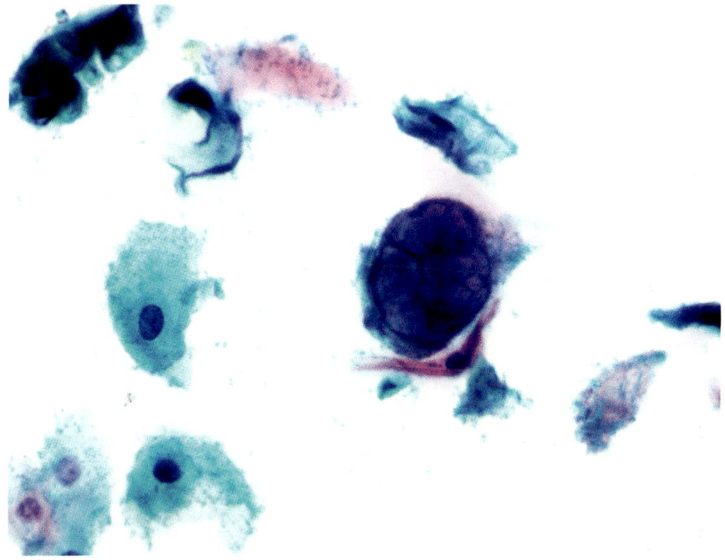

Figure 5.3b — Herpes Simplex Virus, Anal Pap [Pap stain; high power]. In the center of the field, note the enlarged, multinucleated cell with an increased nuclear-to-cytoplasmic ratio, glassy nuclear material, and peripheral margination of the chromatin. Again, these findings are classic for the viral cytopathic effect seen in herpes simplex virus (HSV) infection. In addition, this image shows glandular epithelium at the top left corner and squamous epithelium throughout, indicating that the specimen was obtained from the proper location: the anal transitional zone. It is important to comment on the presence of squamous metaplasia and columnar epithelial cells because clinicians are blindly sampling the anal canal. As a result, this feedback ensures adequate sampling from a clinical perspective.

Figure 5.4a — Low-Grade Squamous Intraepithelial Lesion (LSIL), Anal Pap [Pap stain; high power]. Near the middle of the field, there is a koilocyte with perinuclear clearing and a well-defined peripheral rim of dense cytoplasm. In addition, this cell exhibits nuclear enlargement, nuclear membrane irregularity, and hyperchromasia, features that are classic for the dysplasia typically seen in LSIL of the anus and gynecologic tract. Immediately adjacent and to the left of this cell, there are three additional cells showing LSIL features. Although there are no clear screening guidelines for when to perform anal Pap smears, many clinicians choose to screen patients with an increased risk for anal dysplasia, including patients with AIDS, men who have sex with men (MSM), and women with a history of recurrent cervical, vaginal, or vulvar human papillomavirus–related dysplasia.

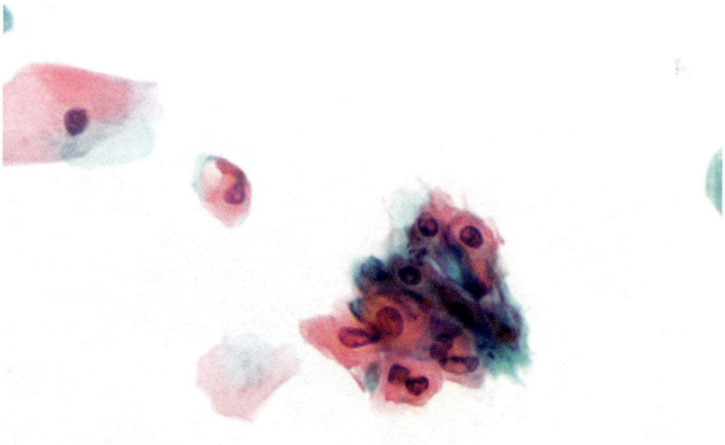

Figure 5.4b — Low-Grade Squamous Intraepithelial Lesion (LSIL), Anal Pap [Pap stain; high power]. This cluster of anal squamous epithelial cells shows several diagnostics features of LSIL, including perinuclear halos at 12 and 1 o'clock, binucleation at 6 o'clock, and nuclear membrane folding/irregularity at 9 o'clock. The presence of dysplasia on anal Pap smears is often subtler than in the cervix. One reason is because anal squamous cells are typically more mature and highly keratinized, as demonstrated in this case.

Figure 5.4c — Low-Grade Squamous Intraepithelial Lesion (LSIL), Anal Pap [Pap stain; high power]. In this image, a binucleate mature squamous cell is seen with densely keratinized [orangeophilic] cytoplasm and a minimal degree of nuclear hyperchromasia. Immediately adjacent to this cell, there is a squamous cell with orangeophilic cytoplasm, nuclear membrane irregularity, and nuclear hyperchromasia. These changes are consistent with human papillomavirus–related viral cytopathic effect like one would see in anal, gynecologic, or penile squamous intraepithelial lesions that arise in the squamo-columnar transitional zone. In anal cytology specimens, however, LSIL is less likely to have perinuclear halos and tends to affect more mature squamous cells. In addition, reactive and reparative changes are observed much less frequently in anal Pap specimens than in cervical specimens. As a result, cytopathologists typically have a lower threshold for dysplasia in anal Pap smears.

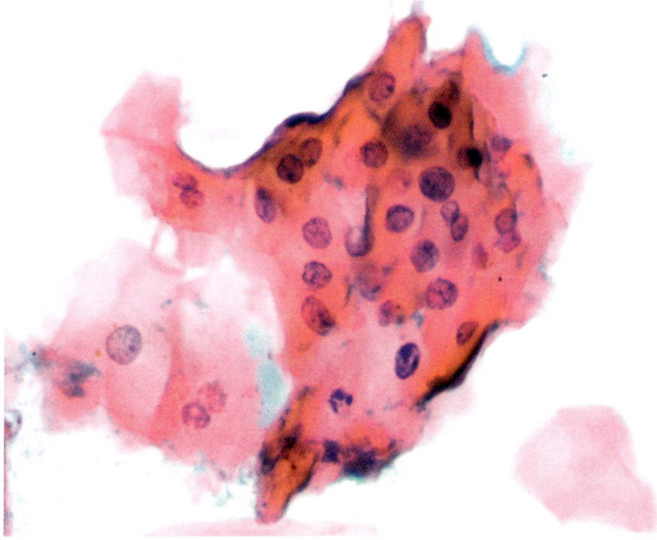

Figure 5.4d — Low-Grade Squamous Intraepithelial Lesion (LSIL), Anal Pap [Pap stain; high power]. This high-power image shows an intact fragment of orangeophilic squamous epithelial cells with occasional binucleation, increased nuclear-to-cytoplasmic ratio, irregular nuclear membranes, and clumpy chromatin. These changes are characteristic of LSIL. It has been shown that as many as 65% of men who have sex with men (MSM) and 95% of MSM with HIV are infected with human papillomavirus (HPV) in the perianal/anal canal at some point in their lifetime. As a result, there is ample evidence that screening this population is important.

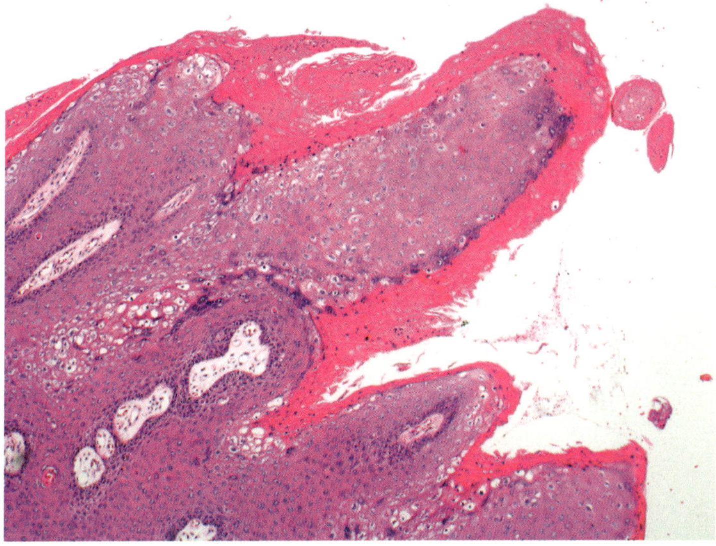

Figure 5.5a — Anal Intraepithelial Neoplasia 1 (AIN1), Tissue Biopsy [H&E stain; low power]. Condyloma acuminata are benign squamous proliferations that display hyperkeratosis and parakeratosis with hyperplastic papillary projections and koilocytic change. The lesions are part of the low-grade squamous intraepithelial lesion (LSIL) [AIN1] category but are hyperplastic and not flat. The differential diagnosis of a hyperplastic proliferation with prominent parakeratosis is verrucous carcinoma, which displays church spire–like growths of parakeratosis, but verrucous carcinoma lacks the koilocytic atypia and has a broad-based, pushing invasion.

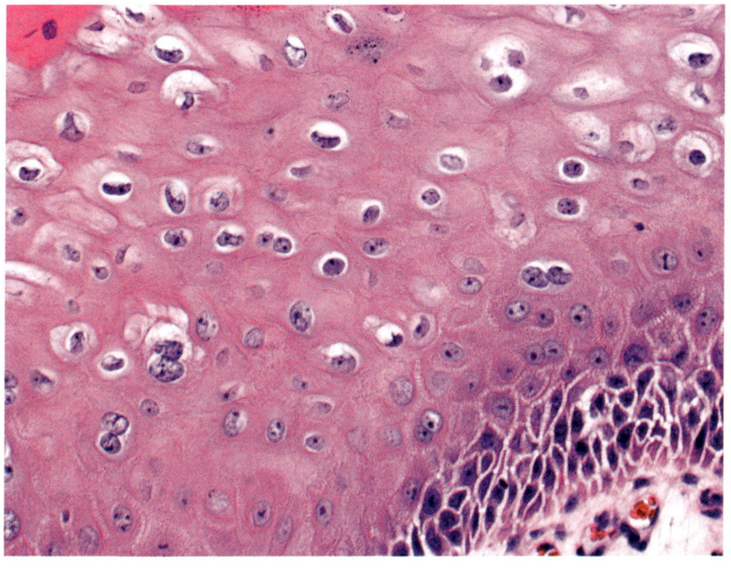

Figure 5.5b — Anal Intraepithelial Neoplasia 1 (AIN1), Tissue Biopsy [H&E stain; high power]. Low-grade squamous intraepithelial lesion (LSIL) of both the anus and the cervix display characteristic koilocytic nuclear features, including binucleation, perinuclear halos, and irregular nuclear borders. The nuclear changes are related to low-risk human papillomavirus (HPV) infection. The category of anal LSIL/AIN1 also includes squamous dysplasia that involves only the lower third of the epithelium.

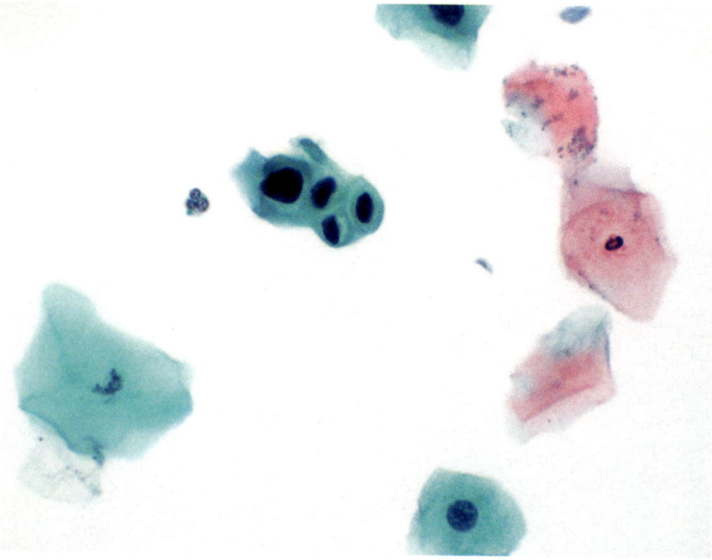

Figure 5.6 — Atypical Squamous Cells, Cannot Exclude High-Grade Squamous Intraepithelial Lesion (HSIL), Anal Pap [Pap stain; high power]. In this high-power view, a cohesive group of four nucleated cells are seen with an increased nuclear-to-cytoplasmic ratio, minimal perinuclear clearing, and hyperchromatic nuclei. The adjacent intermediate and superficial squamous cells appear unremarkable with a minimal amount of bacterial stippling. The hyperchromasia and increased nuclear-to-cytoplasmic ratio raise concerns for HSIL; however, the nuclei do not appear to be greatly enlarged compared with the intermediate cell at the bottom of the field. Furthermore, the atypia was focal. For these reasons, this specimen is best regarded as atypical squamous cells, cannot exclude HSIL.

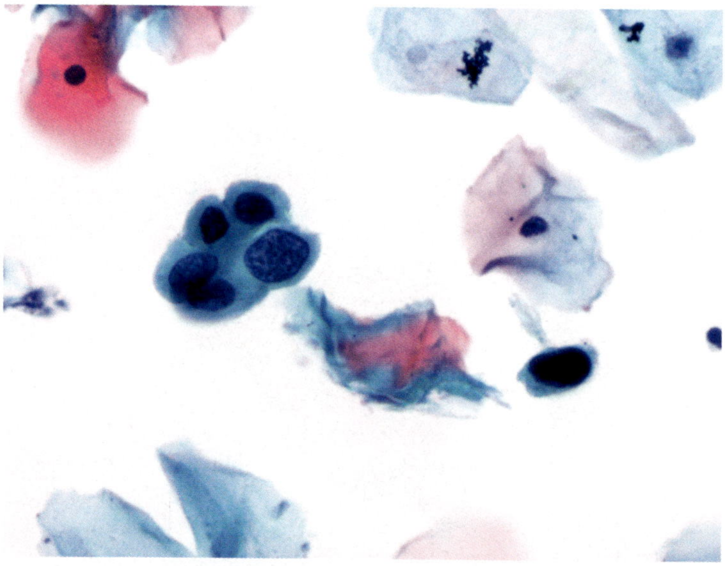

Figure 5.7a — High-Grade Squamous Intraepithelial Lesion (HSIL), Anal Pap [Pap stain; high power]. In this image, a single cluster of atypical epithelial cells can be seen at the left, along with a single atypical cell at the right. The size of these nuclei can be compared with that of the surrounding superficial cells [intermediate cells not present in this view]. The cytoplasm is dense and nuclei are enlarged with a markedly elevated nuclear-to-cytoplasmic ratio, irregular nuclear membranes, and hyperchromasia. In HSIL, cell size can vary, since small basal type cells, metaplastic cells, or larger more superficial cells can all be affected. The key is that regardless of the overall size of the cell, the nuclear-to-cytoplasmic ratio is markedly elevated. HSIL is caused by high-risk human papillomavirus (HPV) genotypes, most commonly types 16 and 18. As a result, there are multiple liquid-based assays, as well as tissue-based based techniques that detect the presence of high-risk HPV genotypes.

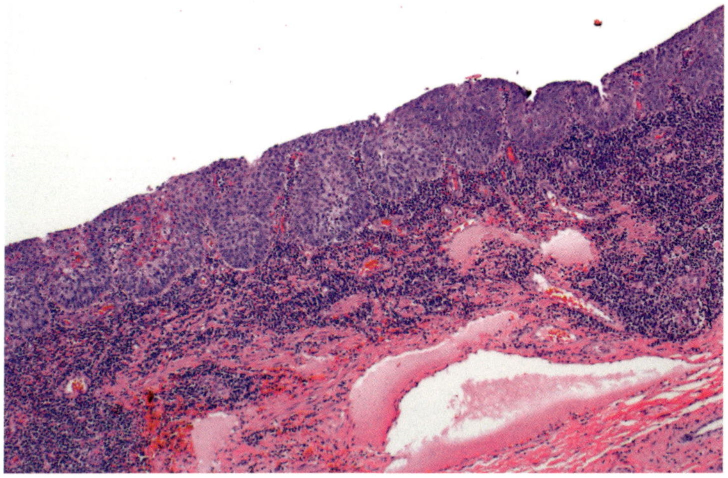

Figure 5.7b — Anal Intraepithelial Neoplasia 3 (AIN3), Hemorrhoidectomy [H&E stain; low power]. Typical hemorrhoidectomy specimens are lined by benign squamous mucosa and display dilated submucosal vessels. However, this field displays an area of squamous epithelium that is basaloid at low power and lacks surface maturation. On anal biopsy, the differential diagnosis at low power includes a high-grade squamous intraepithelial lesion/AIN2-3 and benign anal transition zone. However, in an external hemorrhoidectomy specimen, this is highly concerning for a high-grade squamous intraepithelial lesion/AIN2-3.

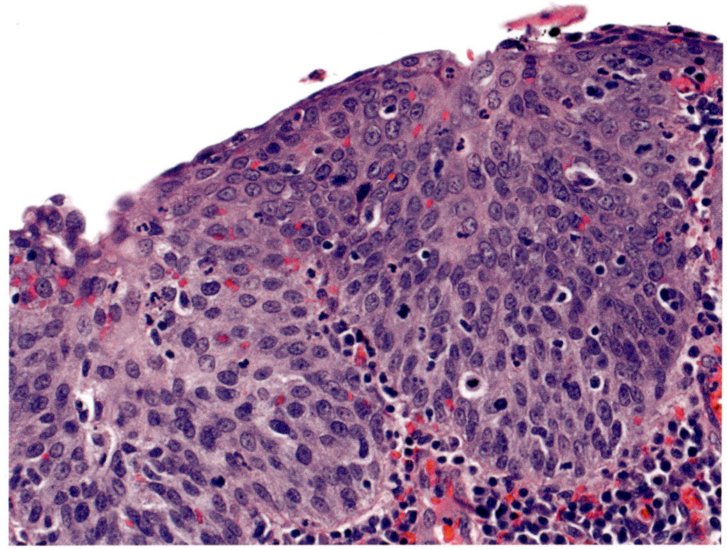

Figure 5.7c — Anal Intraepithelial Neoplasia 3 (AIN3), Hemorrhoidectomy [H&E stain; high power]. High-power examination of this hemorrhoidectomy specimen reveals full-thickness squamous atypia composed of cells with a high nuclear-to-cytoplasmic ratio, a lack of maturation, and mitotic figures extending into the upper third of the epithelium. These changes are characteristic of high-grade squamous intraepithelial lesion/AIN 3/carcinoma in situ. High-grade squamous intraepithelial lesions are due to high-risk human papillomavirus (HPV) infection, and patients with HIV infection are at higher risk of developing them.

6

Respiratory Tract Exfoliative Cytopathology

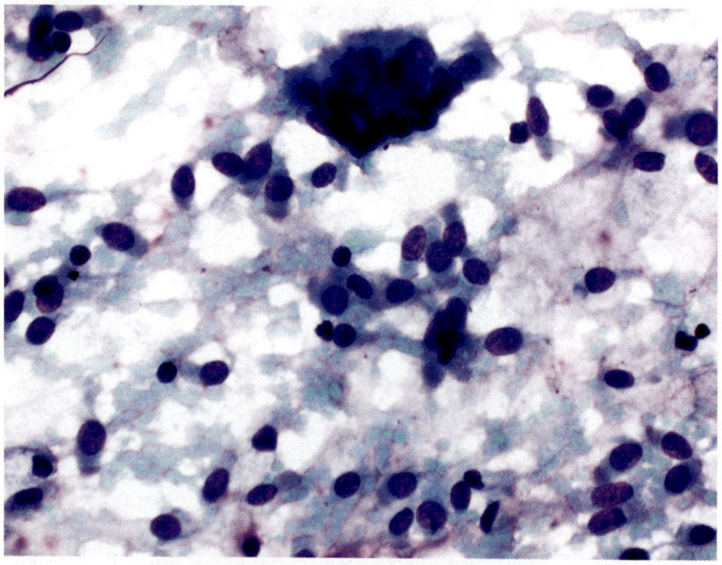

Figure 6.1a — Bronchial Epithelial Cells, Bronchial Washing [Diff-Quik stain; high power]. The field contains predominantly benign ciliated respiratory cells, although only some cells have identifiable cilia, which appear fuzzy and magenta and at the opposite end of the cell as the nucleus. Some cells are present singly, whereas others are present in small fragments. Most cells have a columnar appearance; others appear round or ovoid. Although the presence of cilia confirms cells as benign, cilia may not be seen in some cells, or occasionally any cells, in a given preparation. It is useful to compare the nuclear details; in this case, cells without obvious cilia have similar nuclei as cells with cilia. Thus, one can be reassured that these cells are all benign respiratory cells.

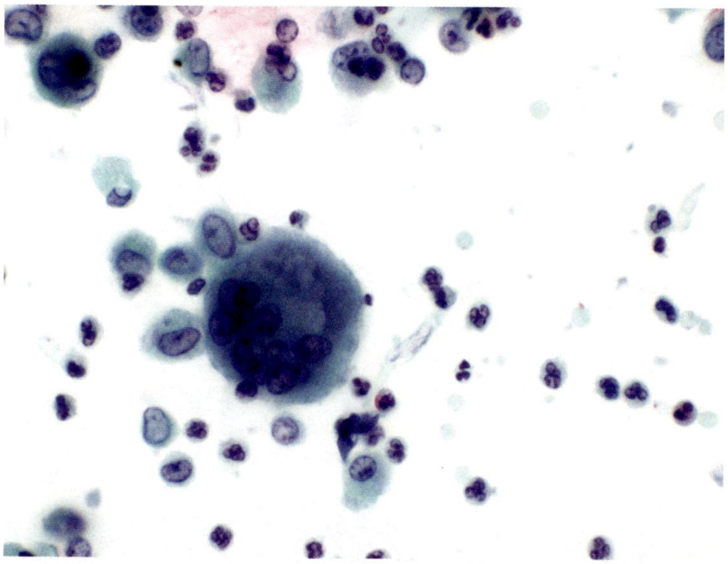

Figure 6.1b — Multinucleated Giant Cell, Bronchoalveolar Lavage [Pap stain; high power]. Multinucleated giant cells can be encountered in the airways and are a nonspecific finding. This giant cell contains numerous nuclei that have the same shape and size, each containing a prominent chromocenter. The nuclei are asymmetrically arranged within the cell's cytoplasm. The background contains mixed inflammations and alveolar macrophages.

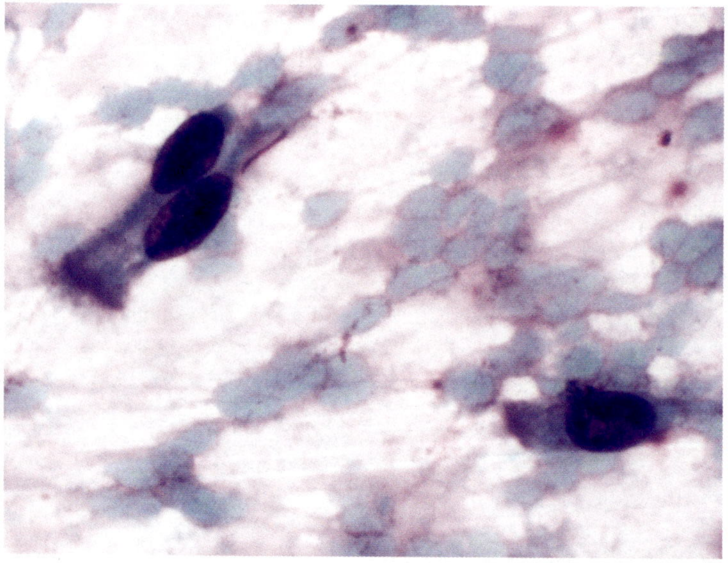

Figure 6.1c — Benign Respiratory Epithelium, Bronchial Washing [Diff-Quik stain; high power]. Several benign respiratory epithelial cells can be seen, which are the cells that line the bronchial mucosa. These cells are columnar with basally oriented nuclei, show a hint of pseudostratification, and contain round to oval nuclei with smooth nuclear contours and evenly distributed chromatin. There are abundant apical cytoplasm and an overlying discrete terminal bar with attached, fine cilia projecting into the airway space, a reassuring finding that these cells are benign. Ciliated respiratory epithelial cells play an important barrier function and provide a mechanism for clearing pathogens and foreign material from the airway through a mucociliary "escalator" mechanism. Diseases associated with ciliary dysfunction, such as primary ciliary dyskinesia, highlight the important role that cilia play in the innate defense of the airway.

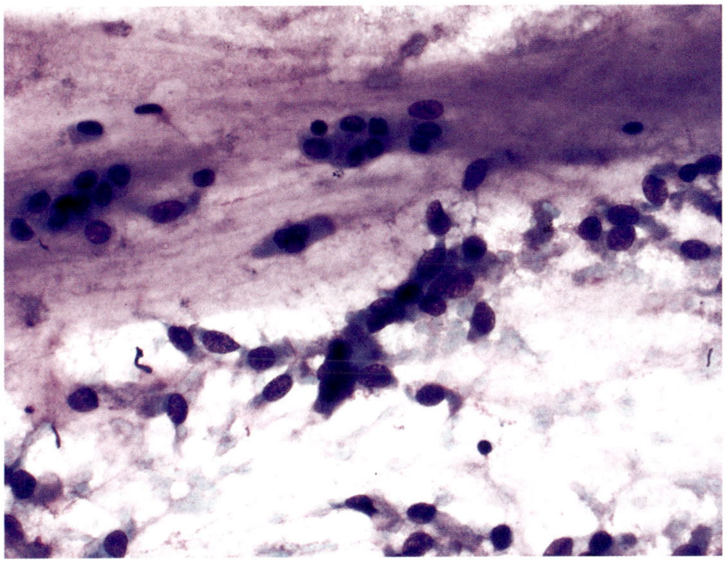

Figure 6.1d — Benign Respiratory Epithelium, Bronchial Washing [Diff-Quik stain; high power]. Small clusters of, as well as scattered single, benign respiratory epithelial cells have round to oval nuclei with smooth nuclear contours and fine, even chromatin. There are abundant cytoplasm and rare apical terminal bars with no obvious cilia present; however, it is not uncommon for reactive or degenerative respiratory epithelial cells to lose cilia and terminal bars. Cilia may also be lost or obscured during specimen retrieval and preparation. Although the presence of cilia may be helpful when significant atypia is present, the absence of cilia does not indicate a malignancy. It is important to take all cytologic features into consideration when making a final diagnosis.

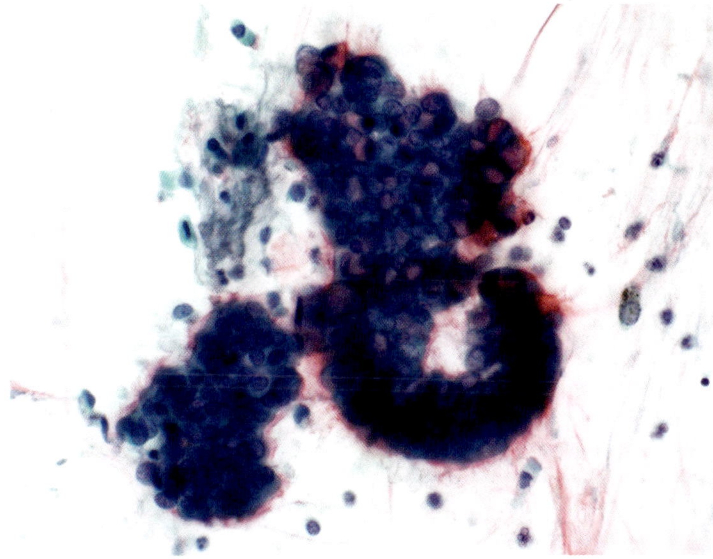

Figure 6.2a — Benign Respiratory Epithelium and Goblet Cells, Bronchial Washing [Pap stain; high power]. This field shows two cellular fragments containing sheets of columnar epithelial cells with abundant cilia along the fragment edges. Within the fragments, scattered goblet cells with metachromatic intracytoplasmic mucin can be identified. The nuclei are small and uniform with smooth borders and unremarkable chromatin. Although there is slight nuclear overlap, the presence of scattered goblet cells and a large population of ciliated epithelium along the surface indicates that the normal architecture is intact. Goblet cells secrete mucin, which provides a protective coating to the cells of the conducting airway.

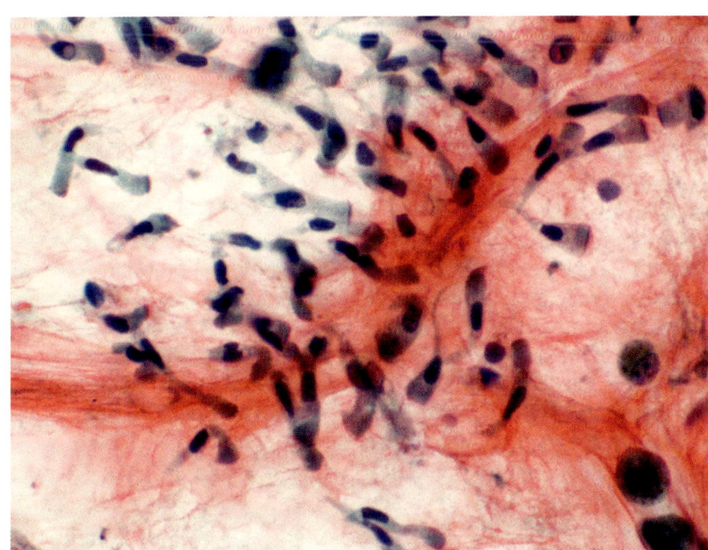

Figure 6.2b — Benign Respiratory Epithelium and Pulmonary Alveolar Macrophages, Sputum [Pap stain; high power]. This view shows multiple scattered, single columnar epithelial cells with basally oriented, oval, and regular nuclei and prominent cytoplasm. In many of the cells, a magenta terminal bar can be seen, suggesting the presence of cilia. The overlying mucus in this sample obscures the ability to definitively identify cilia. At the bottom right portion of the image, three pulmonary alveolar macrophages are seen with foamy cytoplasm containing anthracotic pigment and bean-shaped nuclei. The presence of pulmonary alveolar macrophages is indicative of adequate alveolar sampling.

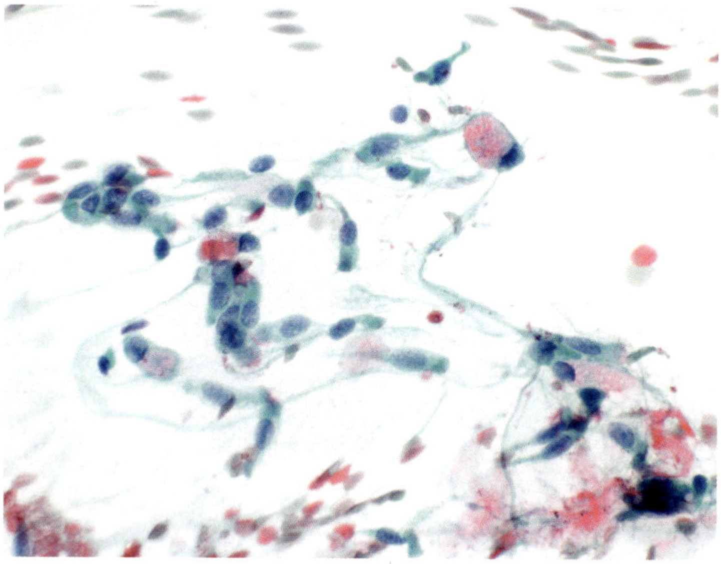

Figure 6.2c — Benign Respiratory Epithelium and Goblet Cells, Bronchial Brushing [Pap stain; high power]. This specimen predominantly consists of single and small clusters of respiratory columnar epithelial cells with abundant elongated, wispy cytoplasm and round to oval nuclei with smooth nuclear contours and pale chromatin with occasional inconspicuous nucleoli. There is a hint of a terminal bar and cilia in a few cells; however, the cytoplasmic elongation seen in this case obscures definitive determination. Overall, the lack of cytologic atypia confirms the benign nature of these cells. In addition, several goblet cells are seen with prominent metachromatic cytoplasmic vacuoles that push the nuclei to the edge of the cell.

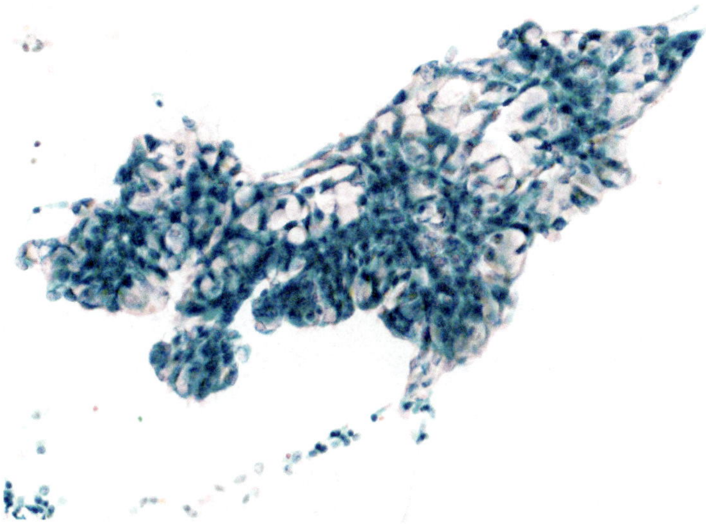

Figure 6.2d — Goblet Cell Hyperplasia, Bronchial Washing [Pap stain; medium power]. A large cellular fragment of epithelial cells containing a prominence of goblet cells with abundant, clear intracytoplasmic vacuoles gives a punched-out appearance. Although the cytologic detail is difficult to appreciate at this power, this pattern showing mixed bland epithelial cells and goblet cells indicates a benign process. More specifically, the numerous goblet cells are indicative of goblet cell hyperplasia, which occurs as a result of chronic bronchial irritation. In general, neoplastic processes are clonal and contain a uniform or pleomorphic population of cells of a singular origin.

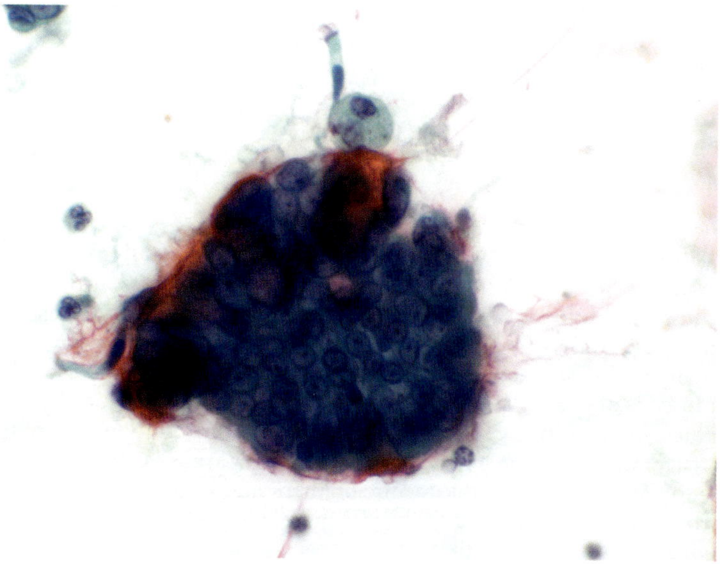

Figure 6.3a — Creola Body, Sputum [Pap stain; high power]. In the center of the field, there is a large, round fragment of reactive respiratory epithelium. The cells in the center show mild nuclear overlap but distinct smooth nuclear contours, round and monomorphic nuclei with pale chromatin, and small, punctate nucleoli. Along the periphery, the cells contain terminal bars and cilia. In this field, there are also a few intact, benign-appearing respiratory epithelia with prominent cilia, a pulmonary alveolar macrophage, and scattered neutrophils. This round, dense cluster is called a *Creola body* and is classically associated with asthma, although in reality, the finding is nonspecific and important enough to be recognized only as a distractor.

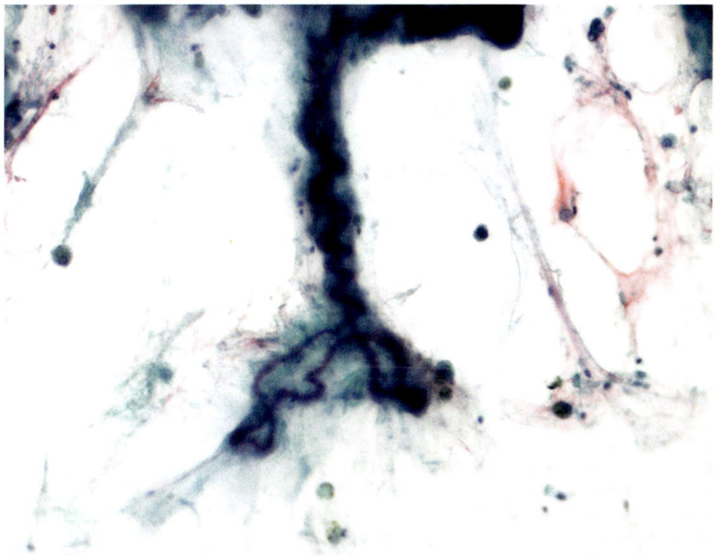

Figure 6.3b — Curschmann Spiral, Sputum [Pap stain; medium power]. In the center of the field, metachromatic material can be seen coiling in on itself. This substance is known as a *Curschmann spiral* and is thought to represent a mucous cast that forms in the subepithelial mucous gland ducts or small bronchi. Curschmann spirals classically had an association with asthma; in these patients, they may be seen alongside Creola bodies and Charcot–Leyden crystals. In practice, structures similar to Curschmann spirals can be identified in gastrointestinal tract and other specimens in which mucin is present. Although obvious in this case due to the size, it is important to recognize these structures because, in some circumstances, they can be confused for fungal organisms. Scattered in this field, multiple small pulmonary alveolar macrophages are present, representing alveolar sampling.

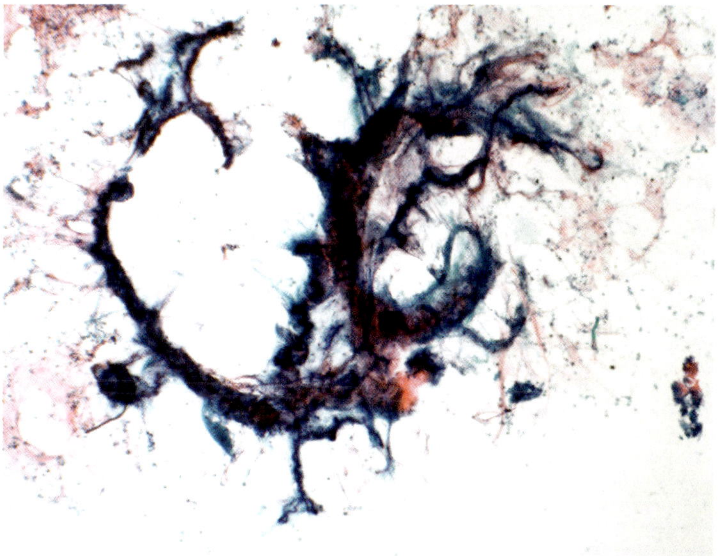

Figure 6.3c — Curschmann Spiral, Sputum [Pap stain; low power]. Exfoliative respiratory tract specimens may contain spiral structures composed of mucin from the airways. In most instances small Curschmann spirals can be seen in the background and are eventually ignored by experienced eyes. This field contains an unusually large Curschmann spiral. In addition to being a distractor, diagnostic cells may stick to these spirals and become obscured.

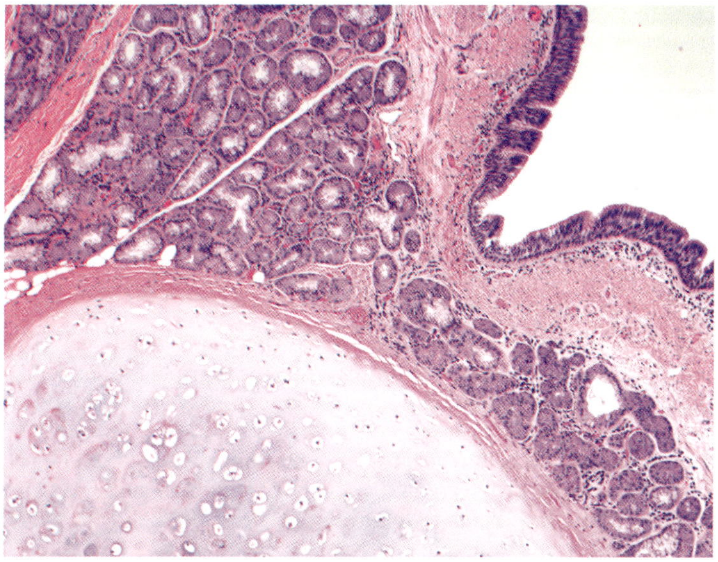

Figure 6.4a — Benign Bronchial Wall, Lobectomy [H&E stain; low power]. There are three defining histologic features for bronchi: the presence of a pseudostratified columnar (respiratory) epithelium, the presence of submucosal seromucinous glands, and the presence of hyaline cartilage. The presence of the hyaline cartilage in the wall of the airway distinguishes bronchi from bronchioles, which lack cartilage in the airway wall. In addition, bronchioles are lined by simple ciliated columnar epithelium rather than pseudostratified columnar epithelium. Neoplasms can arise from any cellular element of the bronchial wall, including the respiratory epithelium, seromucinous salivary-type glands, and cartilage.

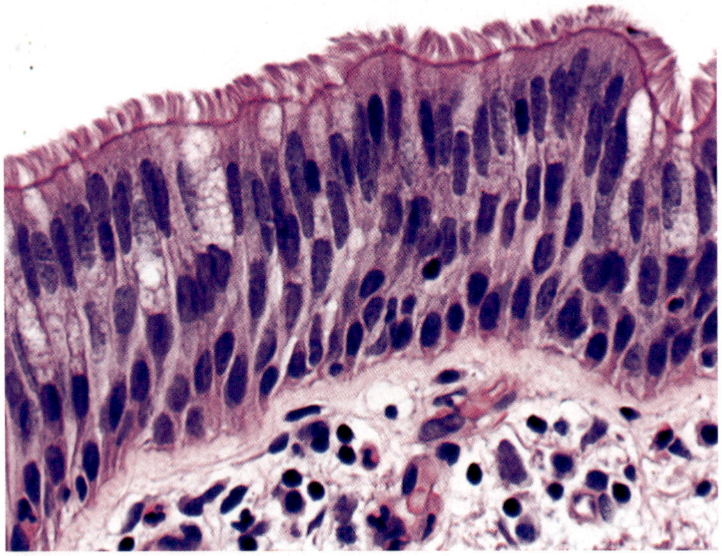

Figure 6.4b — Benign Bronchial Wall, Lobectomy [H&E stain; high power]. The bronchi are lined by pseudostratified, columnar epithelium, also known as *respiratory epithelium*. These columnar cells are ciliated. The ciliated cells do not participate in gas exchange (which occurs across the pneumocytes in the alveolar walls), but rather serve to move mucus and debris along the airway. This epithelium is termed *pseudostratified* because, in contrast to true stratified epithelium (such as squamous epithelium or urothelium), every cell in the pseudostratified epithelial layer touches the basement membrane. In addition to columnar cells, there are also basal cells and mucin-containing goblet cells within the epithelium.

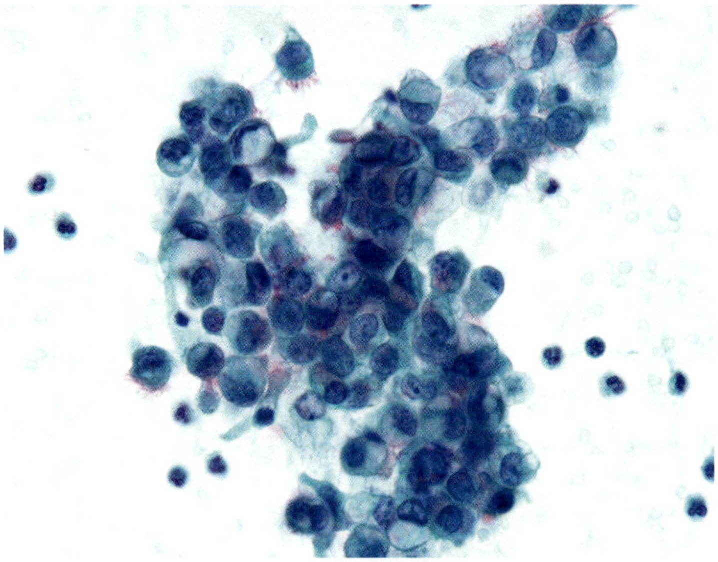

Figure 6.5a — Reactive Bronchial Epithelial Cells, Bronchial Brush [Pap stain; high power]. Reactive bronchial epithelial cells generally retain their cohesiveness and appear in cytologic preparations as three-dimensional clusters of cells with enlarged nuclei, prominent nucleoli, and multinucleation. Cells typically retain a columnar or cuboidal shape with identifiable terminal plates and cilia, as seen in this field.

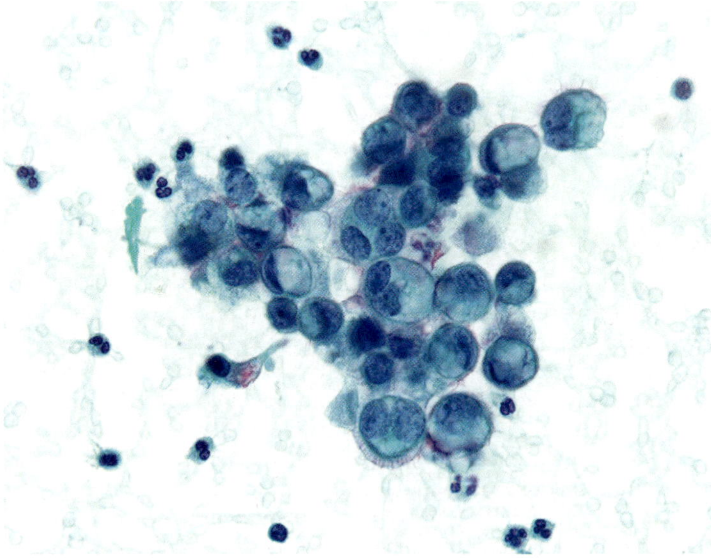

Figure 6.5b — Reactive Bronchial Epithelial Cells, Bronchial Brush [Pap stain; high power]. Multinucleation is a nonspecific response of the bronchial cell to injury. The nuclei tend to be uniform and resemble those of other benign cells in the same specimen. The chromatin should be finely granular and evenly dispersed. Nuclear membrane irregularity is typically absent or minimal, and cilia may be present. Some vacuolated cells without cilia can also be identified in this field; they are benign goblet cells.

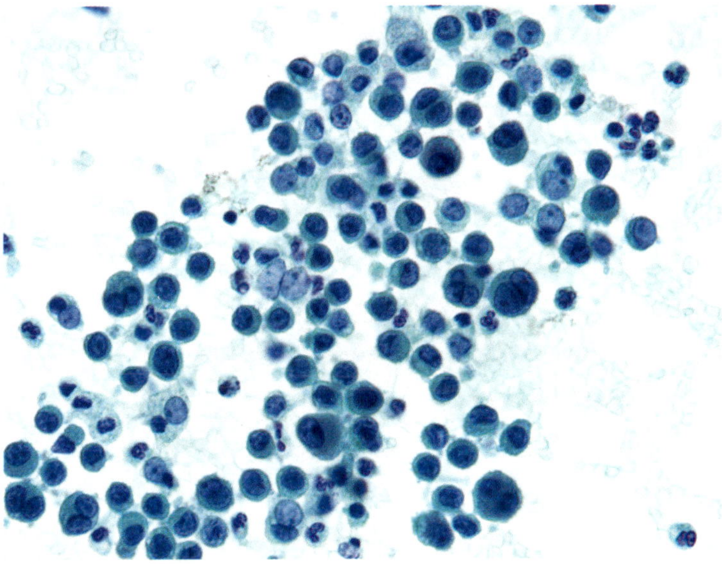

Figure 6.5c — Reactive Bronchial Epithelial Cells and Pulmonary Alveolar Macrophages, Bronchial Brush [Pap stain; medium power]. This field shows discohesive, round, mononucleated and binucleated cells with evenly distributed chromatin, small nucleoli, and densely granular cytoplasm. Although no cilia or terminal bars are present, these features are consistent with reactive and/or regenerating epithelial cells. These cells may represent bronchial epithelium, which tends to shed ciliary tufts as intact units (ciliacytophthoria) in response to injury, or reactive type II pneumocytes. Note the alveolar macrophages with dusky, finely vacuolated cytoplasm in the background. It can be difficult to distinguish these discohesive cells from each other, but this reactive pattern is not specific to any particular etiology.

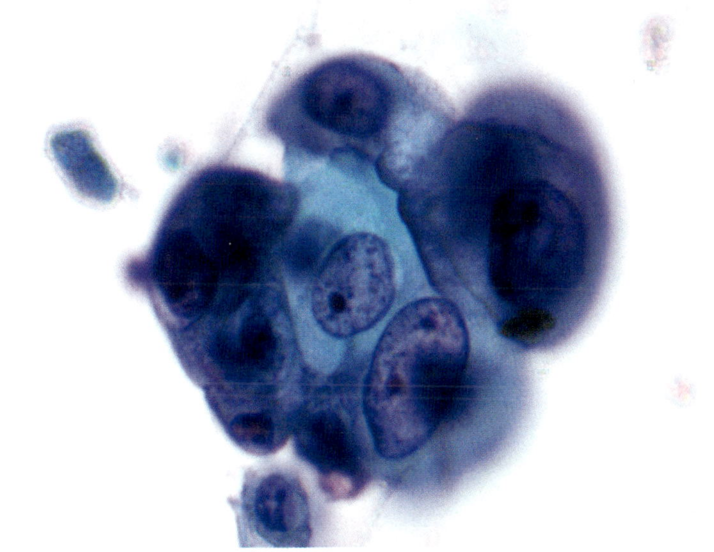

Figure 6.5d — Reactive Bronchial Epithelial Cells, Bronchial Brush [Pap stain; high power]. This field shows large polygonal cells with enlarged nuclei and prominent nucleoli, mimicking malignant glandular cells. Unlike neoplastic cells, however, the nuclear-to-cytoplasmic ratio is preserved and the chromatin is uniformly open, with one or two chromocenters identified per nucleus.

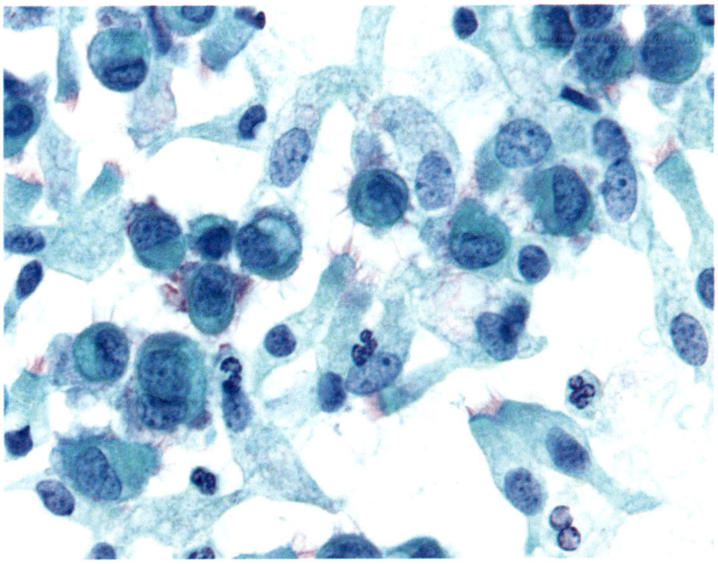

Figure 6.6a — Reactive Bronchial Epithelial Cells, Bronchial Brush [Pap stain; high power]. Reactive bronchial epithelial cells often retain cilia and terminal bars that aid with a benign diagnosis. In the absence of visible cilia or partial loss of cytoplasm, a benign diagnosis can be established by comparing the nuclear shape and chromatin quality of the cells in question with other clusters containing benign ciliated cells or identifying terminal bars. In this field, one can appreciate scattered round, reactive mononuclear cells in a background of columnar cells with preserved nuclear polarity, terminal bars, and cilia.

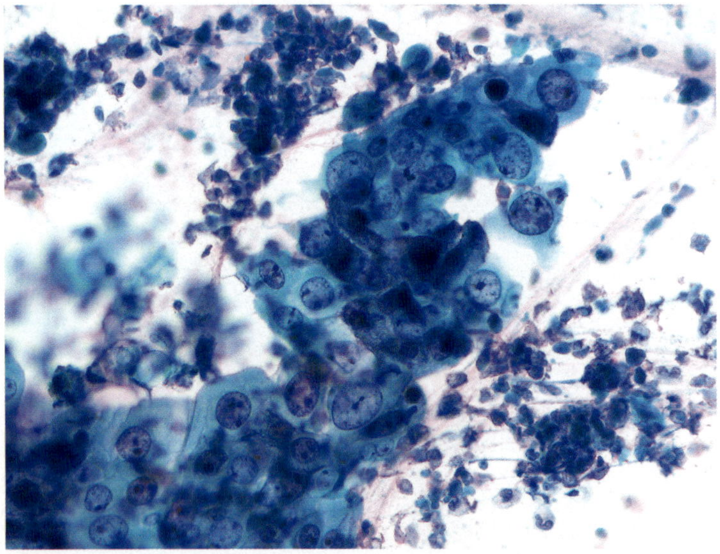

Figure 6.6b — Reactive Bronchial Epithelial Cells, Bronchial Brush [Pap stain; high power]. Reactive changes in the lung are relatively common and nonspecific, occurring as a response to inflammatory and environmental stimuli, including diffuse alveolar damage, pulmonary infarct, chemoradiation, instrumentation, asthma, and infections. Although there can be marked variation in size, macronucleoli, nuclear contour irregularities, and moderately coarse chromatin, irregularly distributed chromatin with condensation under the nuclear membrane should not be seen. In reactive conditions, careful examination usually reveals a spectrum of changes, whereas two distinct populations of normal and clearly malignant changes are typically identified in adenocarcinomas. The cells in this field have predominantly centrally placed nuclei, smooth nuclear contours, small nucleoli, abundant cytoplasm, and mild variation in size. Although this is a dramatic example of reactive atypia, some clues that this is not a malignancy are the abundant cytoplasm, the bland chromatin pattern, and the relative roundness and regularity of the nuclear membranes.

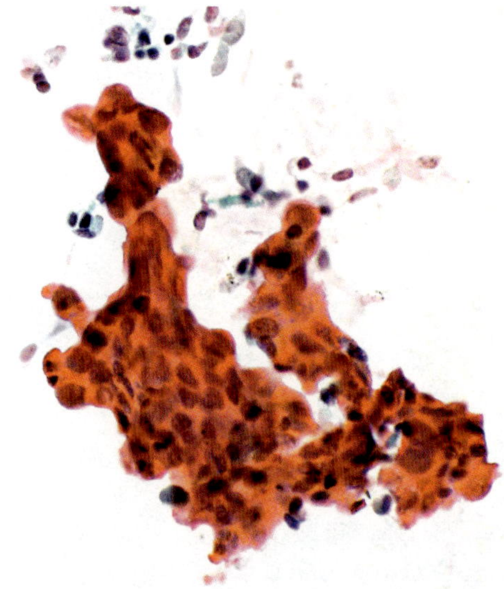

Figure 6.7 — Squamous Metaplasia, Bronchial Washing [Pap stain; high power]. The presence of keratinized squamous fragments should not cause one to assume that the diagnosis is squamous cell carcinoma. Numerous reactive processes can cause squamous metaplasia within the large airways, which can appear as fragments, such as this on pulmonary specimens. Note how the nuclei have regular borders and a bland chromatin pattern, although nuclei in both benign and malignant keratinized squamous cells often stain poorly. The number of metaplastic squamous cells is typically less than what is seen in a neoplastic process such as squamous cell carcinoma.

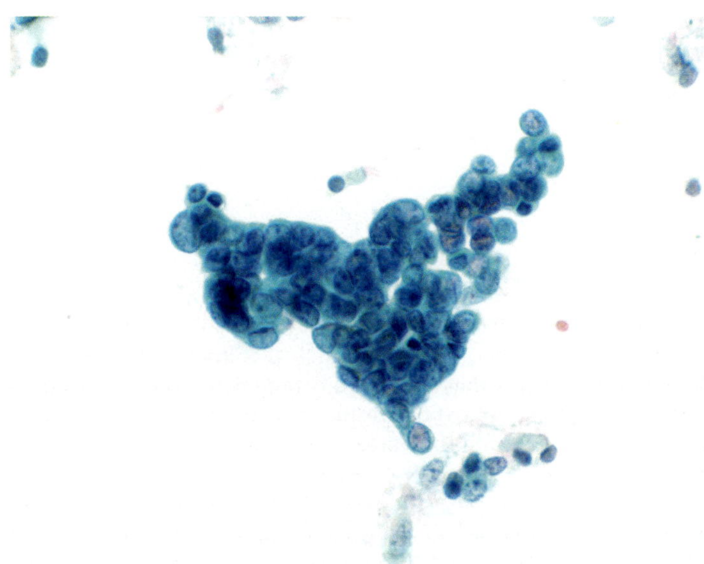

Figure 6.8a — Reserve Cell Hyperplasia, Bronchial Washing [Pap stain; high power]. Reserve cells are found under the ciliated cells of the respiratory tract. Under certain circumstances, reserve cells proliferate to form multiple cellular layers and may be seen as small fragments in respiratory specimens. The cells are small and have high nuclear-to-cytoplasmic ratios. These findings mimic small cell carcinoma. Reserve cell hyperplasia lacks the following features that are associated with small cell carcinoma: necrosis, mitotic figures, increased cellularity, discohesive cells, smearing artifact, and a salt-and-pepper chromatin pattern.

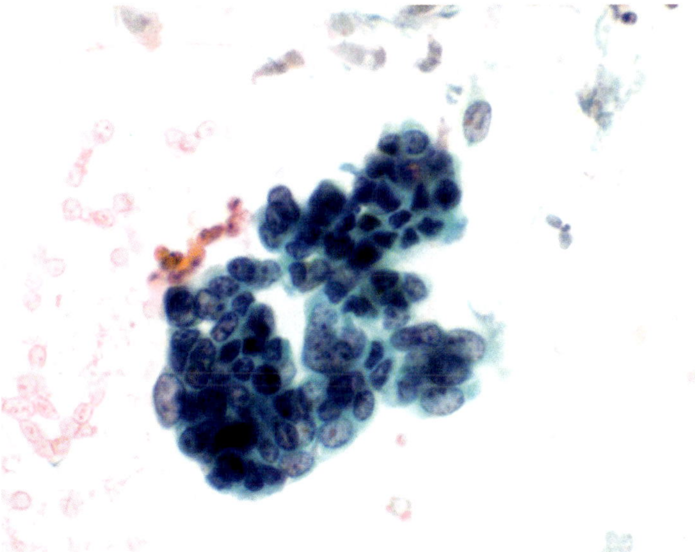

Figure 6.8b — Reserve Cell Hyperplasia, Bronchial Washing [Pap stain; high power]. The reserve cells seen here vary in nuclear size and have moderate nuclear border irregularities. These features, combined with the high nuclear-to-cytoplasmic ratios, raise concern for malignancy. Note the columnar cells at the bottom right corner of the fragment; they represent ciliated respiratory epithelial cells attached to the hyperplastic reserve cells. This finding provides reassurance that the cells within the fragment are benign, consistent with the saying "birds of a feather flock together."

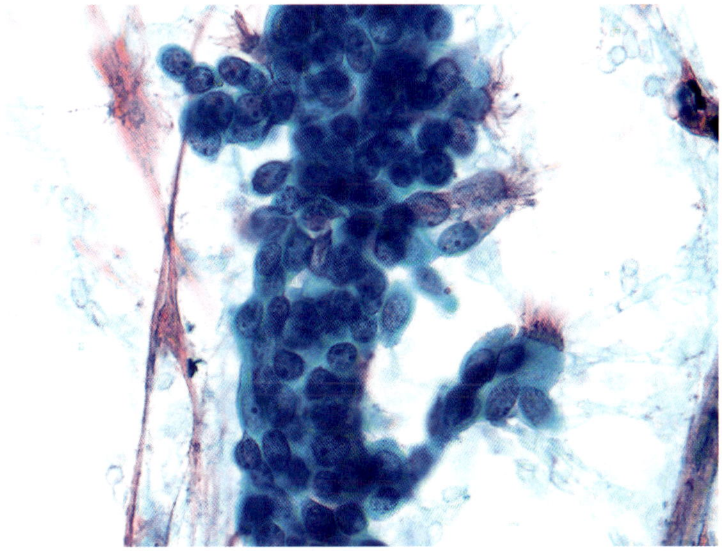

Figure 6.8c — Reserve Cell Hyperplasia, Bronchial Washing [Pap stain; high power]. The benign nature of these reserve cells is indicated by their association with ciliated respiratory epithelial cells. The ciliated cells are benign and not simply artifactually associated with the reserve cells; the reserve cells proliferate below the respiratory cells. In instances in which the ciliated cells are not present or identified, the reserve cells can mimic small cell carcinoma.

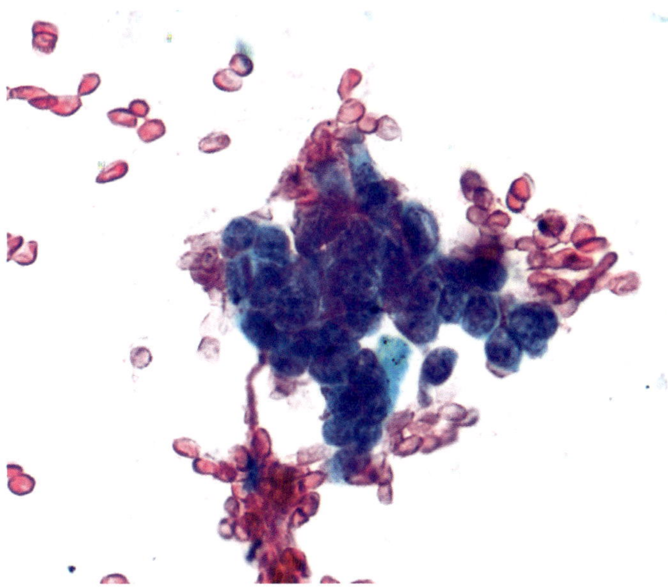

Figure 6.8d — Reserve Cell Hyperplasia, Bronchial Washing [Pap stain; high power]. Here the reserve cells have very little cytoplasm and the nuclei appear to mold against one another, suggestive of small cell carcinoma. Several nuclei have two or three chromocenters, which provide some coarseness to the chromatin and mimic a salt-and-pepper appearance. One clue is the presence of two columnar ciliated respiratory epithelial cells associated with the reserve cells at the top of the fragment; unfortunately, their pink cilia has become obscured by the red blood cells coating the fragment.

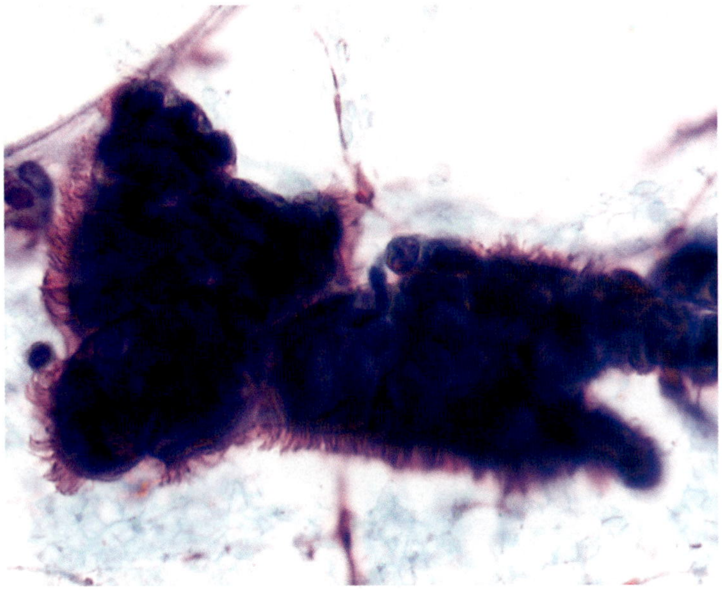

Figure 6.9a — Reserve Cell Hyperplasia, Bronchial Washing [Pap stain; high power]. This fragment is essentially an irregularly shaped Creola body, which typically has a round or oval shape. Here, the fragment forms a papillary-like projection and is distinctly lined by ciliated respiratory epithelial cells. Reserve cells have proliferated under the ciliated cells, giving the fragments a hyperchromatic appearance and making cytoplasmic features difficult to assess. Breaking even benign cytologic findings down into components can help one understand potential dilemmas in other specimens: For instance, the reserve cells would look malignant if it were not for the ciliated cell component.

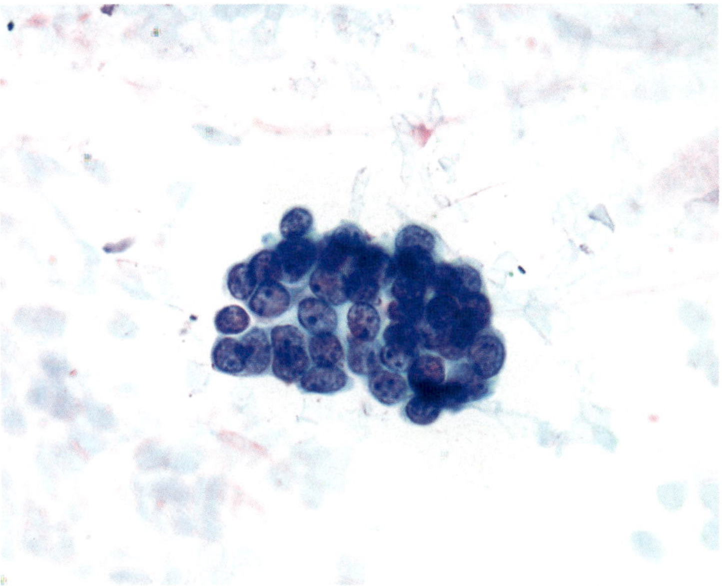

Figure 6.9b — Reserve Cell Hyperplasia, Bronchial Washing [Pap stain; high power]. The reserve cells form a fragment and are not as loosely associated as those seen in small cell carcinoma. Although the nuclei overlap, they do not appear to indent one another or "mold" together, as is seen in small cell carcinoma. Note how several nuclei may contain two or more chromocenters; several cells have only one distinct chromocenter, a feature not seen in small cell carcinoma.

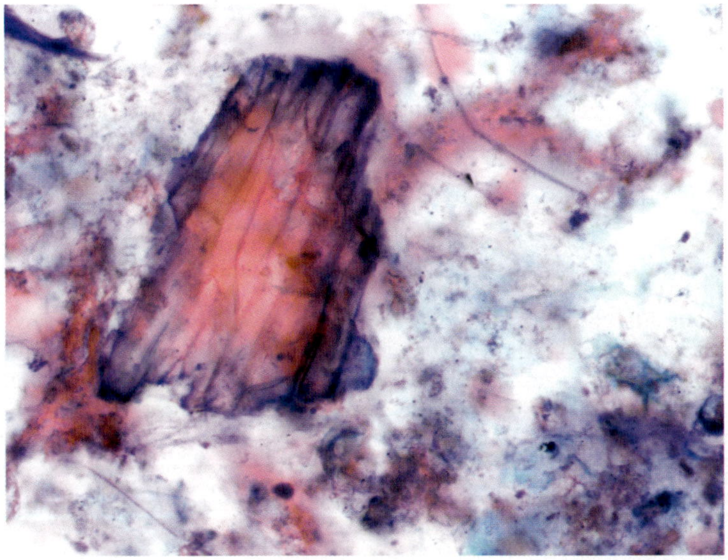

Figure 6.10a — Vegetable Material, Sputum [Pap stain; medium power]. The presence of vegetable material in cytologic specimens should always be interpreted in context. In many instances, vegetable material may simply be contamination (e.g., from the oral cavity). On the other hand, hospitalized patients are often at risk for aspiration, which can lead to pneumonia and radiologic findings indicating that a sputum sample should be collected. Because one cannot be certain, noting vegetable material in respiratory specimens allows the clinical team to make the final determination.

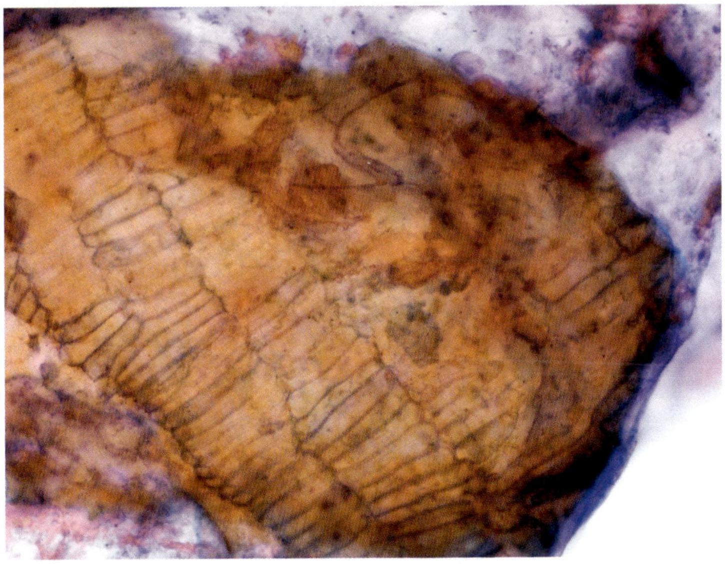

Figure 6.10b — Vegetable Material, Sputum [Pap stain; medium power]. Vegetable material often forms rigid, geometric structures that can otherwise be varied in appearance. They typically have a distinct enough appearance and do not create a differential diagnosis, though some may raise concern for a parasitic infection. In this case, degenerated blood coats the structure, suggesting that it may have been lodged in the airway rather than contaminated from the oral cavity due to a recently eaten meal.

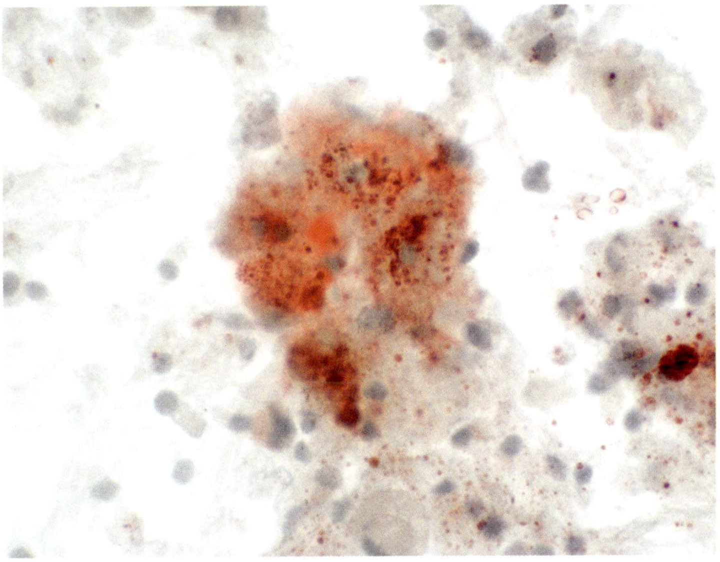

Figure 6.11a — Lipid Aspiration, Bronchoalveolar Lavage [Oil Red O special stain; high power]. The counterstain is weak, demonstrating the nuclei of columnar benign respiratory epithelial cells and round pulmonary alveolar macrophages. The oil red O special stain can be performed to identify lipid on air-dried slides that have not been exposed to alcohol, which would otherwise remove the lipid from them. The test is often performed to identify the presence of lipid in the airway, which is an abnormal finding and indicates aspiration or structural disorders such as tracheoesophageal fistula. The presence of lipid material within alveolar macrophages, seen here as small red spherules of different sizes, is a positive result and requires clinical explanation.

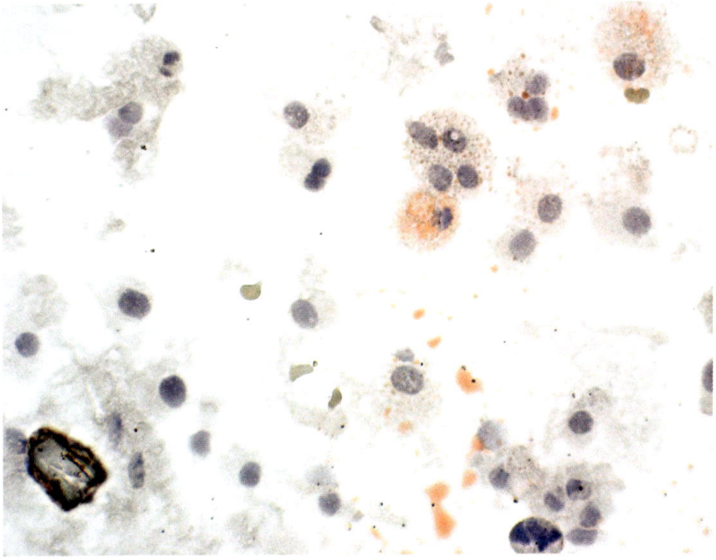

Figure 6.11b — Lipid Aspiration, Bronchoalveolar Lavage [Oil Red O special stain; high power]. The red-orange stippling seen in pulmonary alveolar macrophages is highlighting the presence of lipids that have been phagocytosed. An oil red O special stain can be performed on air-dried, nonalcohol fixed slides. The stain is commonly performed in neonatal patients to identify a tracheoesophageal fistula or in older patients to identify aspiration of lipid-containing particles. Occasionally, cholesterol clefts and foreign body giant cells, in addition to other particles, may be seen. Lipid material does not polarize, but foreign material aspirated with lipid material may polarize, providing a clue to an aspiration event.

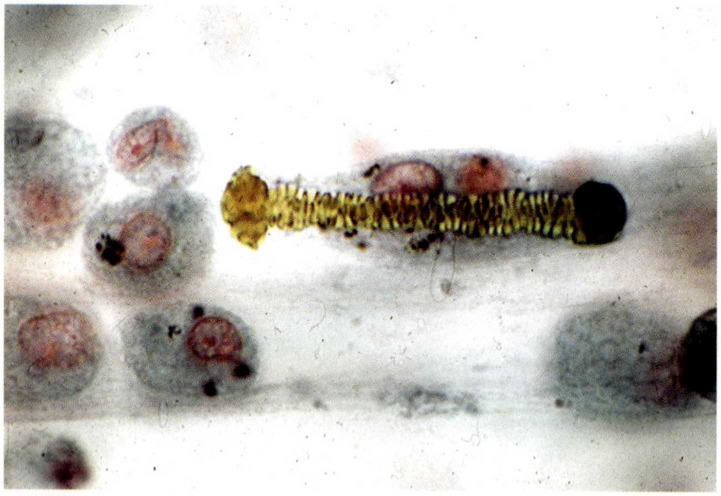

Figure 6.12 — Asbestos [Ferruginous] Body, Bronchial Washing [Pap stain; high power]. This high-power view reveals multiple scattered pulmonary alveolar macrophages and a large, atypical form coursing through a macrophage in the center of the field. This structure represents an asbestos body, also known as a *ferruginous body*. Macrophages phagocytose asbestos fibers; however, they cannot digest the fibers. The color of these bodies is due to the presence of ferritin and hemosiderin coating the outside of the fibers and is not the result of the Pap stain. In some circumstances, the fibers are slender and contain a beaded pattern. In other cases, like this one, the asbestos bodies are coated with more iron-rich materials and form a dumbbell-shaped structure. In addition to being implicated in mesothelioma, asbestos bodies may be associated with interstitial lung disease.

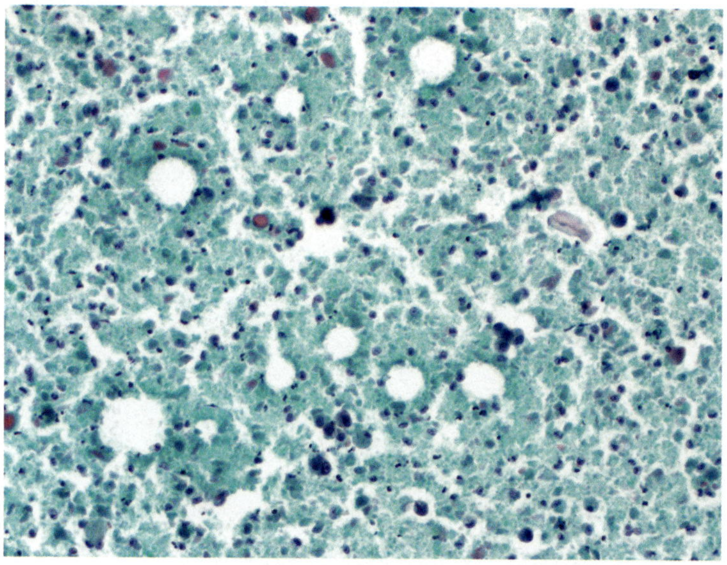

Figure 6.13a — Pulmonary Alveolar Proteinosis [PAP], Bronchoalveolar Lavage [Pap stain; low power]. This specimen contains an abundance of granular debris and scattered pulmonary alveolar macrophages. On low power, it may be difficult to distinguish this granular debris from necrotic debris; however, the presence of smooth magenta globules scattered in the background favor that the material is a protein-rich fluid, as seen in PAP. On higher power, the protein globules are also seen in cytoplasmic vacuoles within pulmonary alveolar macrophages. PAP is typically due to surfactant excess or macrophage dysfunction as a result of autoantibodies against granulocyte–macrophage colony-stimulating factor (GM-CSF); however, it can be seen in a variety of clinical circumstances, such as inhalation injury, immunosuppression, or hematologic malignancies.

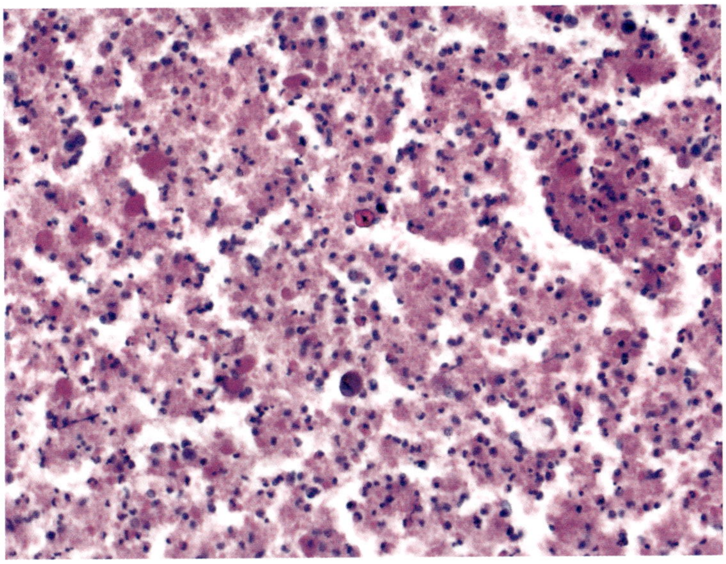

Figure 6.13b — Pulmonary Alveolar Proteinosis (PAP), Bronchoalveolar Lavage [Periodic acid–Schiff (PAS) stain; low power]. In this field, there is an abundance of granular material; larger smooth, pink globules; and pulmonary alveolar macrophages with pink, vacuolated cytoplasm. This PAS stain is highlighting the protein-rich globules and granular debris, confirming the diagnosis of PAP. Patients present in respiratory distress, and on histology, the proteinaceous material is seen filling the alveoli with preserving of the alveolar architecture.

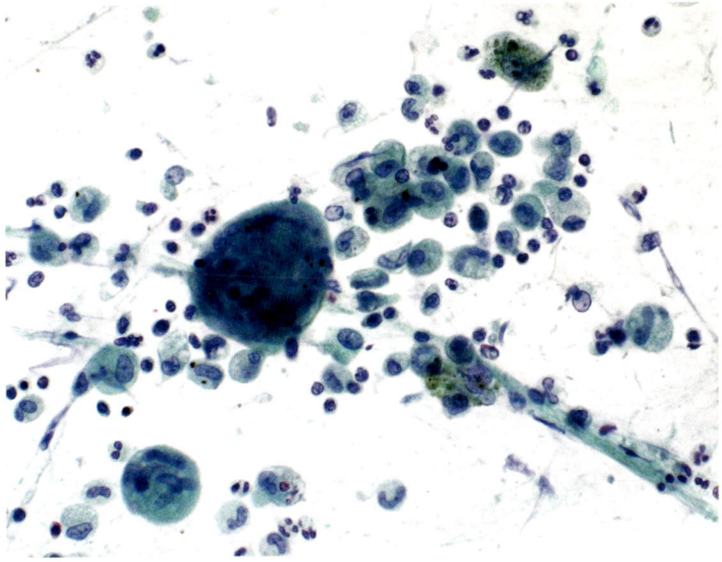

Figure 6.14 — Mixed Inflammatory Infiltrate with Multinucleated Giant Cell Reaction, Bronchial Brushing [Pap stain; high power]. Multiple inflammatory cell types are seen in this image, including neutrophils, eosinophils, lymphocytes, pulmonary alveolar macrophages, and multinucleated giant cells. The presence of multinucleated giant cells is a nonspecific finding with a broad differential diagnosis, including fungal and mycobacterial infections, autoimmune and connective tissue disorders such as sarcoidosis, and aspiration of foreign materials. In all cases, fungal and acid-fast bacillus (AFB) stains should be performed with available material, and clinical correlation is often required to suggest the correct differential diagnosis.

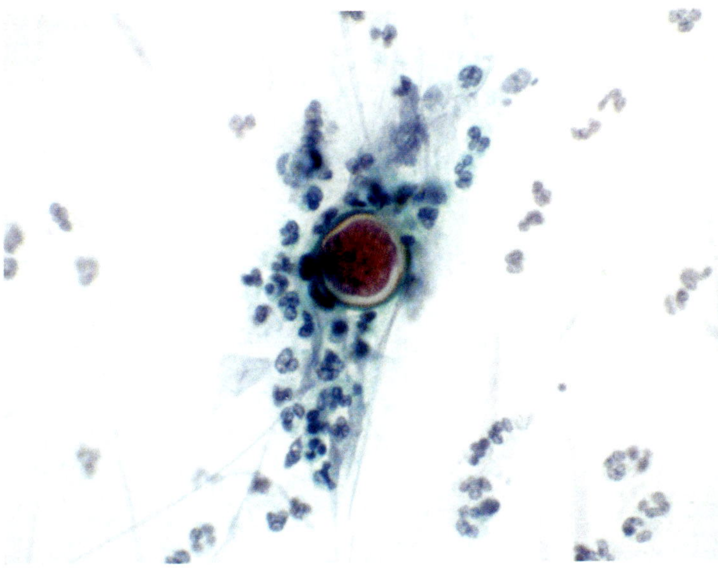

Figure 6.15a — Pulmonary Blastomycosis, Bronchial Washing [Pap stain; high power]. Blastomycosis is acquired through inhalation of microconidia (spores) produced by the dimorphic fungus *Blastomyces dermatitidis*, which is found in moist soil enriched with decomposing acidic organic vegetation. Cytology smears typically contain marked neutrophilic/histiocytic (pyogranulomatous) inflammation and large (15–20 μm), multinucleate yeast cells with refractile, double-contoured cell walls and single broad-based buds. If broad-based buds are not present, *B. dermatitidis* may be confused with *Cryptococcus neoformans*. Unlike *B. dermatitidis*, the mucopolysaccharide capsule of *C. neoformans* stains intensely with mucicarmine. Capsule-deficient forms of *C. neoformans* can be differentiated based on positive staining with melanin stains (*B. dermatitidis* is negative).

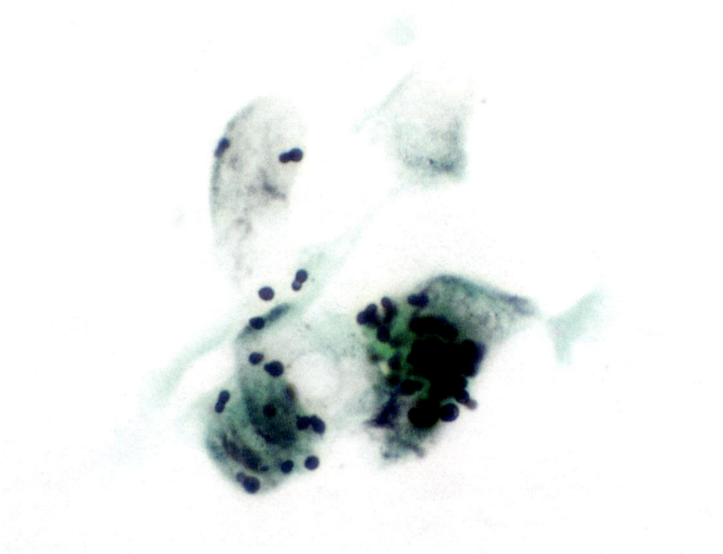

Figure 6.15b — Pulmonary Histoplasmosis, Bronchial Washing [Grocott's methenamine silver [GMS] stain; high power]. Histoplasmosis is caused by inhalation of microconidia of *Histoplasma capsulatum*, which is a dimorphic fungus found in soil rich in bird and bat droppings. The organisms are small, about half the size of a red blood cell (2–5 µm), and uniform in size, with single narrow-based buds (top center). They are typically not visible on H&E, except in disseminated histoplasmosis, in which they cluster within histiocytes. On H&E, the basophilic yeast cytoplasm is often separated from the surrounding tissue by a clear zone corresponding to the cell wall, which can be highlighted with a GMS stain. The appearance is similar to that of leishmaniasis, but positive GMS staining excludes that diagnosis. In cases of acute histoplasmosis, scattered organisms can be found in the central-most necrotic areas of granulomas.

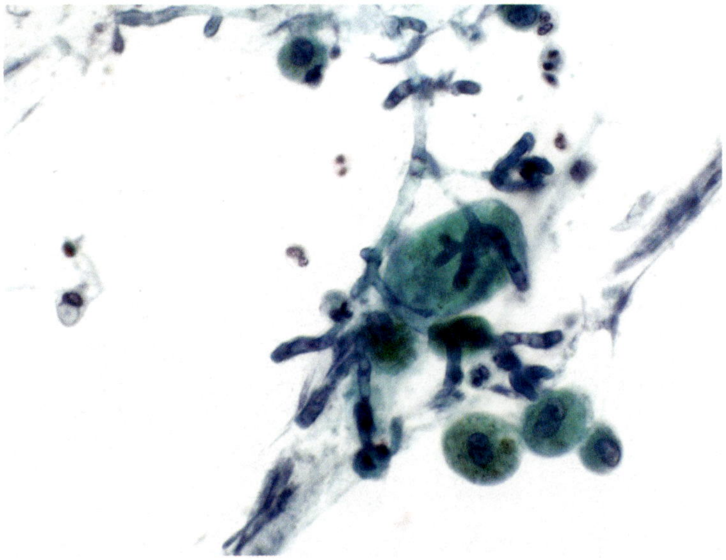

Figure 6.16a — *Aspergillus*, Bronchial Washing [Pap stain; high power]. This high-power view shows several scattered pulmonary alveolar macrophages with metachromatic cytoplasm and bean-shaped nuclei, along with a few neutrophils in the background. Coursing through the center of this field, fungal forms are seen with septate and acute (45°) hyphae with well-defined straight edges. Because this morphology is not specific, it is best to provide a description of the fungal organism and raise the possibility of *Aspergillus*. In immunocompromised patients, such as those with AIDS or undergoing chemotherapy, *Aspergillus* can cause life-threatening pneumonia; however, in a healthy person, the presence of *Aspergillus* on a bronchial washing could be an incidental finding. In this case, though, the presence of neutrophils makes this organism more likely to be of clinical significance.

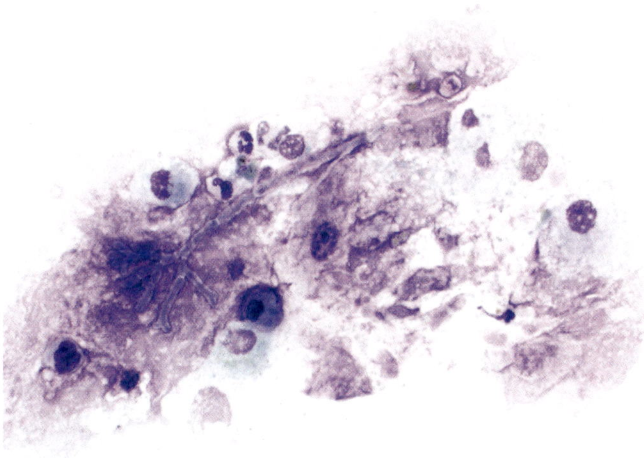

Figure 6.16b — *Aspergillus*, Bronchial Washing [Diff-Quik stain; high power]. This Diff-Quik preparation shows several leukocytes, including pulmonary alveolar macrophages with bubbly cytoplasm, lymphocytes, and neutrophils. A single fungal organism is present with sharp borders, septations, and acute-angle-branching hyphae at one end. These findings raise the possibility of *Aspergillus*; however, correlation with microbiology studies such as culture is required for a definitive diagnosis. The presence of *Aspergillus* in a washing specimen may be incidental finding, indicative of an infectious pneumonia, or a cause of a hypersensitivity reaction called *allergic bronchopulmonary aspergillosis*. Ultimately, clinical correlation is required to determine the significance and the appropriate course of treatment.

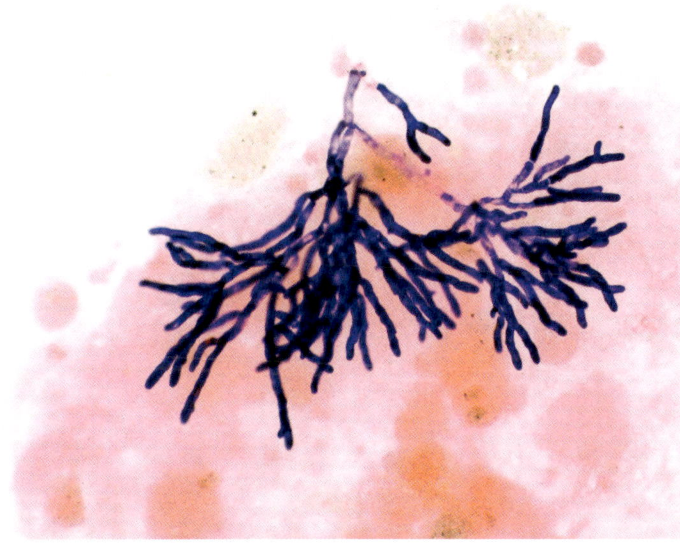

Figure 6.16c — *Aspergillus*, Bronchial Washing [GMS stain; high power]. This GMS stain highlights a fungal organism with septate, acute-angle-branching hyphae with uniform hyphal thickness. These features are consistent with those seen with *Aspergillus*. It is important to attempt to distinguish the features of *Aspergillus* with those seen in the Zygomycetes species, which contain nonseptate hyphae with 90° angle branching and wavy or ribbon-like edges. In addition, the Zygomycetes species have hyphae with a clear or empty appearance.

Figure 6.17a — Necrotizing Granulomatous Inflammation, Wedge Resection [H&E stain; low power]. This wedge resection of lung for a radiographically suspicious mass shows a large granuloma with central necrosis. The periphery of the granuloma contains mixed lymphohistiocytic chronic inflammation with multinucleated giant cells, and the center of the granuloma contains amorphous, coagulative necrosis. The leading differential of necrotizing granulomatous inflammation is an infectious process, namely fungal or mycobacterial infection. Granulomatous inflammation can also be seen in allergic granulomatosis and angiitis (Churg–Strauss syndrome), granulomatous with polyangiitis (Wegener's syndrome), and sarcoidosis. Typically, special stains for microorganisms (such as Grocott's methenamine silver [GMS] and acid-fast bacillus [AFB]) should be performed, as should correlation with microbiologic cultures.

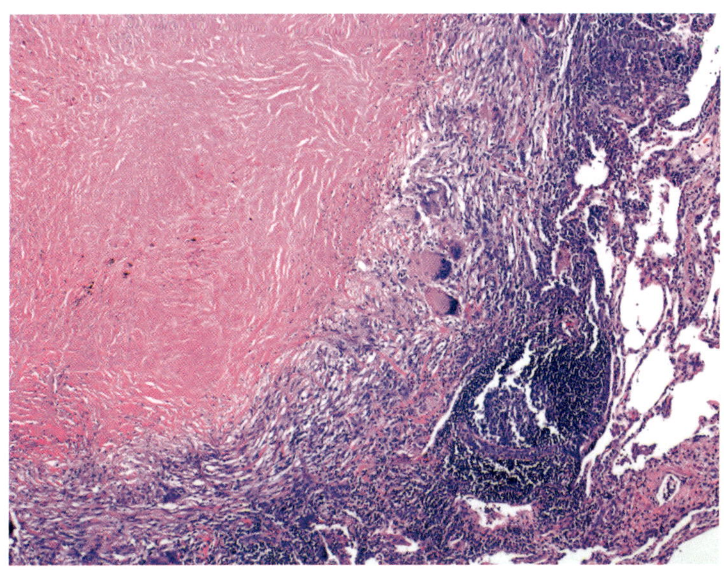

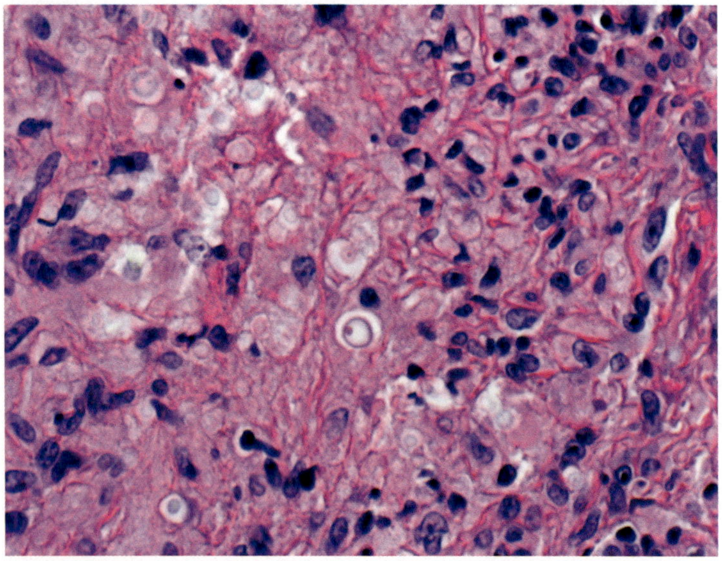

Figure 6.17b — Necrotizing Granulomatous Inflammation, Biopsy [H&E stain; high power]. High-power examination of the cellular component of a necrotizing granuloma reveals a mixed chronic inflammatory infiltrate, including lymphocytes and histiocytes with abundant eosinophilic cytoplasm. In addition, there are structures highly suggestive of microorganisms, consisting of small round, cyst-like structures like those seen in the center of this image. Microorganisms are not always evident on the H&E section, but occasionally, the fungal hyphae of *Aspergillus* or Zygomycetes, as well as the yeast forms of *Cryptococcus* or *Histoplasmosis*, may be evident. Special stains such as Grocott's methenamine silver (GMS) and periodic acid–Schiff (PAS) are recommended for further evaluation and confirmation of the presence of microorganisms.

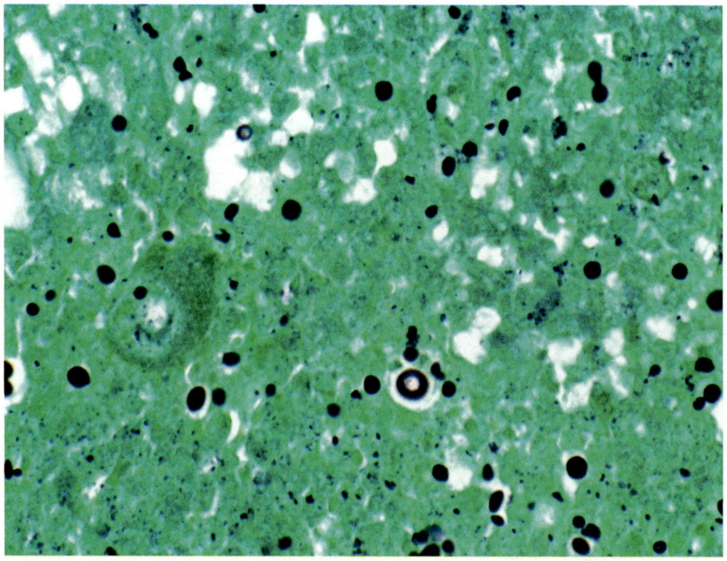

Figure 6.17c — Necrotizing Granulomatous Inflammation, Biopsy [Grocott's methenamine silver (GMS) stain; high power]. A GMS stain highlights the yeast forms detected on the H&E of this necrotizing granuloma, as well as additional scattered budding yeast forms. This is an example of *Cryptococcus* infection, and the *Cryptococcus* yeast forms have a capsule that is also positive on the mucicarmine and Fontana–Masson special stains. Correlation with serology and microbiologic cultures is still recommended. Cryptococcal infection can cause mild pulmonary infection in immunocompetent patients but can also lead to severe infections in immunocompromised patients.

Figure 6.17d — Pulmonary Sarcoidosis, Biopsy [H&E stain; intermediate power]. This lung biopsy reveals scattered non-necrotizing granulomas composed of epithelioid histiocytes with scattered lymphocytes. In this field, the granulomas involve the interstitial space between the alveoli; in other fields, the granulomata are centered around airways and vessels, but there is no vasculitis. The findings could be compatible with sarcoidosis, but sarcoidosis is a diagnosis of exclusion. Infectious processes such as fungal or mycobacterial infection should be excluded before making a diagnosis of sarcoidosis. The granulomas in sarcoidosis very rarely have central necrosis.

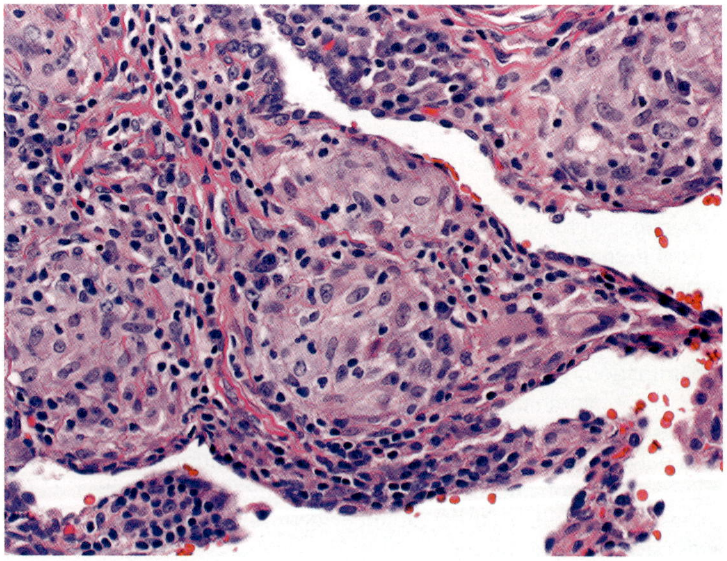

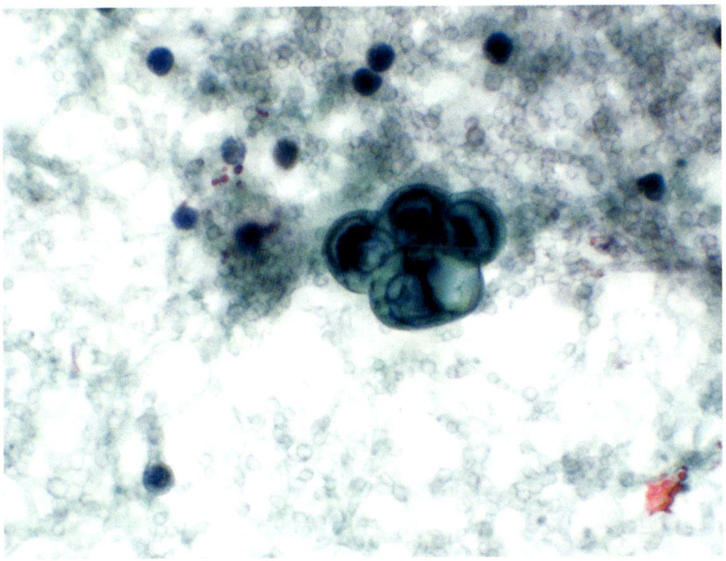

Figure 6.18a — Cytomegalovirus [CMV] Pneumonitis, Bronchoalveolar Lavage [Pap stain; high power]. As in histologic sections, CMV-infected cells are enlarged, with single, large, basophilic intranuclear (Cowdry type A) inclusions that are much larger than a nucleolus. The nuclear chromatin is typically marginated, leaving a clear space between the inclusion and the nuclear membrane. Characteristic intracytoplasmic inclusions as originally defined on other preparations are not often seen in Pap-stained material.

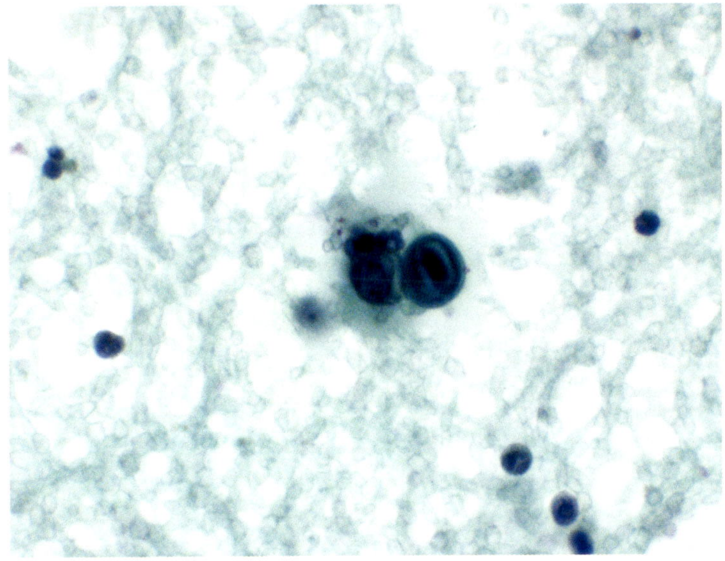

Figure 6.18b — Cytomegalovirus (CMV) Pneumonitis, Bronchoalveolar Lavage [Pap stain; high power]. CMV-cytopathic changes can be found in bronchial cells, type II pneumocytes, alveolar macrophages, and endothelial cells of septal capillaries. Detection of viral cytopathic effect may represent the earliest manifestation of a developing pneumonitis or the detection of a latent virus and indicate CMV infection without clinical or pathologic disease. On such occasions, the decision to treat would depend on the clinical situation. In lung transplant patients, significant CMV disease occurs most commonly in the first 2 to 3 months post-transplant, although it may also occur anytime thereafter. Recipients who are serologically negative and receive serologically positive lung transplants (R^-/D^+) are at the highest risk for CMV disease and require the most aggressive anti-CMV prophylactic regimen.

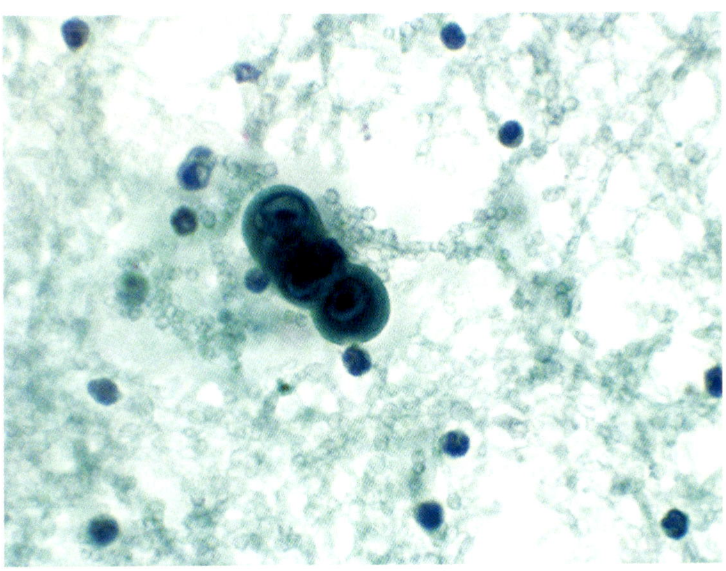

Figure 6.18c — Cytomegalovirus (CMV) Pneumonitis, Bronchoalveolar Lavage Fluid [Pap stain; high power]. When viral cytopathic effect is suspected, a panel of immunohistochemical stains for CMV and other herpesviruses, such as herpes simplex viruses (HSV) 1 and 2, varicella-zoster virus (VZV), and adenovirus, should be performed. Classic Cowdry type A nuclear inclusions can also be seen in VZV and HSV, but only HSV shows multinucleation.

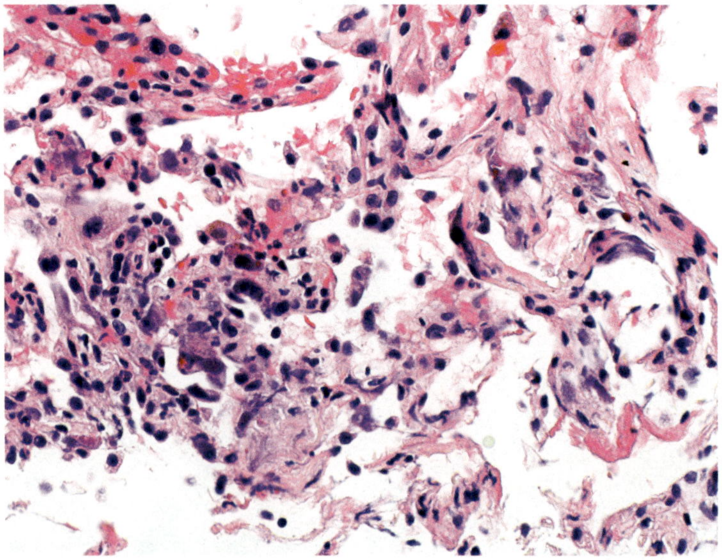

Figure 6.19a — Cytomegalovirus (CMV) Pneumonitis, Biopsy [H&E stain; intermediate power]. Transbronchial biopsy of this patient's lung for post–lung transplant surveillance shows no evidence of acute rejection but displays scattered hyperchromatic and enlarged nuclei. There are scattered neutrophils, but the inflammatory response is not prominent. The differential diagnosis of hyperchromatic nuclei at low power includes reactive pneumocytes and viral inclusions, including CMV, herpes simplex virus (HSV), and adenovirus. Higher-power examination of the nuclei, as well as confirmatory viral immunostains, can aid in the diagnosis. CMV pneumonitis is a major source of morbidity in immunocompromised patients, including lung transplant recipients.

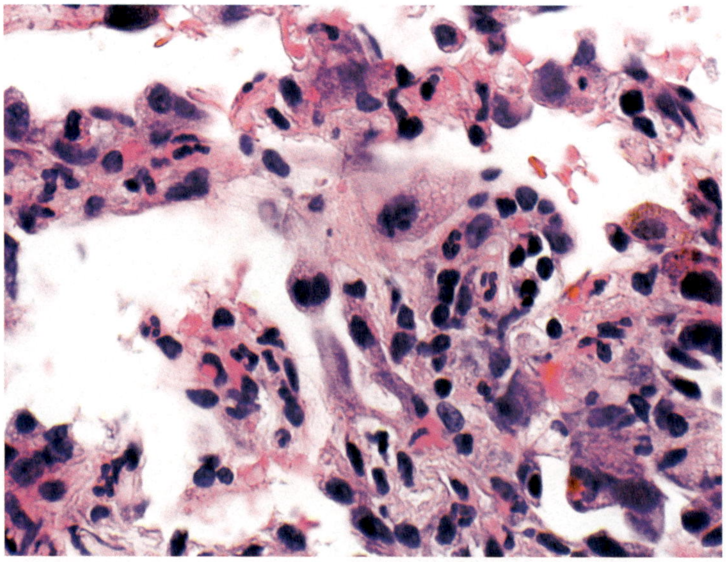

Figure 6.19b — Cytomegalovirus (CMV) Pneumonitis, Biopsy [H&E stain; high power]. Higher-power examination of this transbronchial biopsy reveals numerous pneumocytes with intranuclear inclusions, some of which are basophilic and binucleate in appearance and some of which are bright red. The classic CMV inclusion is the "owl's eye" inclusion with a peripheral halo. CMV can infect the endothelial cells, the pneumocytes, and the alveolar macrophages. The differential diagnosis includes herpes simplex virus (HSV), which results in nuclear molding and Cowdry A and B inclusions, as well as adenovirus, which results in basophilic inclusions.

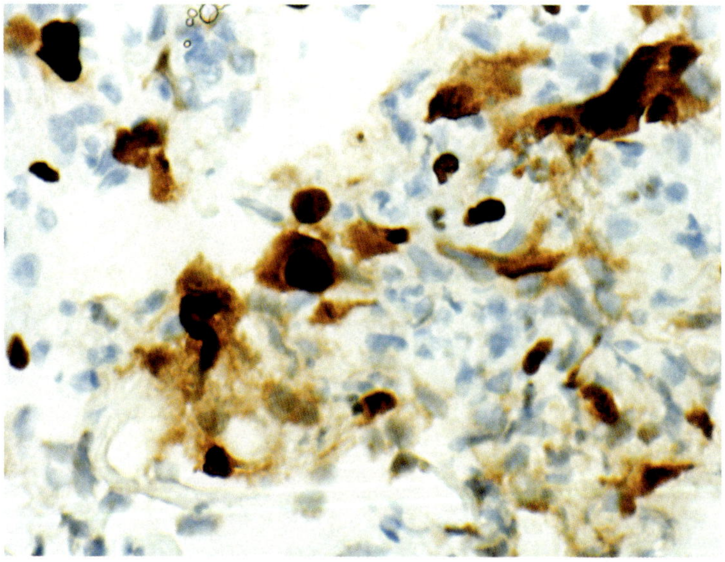

Figure 6.19c — Cytomegalovirus (CMV) Pneumonitis, Biopsy [CMV immunostain; high power]. A CMV immunostain highlights the numerous viral inclusions in this transbronchial biopsy and is confirmatory for a diagnosis of CMV pneumonitis. A low threshold for performing viral stains is warranted in immunosuppressed patients because there may be minimal inflammatory infiltrate associated with infection.

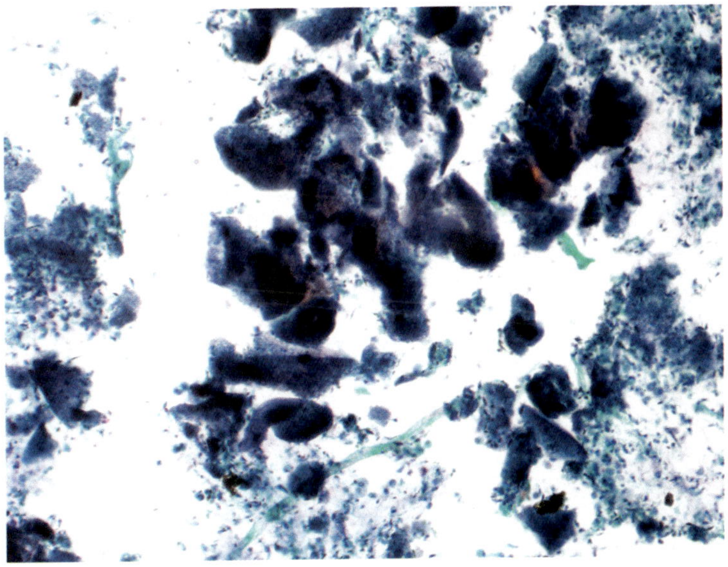

Figure 6.20a — *Pneumocystis jirovecii*, Bronchoalveolar Lavage [Pap stain; low power]. At low power, the Pneumocystis organisms form numerous granular-appearing clumps. In many specimens, the granular material may be a focal finding. Examination at higher power usually confirms the diagnosis on cytomorphology alone. Although classically associated with AIDS, this organism can also be found in other immunocompromised patients, such as transplant patients. Even though modern testing performed in the microbiology laboratory has excellent sensitivity, this finding should never be missed on a cytologic specimen.

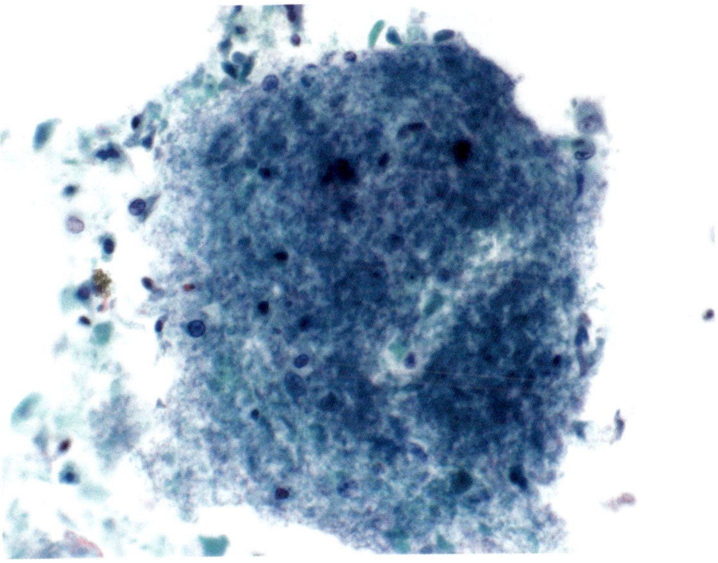

Figure 6.20b — *Pneumocystis jirovecii*, Bronchoalveolar Lavage [Pap stain; high power]. High-power examination reveals the cohesive granular debris ("foamy alveolar casts") classically associated with *Pneumocystis* pneumonia. Focusing in and out of the field often allows the negative staining image of sporozoite structures to be seen. If necessary, a silver stain can be performed on additional material to verify the presence of this yeast-like fungus.

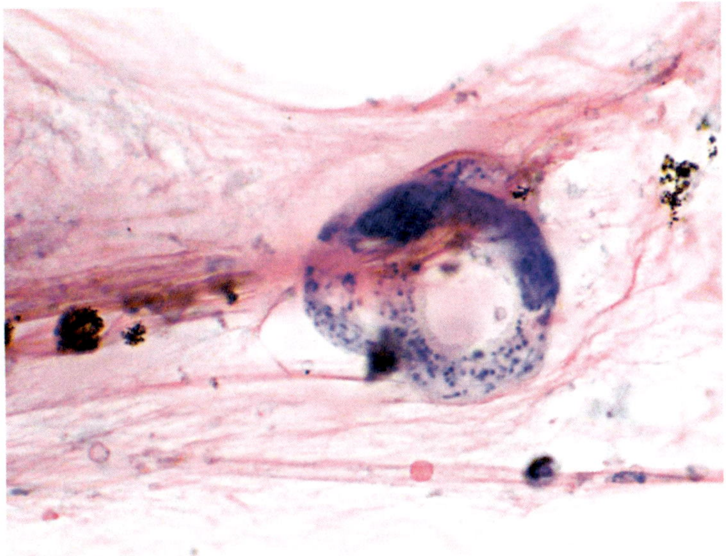

Figure 6.21 — *Strongyloides stercoralis*, Bronchial Washing [H&E stain; high power]. This is a filariform larva of *S. stercoralis.* Other parasites in the differential diagnosis include *Ascaris lumbricoides* and hookworms (both *Necator americanus* and *Ancylostoma duodenale*). The organism is identified as being *S. stercoralis* because of its length (400–500 μm) and blunt tail. This is in contrast to the tails of the filariform larvae in both hookworm and *Ascaris*, which are sharply pointed. Respiratory infection with *S. stercoralis* is found mostly in patients receiving high-dose steroid therapy. Pulmonary infection is produced when the filariform larvae migrate through the intestinal wall into the blood and penetrate into alveolar spaces.

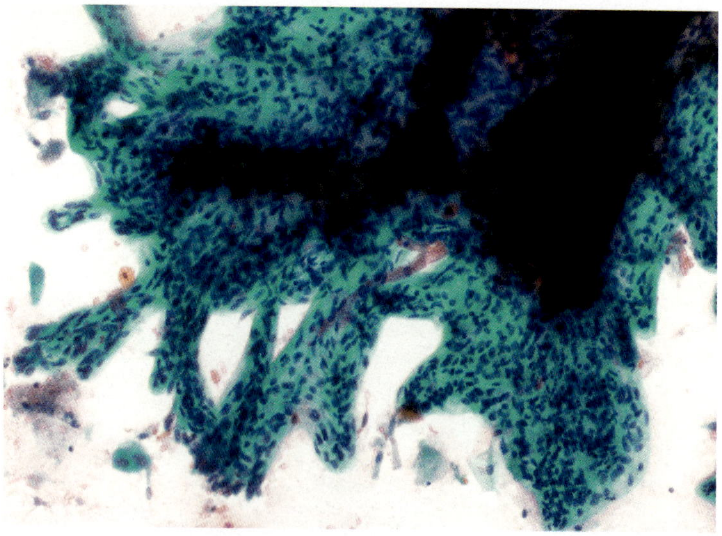

Figure 6.22a — Laryngeal Papillomatosis, Laryngeal Brushing [Pap stain; low power]. Although technically a head and neck specimen, neoplasms within the larynx may be encountered during procedures investigating the respiratory tract. Furthermore, papillomas may involve the tracheobronchial tree. In this case, numerous papillae formed from monotonous cells with bland cytomorphologic features. The cellular nature of this specimen suggests a neoplasm, and cell block material can be obtained for human papillomavirus (HPV) testing (types 6 and 11 in most instances). These lesions have the propensity to recur, and recurrences may become dysplastic or malignant; high-grade dysplasia cannot be reliably distinguished from carcinoma on a brushing specimen alone.

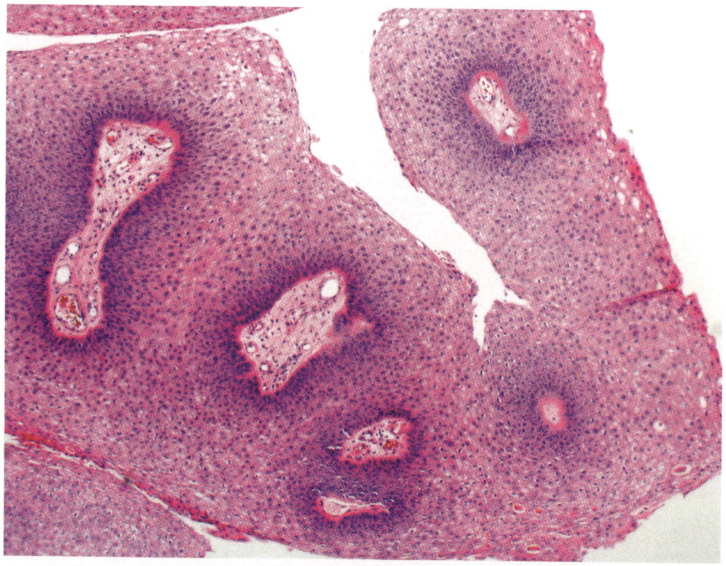

Figure 6.22b — Laryngeal Squamous Papilloma, Biopsy [H&E stain; low power]. Low-power examination of this laryngeal biopsy reveals a papillary proliferation characterized by central fibrovascular cores lined by variably thickened squamous epithelium. The differential diagnosis includes a benign squamous papilloma and a papillary squamous cell carcinoma, but at this power there is no atypia to suggest carcinoma. A vocal cord polyp may also be in the clinical and histologic differential; however, these are characterized by expanded and edematous submucosa with overlying bland squamous epithelium.

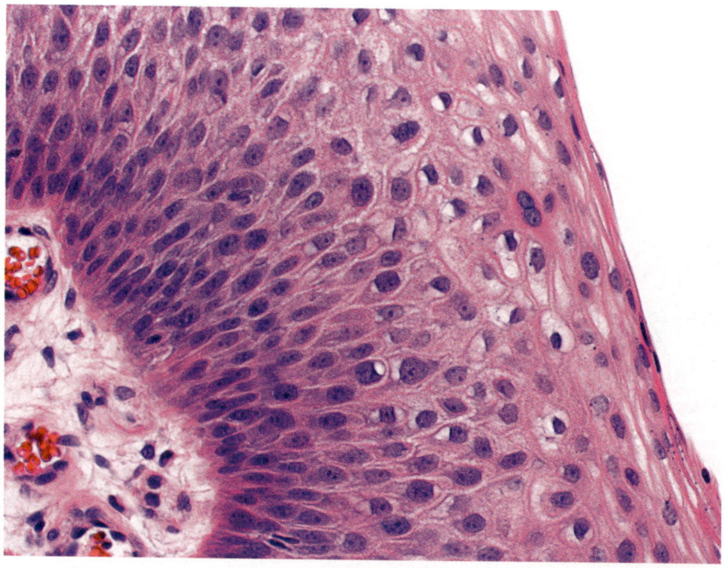

Figure 6.22c — Laryngeal Squamous Papilloma, Biopsy [H&E stain; intermediate power]. Higher-power examination of this vocal cord lesion reveals maturing squamous epithelium that displays koilocytic change and atypia, namely the presence of scattered binucleate forms, nuclear clearing, and enlarged nuclei with irregular nuclear contours. Indeed, most laryngeal squamous papillomas are related to low-risk human papilloma virus (HPV) infection, particularly types 6 and 11. Laryngeal squamous papillomas are commonly multiple and frequently recur, but they are benign.

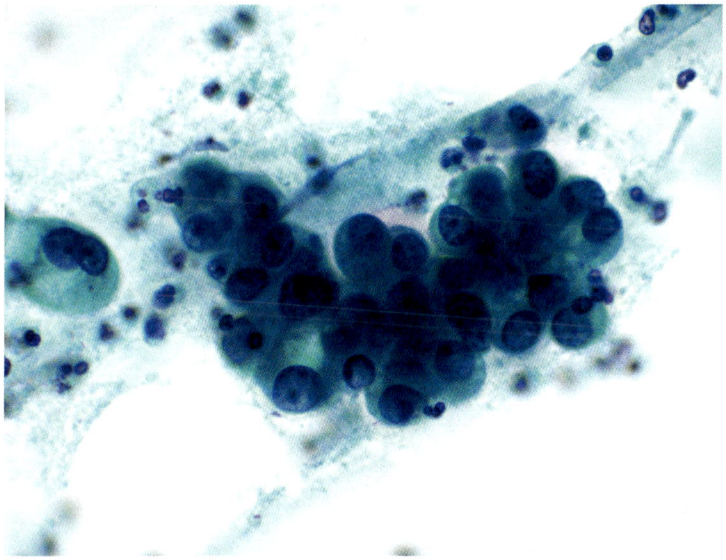

Figure 6.23a — Well-Differentiated Adenocarcinoma, Bronchoalveolar Lavage [Pap stain; high power]. Numerous, loosely cohesive cells can be seen with eccentrically placed nuclei and abundant cytoplasm. Two clues to the neoplastic nature of these cells are their prominent nucleoli and the absence of cilia. In the past, many well-differentiated adenocarcinomas were known as *bronchioloalveolar carcinoma* and were associated with a lepidic growth pattern on histology. These neoplasms had distinctive cytologic features such as finely granular chromatin, nuclear grooves, and a clean background.

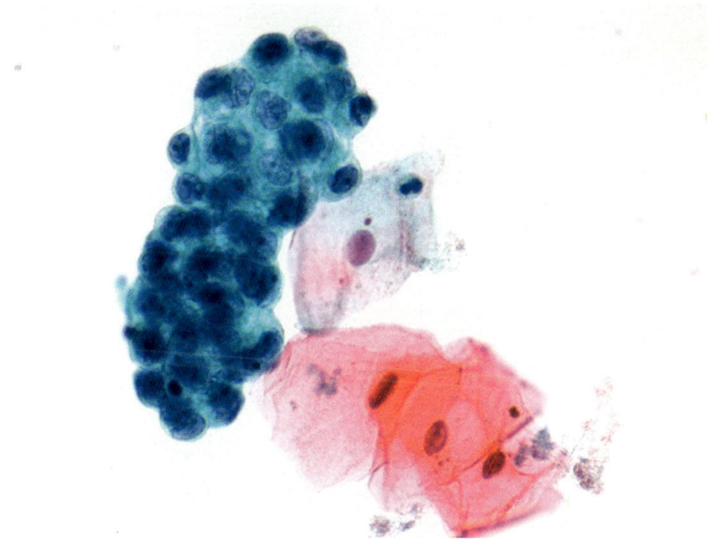

Figure 6.23b — Well-Differentiated Adenocarcinoma with Bronchioloalveolar Features, Sputum [Pap stain; high power]. The small glandular cluster shown here consists of regular small cells with vesicular cytoplasm, high nuclear-to-cytoplasmic ratio, small nucleoli, granular chromatin, and occasional nuclear grooves. The nuclei are eccentrically placed and the outer contour of the cluster is relatively smooth with peripherally protruding nuclei, giving the cells a so-called hobnail appearance. Note the lack of identifiable terminal bars or cilia.

Figure 6.23c — Well-Differentiated Adenocarcinoma with Bronchioloalveolar Features, Sputum [Pap stain; high power]. This is a somewhat cohesive fragment of cells with eccentric hyperchromatic nuclei, prominent nucleoli, and vacuolated cytoplasm. Note the clean mucoid background and occasional prominent cytoplasmic vacuoles. The nuclei of well-differentiated adenocarcinomas tend to be relatively uniform with smooth nuclear contours, making them difficult to differentiate from hyperplastic goblet cells, reactive alveolar/bronchial cells, and alveolar macrophages. The presence of papillary fronds with fibrovascular septa and/or psammoma bodies would suggest the diagnosis of a neoplasm. In this case, the disordered arrangement of cells, nuclear crowding, and eccentric nuclei are telltale signs of malignancy.

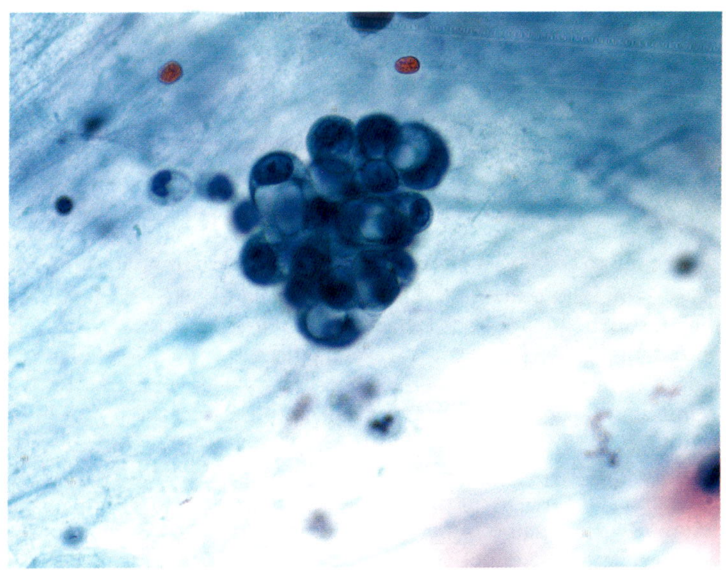

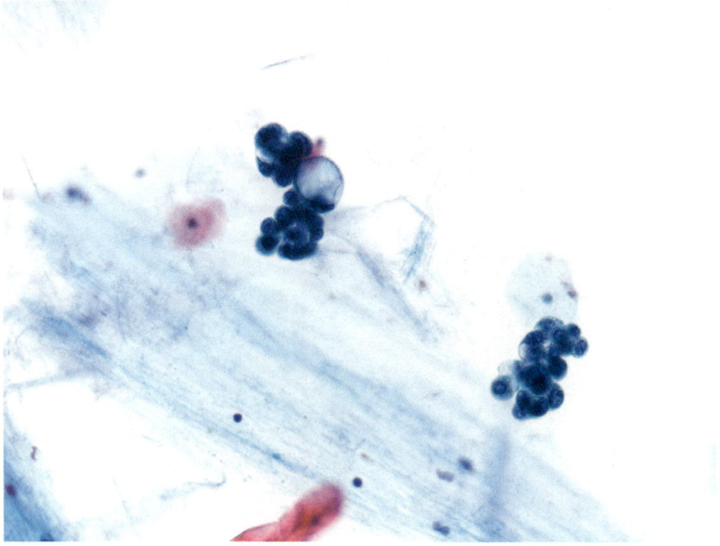

Figure 6.23d — Well-Differentiated Adenocarcinoma, Sputum [Pap stain; high power]. Two clusters of atypical epithelial cells are seen with several features diagnostic of adenocarcinoma. First, the clusters are disorganized, showing a lack of preserved honeycomb architecture. Second, the nuclei are enlarged and most cells have a high nuclear-to-cytoplasmic ratio. Most of the cells with more abundant cytoplasm have small intracytoplasmic mucin vacuoles; note one cell with a large, single mucin vacuole. Finally, the chromatin is clumpy, and the nucleoli, albeit small, are clearly present. On surgical follow-up, this case was shown to have a predominant lepidic growth pattern. Adenocarcinomas showing a predominant histologic lepidic pattern may show nuclear pseudoinclusions and nuclear groves; however, these features are not obvious in this case.

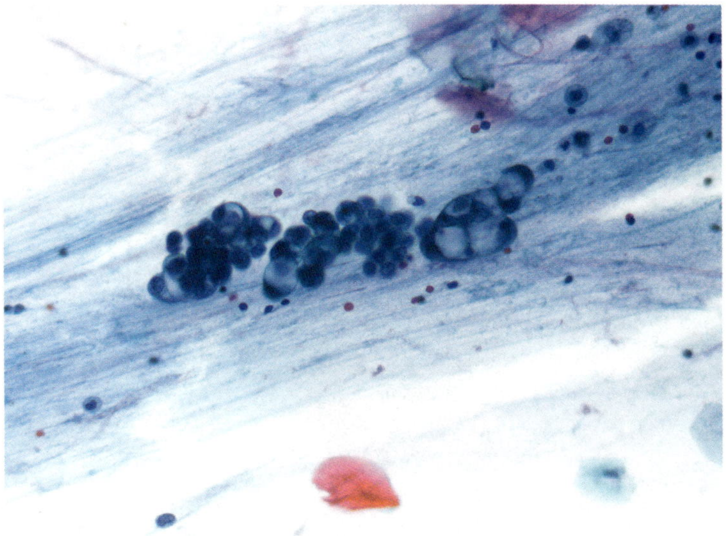

Figure 6.24a — Well-Differentiated Adenocarcinoma, Sputum [Pap stain; high power]. The cells in this field are disorganized and display a degree of three-dimensionality, a feature suggestive of malignancy. The cells contain eccentrically placed hyperchromatic, indented nuclei with large intracytoplasmic mucin vacuoles, a feature suggestive of adenocarcinoma. The cells without obvious intracytoplasmic mucin have a high nuclear-to-cytoplasmic ratio and punctate nucleoli. Taken together, these findings are diagnostic for a well-differentiated adenocarcinoma. Pulmonary adenocarcinoma likely originates from bronchial epithelial cell and pneumocyte precursors and is the most common type of lung cancer among young patients and those without a smoking history.

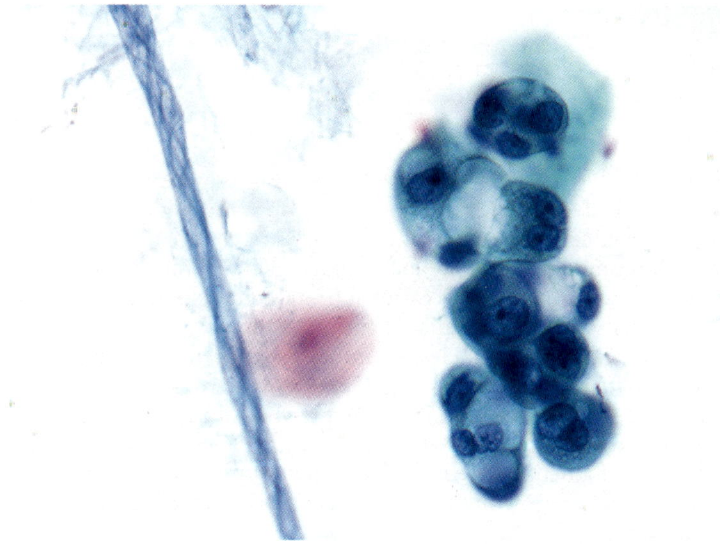

Figure 6.24b — Well-Differentiated Adenocarcinoma, Sputum [Pap stain; high power]. The cells in this field display several features diagnostic for adenocarcinoma: three-dimensional cluster formation, abundant foamy cytoplasm with small and large mucin vacuoles, enlarged, pleomorphic nuclei with irregular nuclear borders, and prominent nucleoli. Compare the enlarged nuclei of the tumor cells to the squamous cell that is slightly out of focus in this field. In sputum samples, oral mucosal squamous cells are a common finding. It is important not to overinterpret their presence as a keratinizing lesion such as squamous metaplasia, squamous cell carcinoma, or adenosquamous carcinoma.

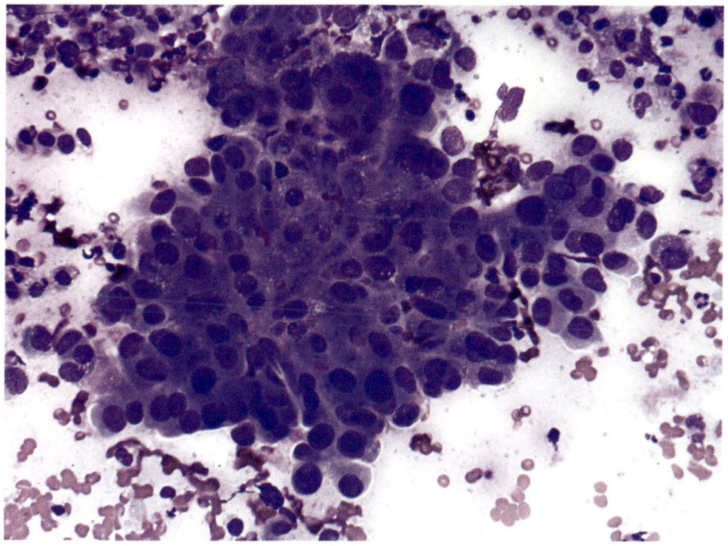

Figure 6.24c — Adenocarcinoma, Bronchial Washing [Diff-Quik stain; high power]. This fragment shows tumor cells with enlarged pleomorphic nuclei and variably prominent nucleoli. Vague glandular structures can be appreciated in the center of the field. In many cases, Diff-Quik stain can help distinguish adenocarcinoma from squamous cell carcinoma as a result of the softer vacuolated cytoplasm of the former and the dense robin's egg blue cytoplasm of the latter.

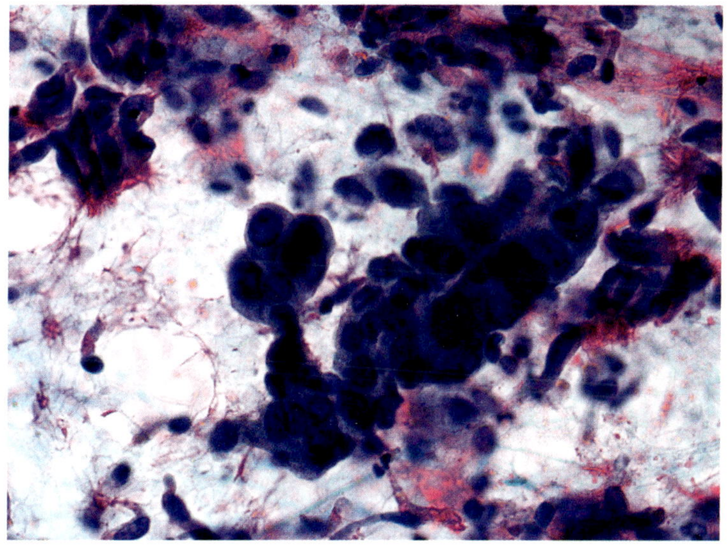

Figure 6.24d — Adenocarcinoma, Bronchial Washing [Pap stain; high power]. A fragment of tumor cells shows high nuclear-to-cytoplasmic ratios, nuclear hyperchromasia, marked nuclear pleomorphism, and glandular arrangement. Compared with the abundant benign bronchial epithelium in the background, the nuclei are markedly enlarged and lack nuclear polarity. One can readily appreciate the two different cell types in this field.

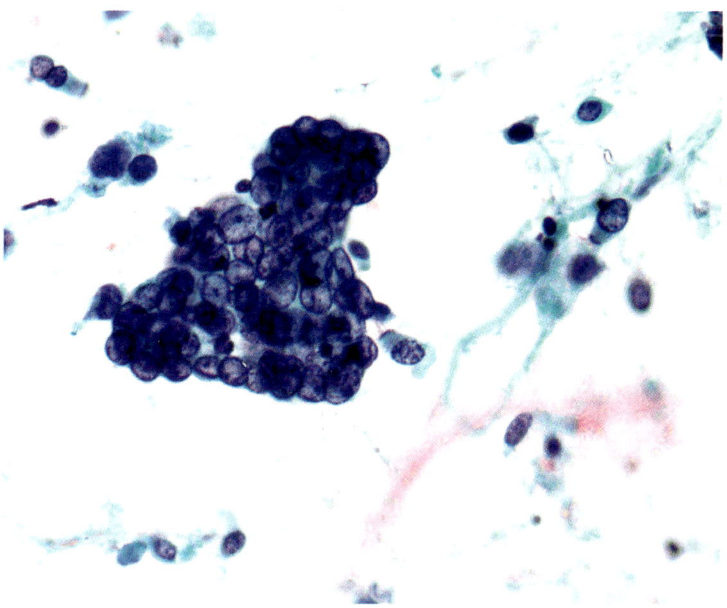

Figure 6.25a — Adenocarcinoma, Bronchial Washing [Pap stain; high power]. This is a three-dimensional fragment of adenocarcinoma demonstrating prominent nucleoli, powdery to open chromatin, nuclear crowding, and glandular structures. The nuclear-to-cytoplasmic ratio is markedly increased here, with only scant cytoplasm bringing small cell carcinoma into the differential diagnosis. Unlike small cell carcinoma, however, there is no chromatin clumping, nuclear molding, or necrosis in the background.

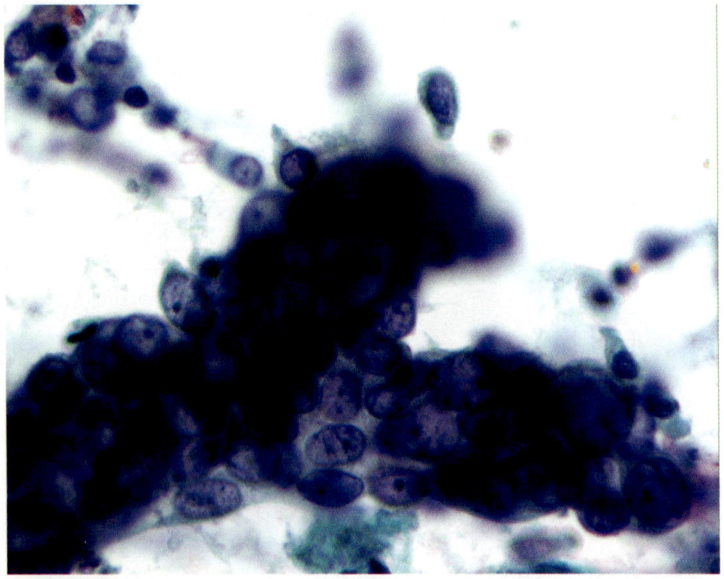

Figure 6.25b — Adenocarcinoma, Bronchial Washing [Pap stain; high power]. This papillary fragment consists of tumor cells with large, eccentric, and pleomorphic nuclei and prominent nucleoli. Occasional nuclear grooves are also present. These features are characteristic of lung adenocarcinomas and should prompt the pathologist to order broad molecular profiling, if possible. One limitation of exfoliative cytology specimens is that they rarely obtain enough material for ancillary studies.

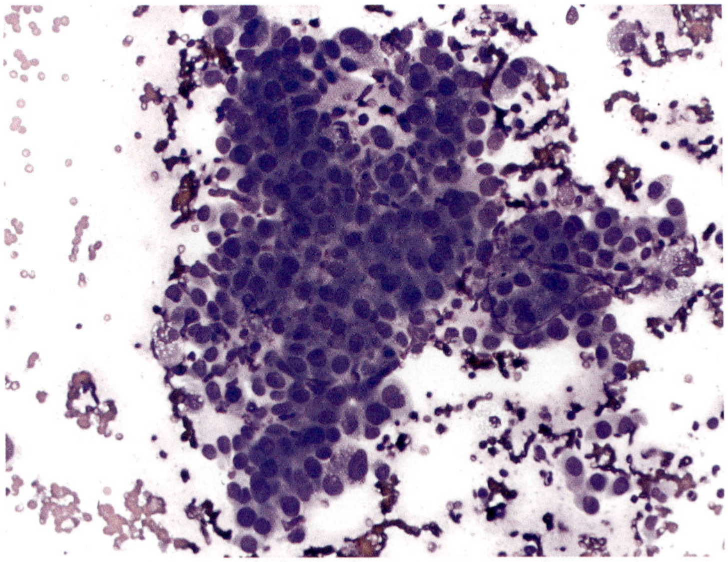

Figure 6.25c — Moderately Differentiated Adenocarcinoma, Bronchial Brush [Diff-Quik stain; medium power]. The tumor cells do not form distinct glands but rather form a three-dimensional sheet. Features that suggest an adenocarcinoma and not squamous cell carcinoma include an eccentrically placed nucleus, prominent nucleoli, and cytoplasmic vacuolization. Because bronchial brushings do not typically gather enough material for confirmatory immunostains, it is important to first determine this to be a non–small cell carcinoma and then at least favor an adenocarcinoma. If necessary, additional sampling might be necessary to gather enough cells for molecular testing.

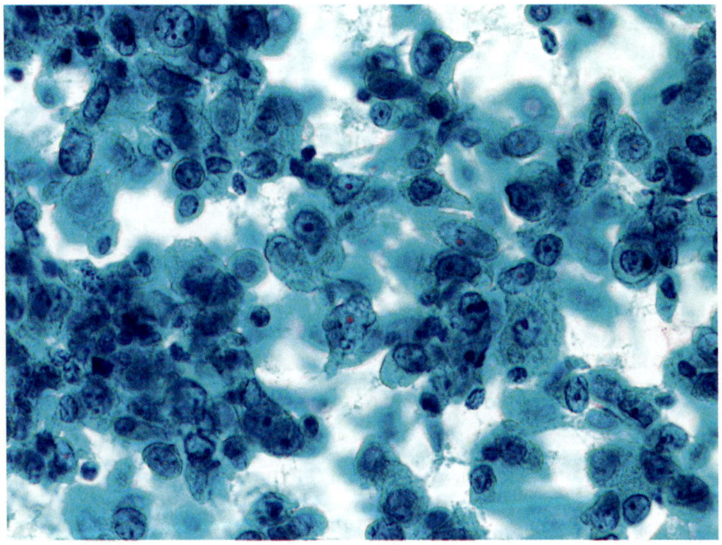

Figure 6.25d — Moderate to Poorly Differentiated Adenocarcinoma, Bronchial Washing [Pap stain; high power]. This view shows highly atypical cells with significant nuclear atypia in a background of necrosis. The nuclei are enlarged with greater than a 4:1 variation in size, nuclear membrane irregularity, and prominent nucleoli. Many cells contain abundant foamy cytoplasm that supports a diagnosis of adenocarcinoma. It is important to remember than the lung is a common site of nonpulmonary metastasis. In particular, colonic adenocarcinomas often contain abundant necrosis and cytomorphologic features indistinguishable from a lung primary tumor. In most cases, combining the patient's clinical history with a panel of immunohistochemical stains can determine whether the adenocarcinoma is of a pulmonary or nonpulmonary origin.

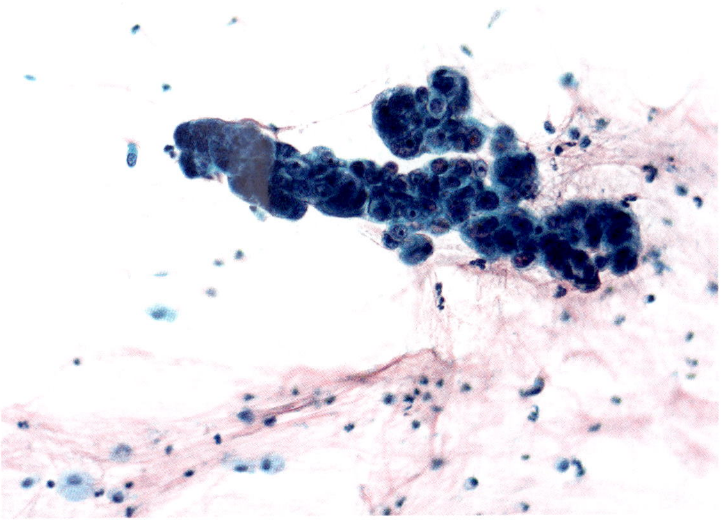

Figure 6.26a — Poorly Differentiated Adenocarcinoma, Bronchial Washing [Pap stain; high power]. The cells have very little cytoplasm and prominent nucleoli. Based on these cytomorphologic features alone, it would be difficult to determine that this is an adenocarcinoma. However, the cells form a rigid, three-dimensional structure with papillary-like projections, strongly indicating that this is an adenocarcinoma. Since this is a poorly differentiated adenocarcinoma, it likely does not correspond, on histology, to an adenocarcinoma with a papillary growth pattern, which typically has a well-differentiated cytomorphologic appearance.

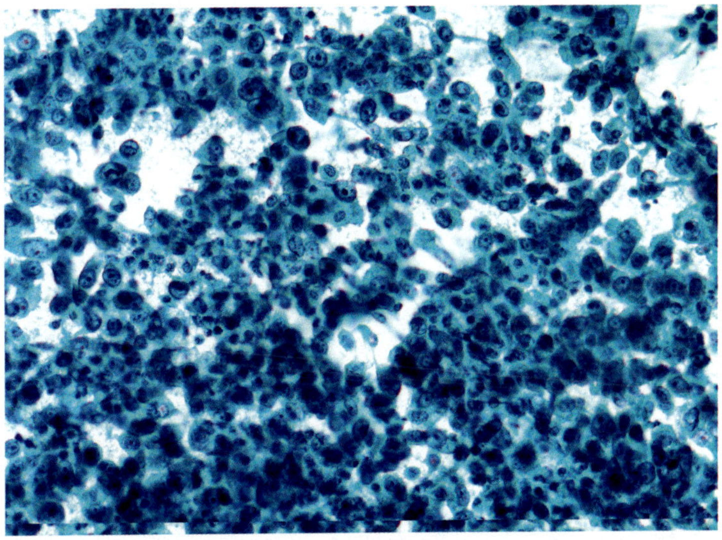

Figure 6.26b — Poorly Differentiated Adenocarcinoma, Bronchial Washing [Pap stain; medium power]. There are numerous highly atypical cells and cells in overlapping sheets in an abundantly necrotic background. The epithelial cells have enlarged nuclei with significant, greater than 4:1, nuclear size variation. The presence of mitotic figures and prominent nucleoli further confirms a diagnosis of malignancy. Some cells have abundant foamy cytoplasm; however, the degree of degeneration and necrotic granular debris obscures the field and makes the determination of adenocarcinoma difficult at this magnification. These features are most suggestive of a poorly differentiated adenocarcinoma, although a diagnosis of non–small cell lung carcinoma may be most prudent in this case if a cell block is not available for confirmatory immunohistochemical studies and the patient does not have a history of prior malignancy.

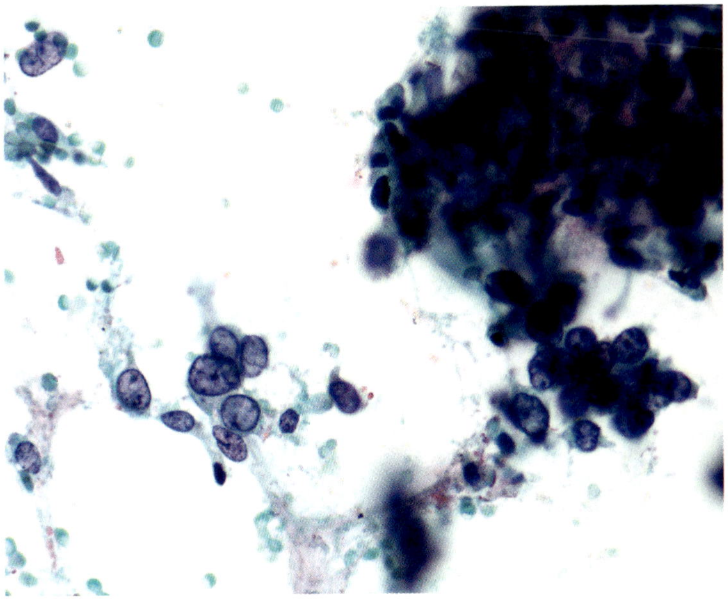

Figure 6.26c — Poorly Differentiated Adenocarcinoma, Bronchial Washing [Pap stain; high power]. A three-dimensional cluster of hyperchromatic, irregular cells is seen at the top right portion of the field. In other areas, the individual cytologic features are more easily assessed. The nuclei are enlarged, with an increased nuclear-to-cytoplasmic ratio, nuclear membrane irregularity, and occasional nuclear grooves. In this field, no gland-like arrangements can be appreciated, making a definitive diagnosis of adenocarcinoma difficult. When possible, perform immunostains to confirm the tumor subtype and preserve as much material as possible for molecular testing. It is helpful to make a more definitive diagnosis than non–small cell lung carcinoma because new targeted therapies are available for patients with adenocarcinoma, not squamous cell carcinoma, with *EFGR*, *ALK*, or *ROS1* mutations.

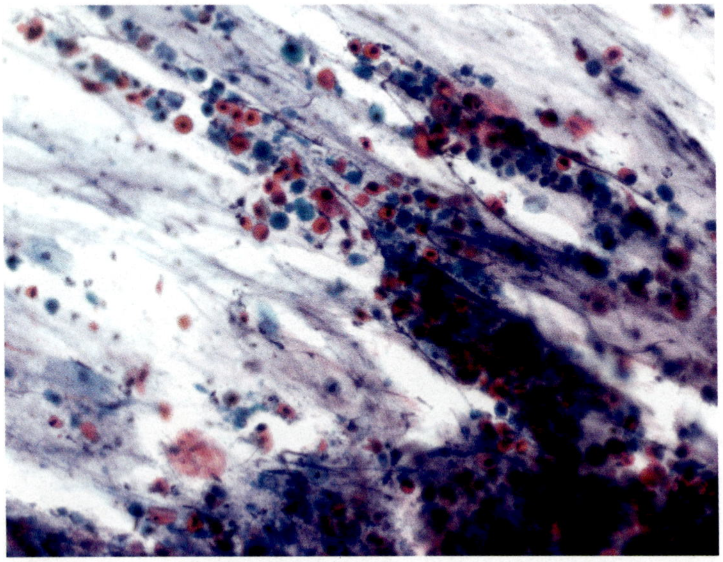

Figure 6.27a — Keratinizing Squamous Cell Carcinoma, Sputum [Pap stain; low power]. Numerous discohesive tumor cells are found within strands of mucin. The tumor cells have dense cytoplasm, some pink and others cyan. Although the nuclear detail is difficult to discern at this power, many nuclei appear dark and pyknotic. Many nonmalignant processes in the lung can cause squamous metaplasia and keratinization of the large airways. However, in this case the malignant cells are too numerous and atypical to be ignored.

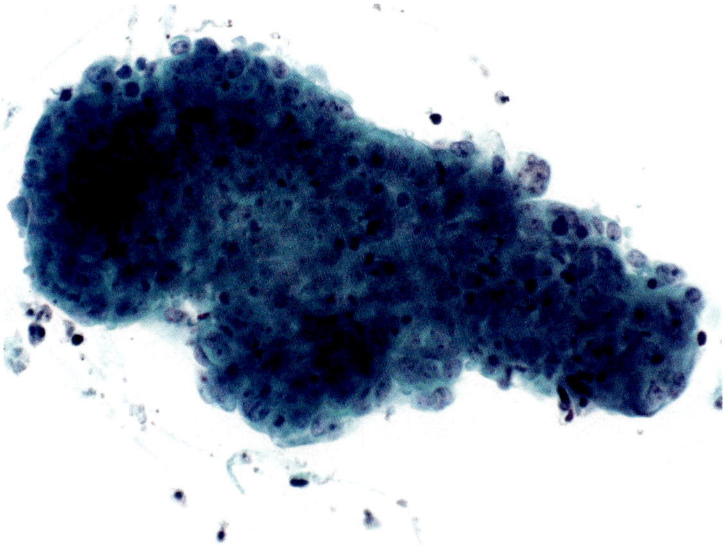

Figure 6.27b — Nonkeratinizing Squamous Cell Carcinoma, Bronchial Brush [Pap stain; high power]. Centrally located squamous cell carcinomas are accessible to the bronchial brush and can be sampled with high diagnostic yield. Unlike well-differentiated squamous cell carcinomas, moderate to poorly differentiated squamous cell carcinomas commonly exhibit little to no keratinization. Cells may be isolated, in loosely cohesive groups, or in syncytial tissue fragments without any architectural pattern. The cell size can be variable with large nuclei and high nuclear-to-cytoplasmic ratios. The only indication leading to squamous differentiation may be an occasional anucleated cell with dense cytoplasm and well-defined rigid cell borders or an intracytoplasmic keratin blob. Because any given squamous carcinoma may present different grades in different parts of the tumor, it is not uncommon to come across a spectrum of morphologic patterns in a given sample. However, without ancillary stains (e.g., p40, thyroid transcription factor-1 [TTF-1], Napsin A, mucicarmine), most cases are difficult to accurately subclassify into squamous cell carcinoma versus poorly differentiated adenocarcinoma.

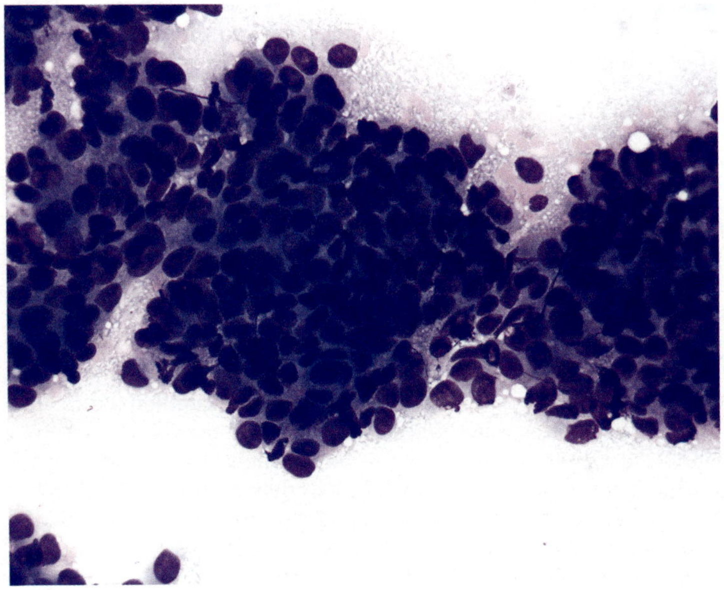

Figure 6.28a — Poorly Differentiated Squamous Cell Carcinoma, Bronchial Brush [Diff-Quik stain; high power]. As the differentiation of squamous cell carcinoma decreases, nuclear and cytoplasmic features of squamous differentiation are less apparent. The only cytologic evidence of squamous differentiation may lie in the tendency of the cells to form a monolayer, as shown here, or the sharp appearance of cytoplasmic borders. A few prominent nucleoli can be seen along the bottom of the fragment.

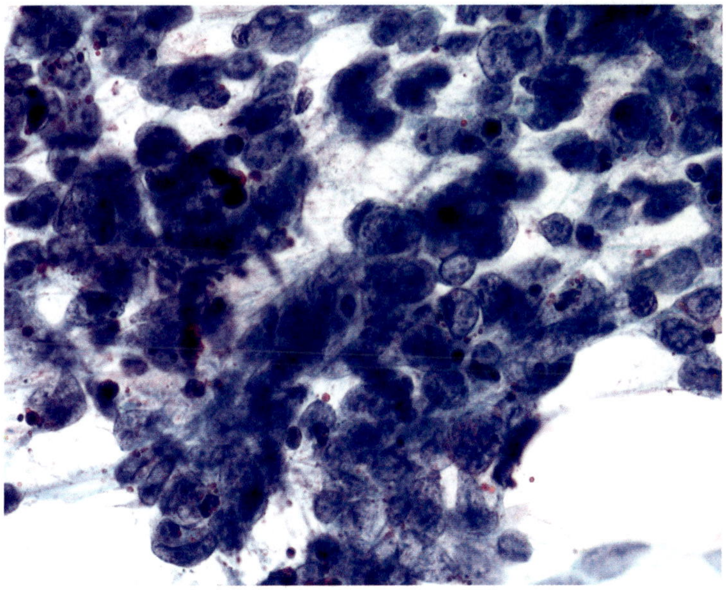

Figure 6.28b — Poorly Differentiated Squamous Cell Carcinoma, Bronchial Brush [Pap stain; high power]. Marked cellular pleomorphism is characteristic of squamous cell carcinomas. The cells in this image also demonstrate chromatin clumping, prominent nucleoli, irregular nuclear borders, and rare keratinized (deep orange-red) tissue fragments.

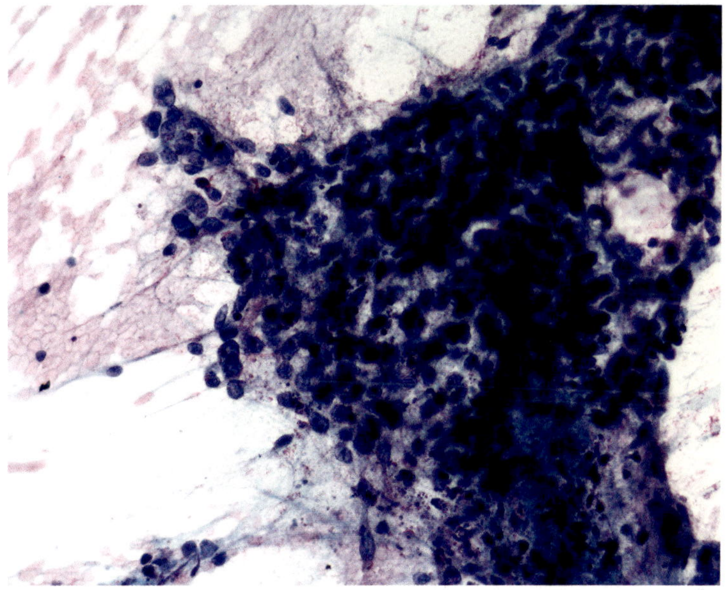

Figure 6.28c — Poorly Differentiated Squamous Cell Carcinoma, Bronchial Brush [Pap stain; medium power]. At lower power, abundant keratinaceous debris can be readily identified throughout the large tissue fragment. Marked nuclear pleomorphism with spindled cells (bottom of the image) are also present. Recently, molecular profiling of squamous cell carcinoma of the lung has revealed potentially targetable alterations, among them, *FGFR* amplifications, *DDR2* mutations, *MET* mutation/amplification, and *PIK3CA* mutations/amplifications.

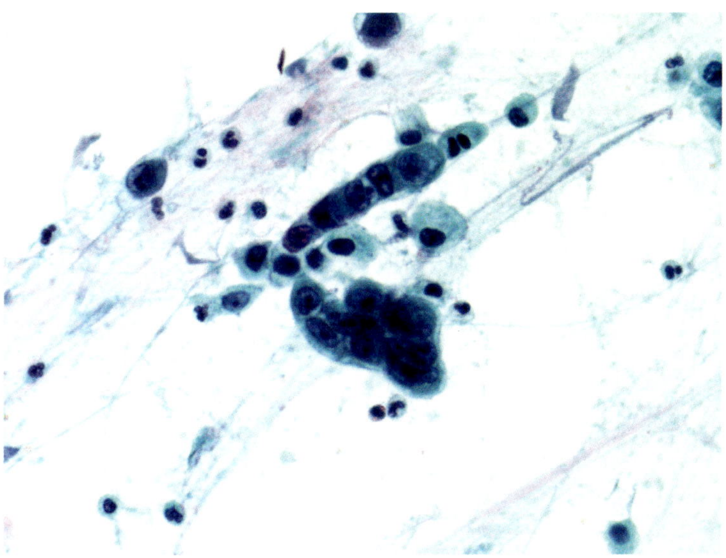

Figure 6.28d — Poorly Differentiated Squamous Cell Carcinoma, Sputum [Pap stain; high power]. Note the sharply defined cytoplasmic borders of the tumor cells, intercellular bridges, dense cyanophilic cytoplasm, macronucleoli, and nuclear pleomorphism. Squamous cell carcinomas of the lung are usually centrally located and associated with a large airway. A metastasis from another organ site, such as the esophagus, is frequently difficult to exclude. If a previous history of squamous cell carcinoma related to human papillomavirus (HPV) is known, high-risk HPV in situ hybridization or p16 immunohistochemistry can be useful in differentiating primary versus secondary malignancies.

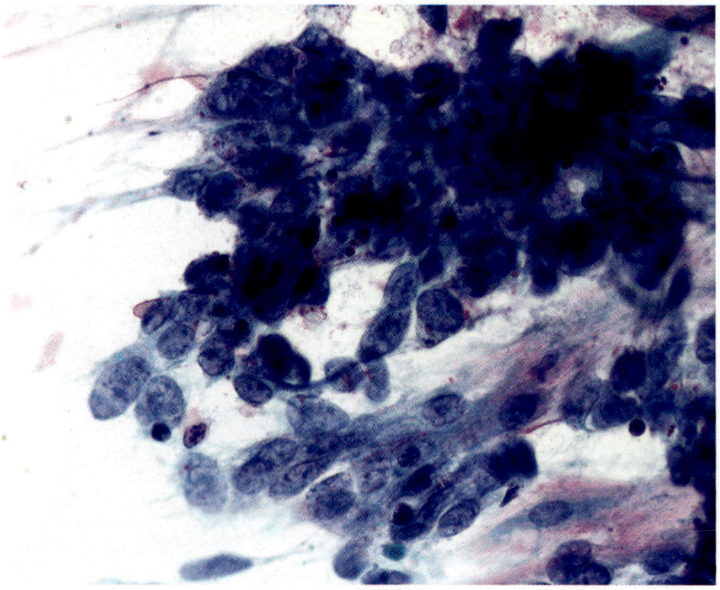

Figure 6.29a — Poorly Differentiated Squamous Cell Carcinoma, Bronchial Brushing [Pap stain; high power]. Cells are seen clumping together with marked atypia in a necrotic background. The nuclei are enlarged with chunky chromatin. The Orange G component of the Pap stain highlights focal orangeophilic cytoplasm, indicating a keratinizing component to this tumor. Otherwise, typical features suggestive of a squamous cell origin, such as intercellular bridges, are absent. Most squamous cell carcinomas are due to smoking and are more likely to present as central versus peripheral lesions.

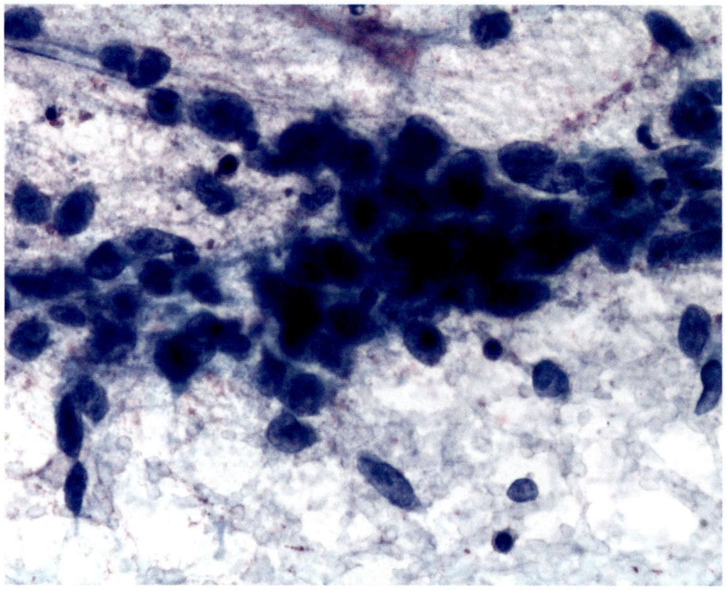

Figure 6.29b — Poorly Differentiated Squamous Cell Carcinoma, Bronchial Brushing [Pap stain; high power]. This smear shows single and clustered, highly atypical cells with enlarged, hyperchromatic nuclei and a high nuclear-to-cytoplasmic ratio is a necrotic granular background. Several mitotic figures are seen; however, no definitive evidence of mucin production or keratinization can be seen. Furthermore, the cells do not show molding, and the nuclei do not contain salt-and-pepper chromatin, which puts this case in the non–small cell lung carcinoma category. On immunohistochemical stains, this case was shown to be positive for p40 and negative for thyroid transcription factor-1 (TTF-1) and Napsin A, confirming a diagnosis of poorly differentiated squamous cell carcinoma.

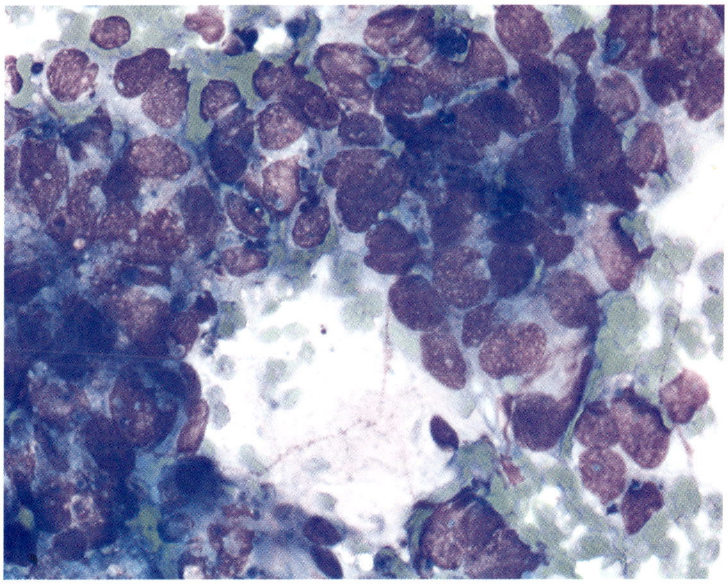

Figure 6.29c — Poorly Differentiated Squamous Cell Carcinoma, Bronchial Brushing [Diff-Quik stain; high power]. This high-power view shows a sheet of disorganized, pleomorphic cells with "hard," light blue cytoplasm and enlarged, irregular nuclei. Keratinization on Diff-Quik stain is a characteristic robin's egg blue with a glassy/hard appearance, as seen on the left side of the image. Definitive confirmation with immunostaining is warranted, however, because incorrect classification as squamous cell carcinoma will prevent or delay molecular testing for targeted therapies.

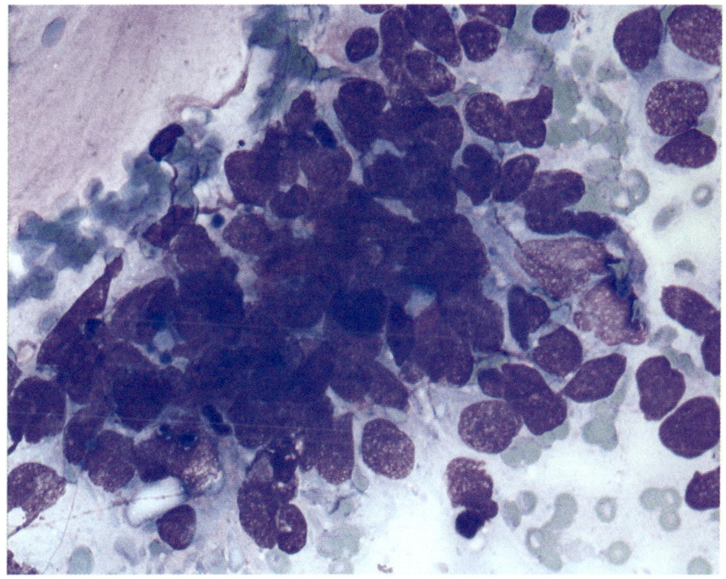

Figure 6.29d — Poorly Differentiated Squamous Cell Carcinoma, Bronchial Brushing [Diff-Quik stain; high power]. This field contains an irregular cluster of epithelial cells with enlarged, hyperchromatic nuclei with irregular nuclear contours and very little basophilic cytoplasm. The cells are notable for a lack of mucin vacuolization and lack of prominent nucleoli. In this patient with a history of squamous cell carcinoma, these features are consistent with a reccurrence, although definitive classification with immunostaining because of the lack of overt keratinization is warranted. The prognosis for pulmonary squamous cell carcinoma is generally poor and most closely related to the stage at the time of diagnosis. However, immunotherapy with anti-PDL1 antibodies has shown favorable results for a subset of patients.

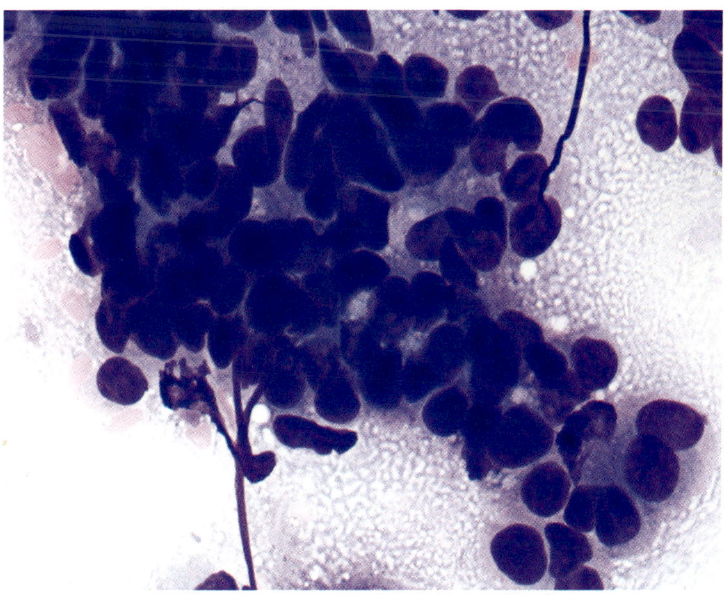

Figure 6.29e — Poorly Differentiated Squamous Cell Carcinoma, Bronchial Brushing [Diff-Quik stain; high power]. This overlapping sheet of epithelial cells shows enlarged, hyperchromatic nuclei with a high nuclear-to-cytoplasmic ratio and crush artifact. There is a notable lack of nuclear molding, salt-and-pepper chromatin, and prominent nucleoli, which make a diagnosis of small cell carcinoma and large cell neuroendocrine carcinoma less likely. Furthermore, no overt mucin production or keratinization is seen. Ultimately, a broad immunopanel should be ordered on this type of poorly differentiated carcinoma to ensure the correct diagnosis. In this case, p40 was diffusely positive, and thyroid transcription factor-1 (TTF-1), Napsin A, synaptophysin, chromogranin, and CD56 were negative.

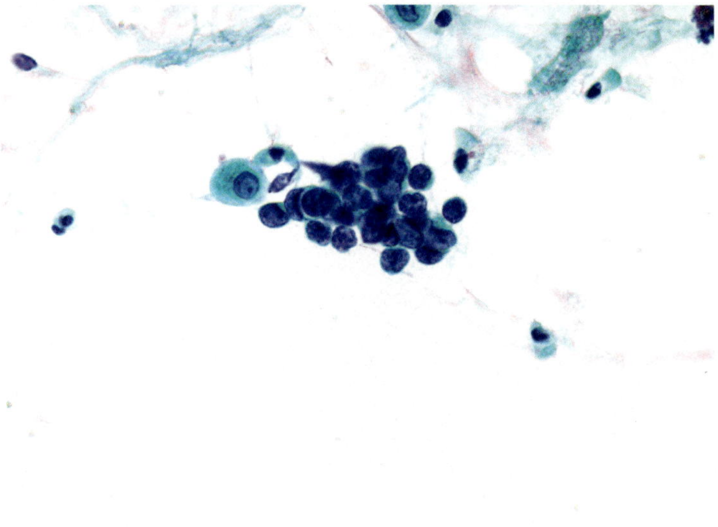

Figure 6.30a — Small Cell Carcinoma, Bronchial Washing [Pap stain; high power]. The cells in the center of the field demonstrate some prominent features of small cell carcinoma: small size, mostly by merit of the lack of cytoplasm (compare the nuclear size to that of the adjacent macrophage); a salt-and-pepper chromatin pattern, as demonstrated by both fine and coarse granules; and nuclear molding, in which some nuclei have a compressed shaped because of forces exerted by an adjacent carcinoma cell. If present focally, their small size may cause them to be mistaken for lymphocytes; reserve cell hyperplasia is also an important entity in the differential diagnosis.

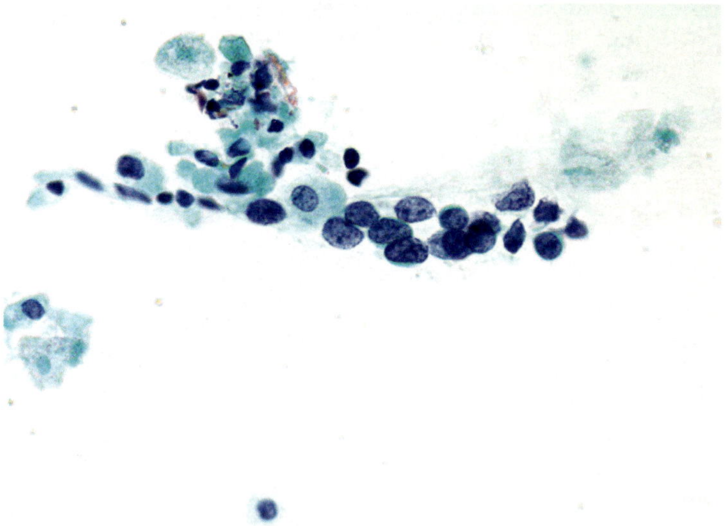

Figure 6.30b — Small Cell Carcinoma, Bronchial Washing [Pap stain; high power]. At close examination, it should not be challenging to differentiate small cell carcinoma cells form lymphocytes. Lymphocytes typically have a thin rim of cytoplasm and a uniform appearance. Here, some carcinoma cells have no identifiable cytoplasm and the nuclei demonstrate size variation. Furthermore, the malignant cells have irregular nuclear borders beyond the extent that would be seen in reactive lymphocytes.

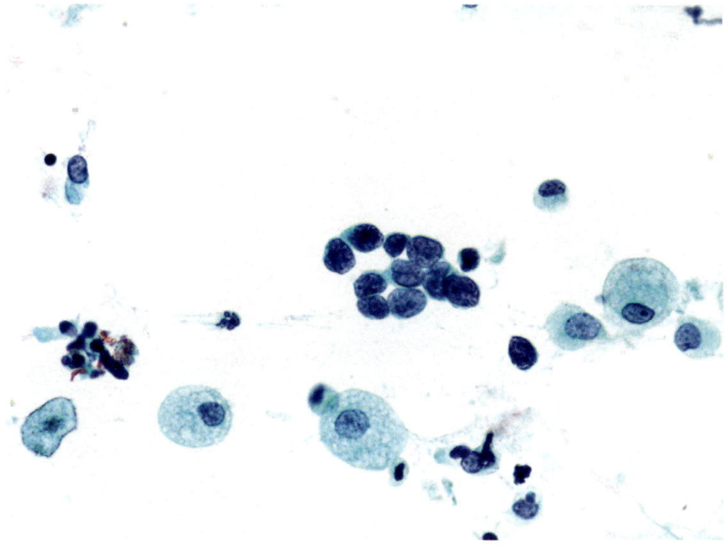

Figure 6.30c — Small Cell Carcinoma, Bronchial Washing [Pap stain; high power]. Features seen on histology that are often lacking in certain exfoliative cytology specimens include nuclear smearing artifact and necrosis. Nuclear molding may also be less dramatic. Specimens that are smeared on a slide (such as sputum) may preserve some of these morphologic features, whereas specimens collected in liquid (lavages and washings) reduce smearing artifact and allow necrotic material to be diluted and washed away during subsequent concentration steps.

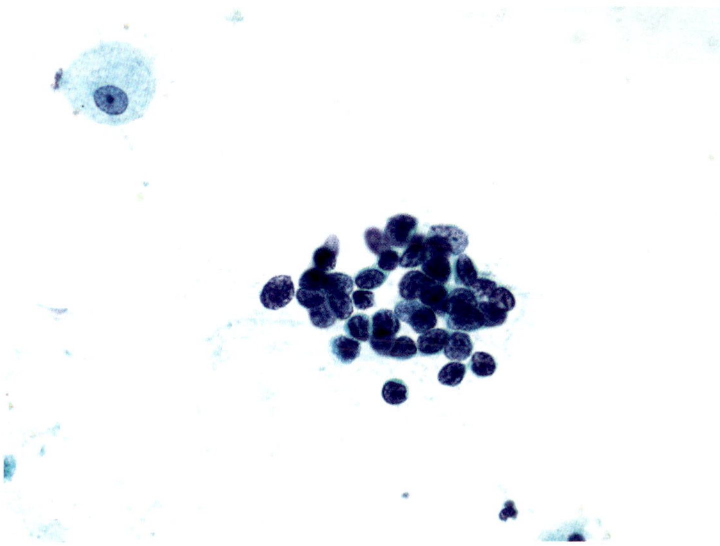

Figure 6.30d — Small Cell Carcinoma, Bronchial Washing [Pap stain; high power]. Although reserve cell hyperplasia usually does not produce highly cellular specimens, small cell carcinoma is a neoplastic process and may result in greater cellularity. However, small cell carcinoma cells may only be seen focally, which makes a definitive diagnosis more challenging. Here, the neuroendocrine pattern of the chromatin strongly suggests small cell carcinoma.

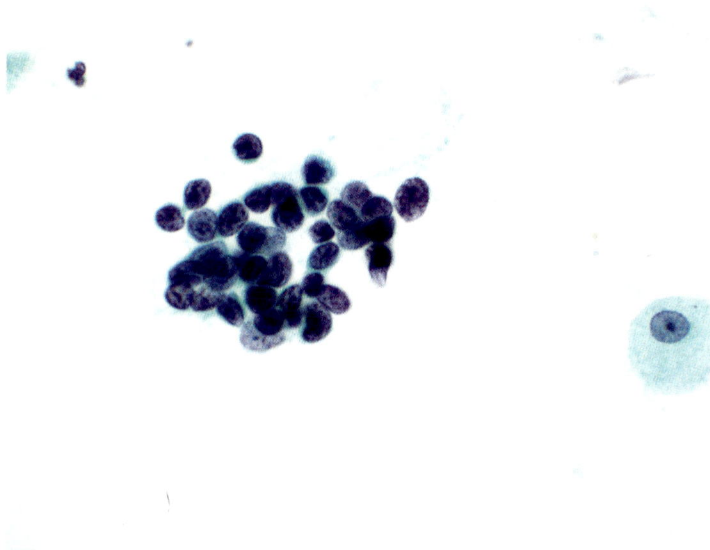

Figure 6.30e — Small Cell Carcinoma, Sputum [Pap stain; high power]. Compare the size of these cells to the adjacent pulmonary alveolar macrophage (right side of the field). The nuclei are about the same size, but the cells contain very little cytoplasm, even less cytoplasm than seen in reserve cell hyperplasia. The cells are a bit more uniform than reserve cells and are less cohesive. The salt-and-pepper appearance of the chromatin, or chromatin that is a mixture of fine and coarse, is distinctive on a good Pap-stained preparation.

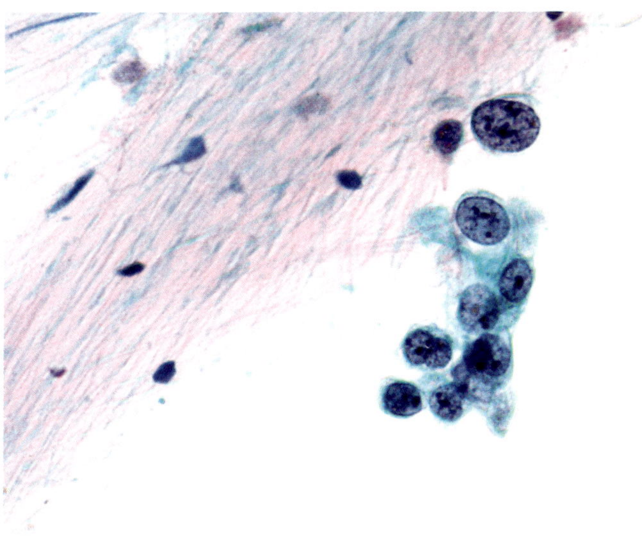

Figure 6.31a — Large Cell Lung Carcinoma, Sputum [Pap stain; high power]. This field contains large, pleomorphic cells with clumpy chromatin and nucleoli without evidence of mucin, keratin production, or neuroendocrine features. On surgical resection, this tumor was positive for CAM5.2 but negative for all other markers of glandular, squamous, or neuroendocrine differentiation. Large cell lung carcinoma is a rare non–small cell lung carcinoma subtype that is necessarily a diagnosis of exclusion. Multinucleation, bizarre atypia, atypical mitoses, and sarcomatoid features may be seen. Overall, the prognosis is poor.

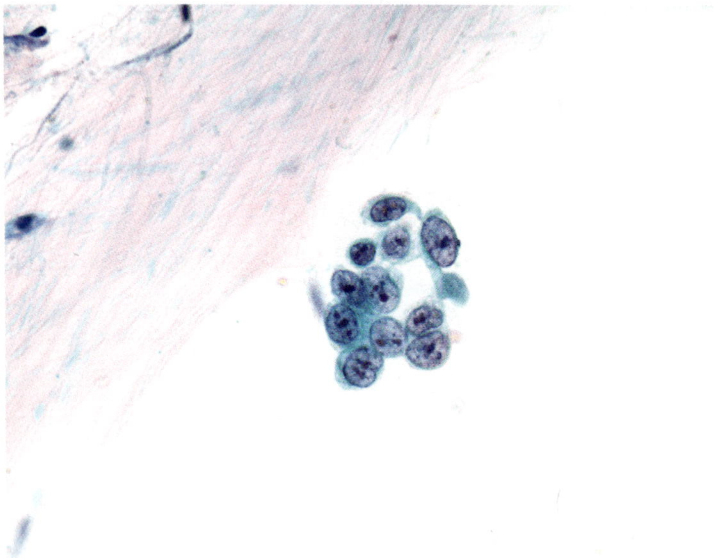

Figure 6.31b — Large Cell Lung Carcinoma, Sputum [Pap stain; high power]. Compared with the inflammatory cell in the upper left corner of the image, the malignant cells have significantly enlarged nuclei with little cytoplasm and clumpy chromatin. There are significant size variation and no obvious signs of glandular, squamous, or neuroendocrine differentiation. On surgical resection, this tumor was diagnosed as a large cell lung carcinoma with pan-negative markers for differentiation; however, on cytologic evaluation, this case is best classified as non–small cell lung carcinoma.

Index